Lecture Notes in Computer Science 15249

The series Lecture Notes in Computer Science (LNCS), including its subseries Lecture Notes in Artificial Intelligence (LNAI) and Lecture Notes in Bioinformatics (LNBI), has established itself as a medium for the publication of new developments in computer science and information technology research, teaching, and education.

LNCS enjoys close cooperation with the computer science R & D community, the series counts many renowned academics among its volume editors and paper authors, and collaborates with prestigious societies. Its mission is to serve this international community by providing an invaluable service, mainly focused on the publication of conference and workshop proceedings and postproceedings. LNCS commenced publication in 1973.

Marc Modat · Ivor Simpson · Žiga Špiclin ·
Wietske Bastiaansen · Alessa Hering ·
Tony C. W. Mok
Editors

Biomedical Image Registration

11th International Workshop, WBIR 2024
Held in Conjunction with MICCAI 2024
Marrakesh, Morocco, October 6, 2024
Proceedings

Springer

Editors
Marc Modat
King's College London
London, UK

Ivor Simpson
University of Sussex
Brighton, UK

Žiga Špiclin
University of Ljubljana
Ljubljana, Slovenia

Wietske Bastiaansen
Erasmus MC
Rotterdam, The Netherlands

Alessa Hering
Radboud University Medical Center
Nijmegen, The Netherlands

Tony C. W. Mok
Hong Kong University of Science
and Technology
Hong Kong, China

ISSN 0302-9743 ISSN 1611-3349 (electronic)
Lecture Notes in Computer Science
ISBN 978-3-031-73479-3 ISBN 978-3-031-73480-9 (eBook)
https://doi.org/10.1007/978-3-031-73480-9

This Springer imprint is published by the registered company Springer Nature Switzerland AG
The registered company address is: Gewerbestrasse 11, 6330 Cham, Switzerland

Preface

The 11th International Workshop on Biomedical Image Registration (WBIR 2024, https://wbir.info) was held in Marrakesh, Morocco on October 6, 2024. To reaffirm the important role of biomedical image registration in the broader medical imaging community the WBIR 2024 meeting was organized as a half-day satellite workshop held in conjunction with the 27th International Conference on Medical Image Computing and Computer Assisted Intervention (MICCAI 2024), organized by the MICCAI Society (https://miccai.org). The co-located venue and temporal proximity to the MICCAI 2024 conference enabled the interested researchers and practitioners to indulge in the topics focused solely on biomedical image registration.

Preceding editions of WBIR have run mostly standalone as two-day workshops at various locations: Bled, Slovenia (1999); Philadelphia, USA (2003); Utrecht, The Netherlands (2006); Lübeck, Germany (2010); Nashville, USA (2012); London, UK (2014); Las Vegas, USA (2016), Leiden, The Netherlands (2018), Portorož, Slovenia (2020/virtually) and Munich, Germany (2022/hybrid). As with previous editions, the major appeal of the WBIR 2024 meeting was bringing together researchers and practitioners with various backgrounds and at different stages of academic careers, who all share a great interest in image registration. The half-day workshop program included in-person oral and poster presentations, with ample opportunities for discussion among peers. As everyone's mindset was on image registration, it made it easier for students to approach and meet their distinguished colleagues.

WBIR 2024 was co-located and held concurrently with the 5th edition of the Learn2Reg challenge (https://learn2reg.grand-challenge.org/), an established venue for objective and comparative evaluation of image registration algorithms. The Learn2Reg challenge organizers provided the data and evaluation platform with online leaderboard for four exciting image registration tasks: (1) brain resection multimodal registration, (2) large-scale unsupervised brain MRI image registration, (3) histopathology tissue section registration using second-harmonic generation microscopy and bright field (BF) microscopy images of cancer tissue and (4) multimodal microscopy 3D image registration of cellular level volume electron microscopy and light microscopy images. All Learn2Reg participants were encouraged to submit full papers to WBIR 2024. Those Learn2Reg participants that were among the top 10 on the leaderboard for any of the four tasks were invited to present a poster during the co-located WBIR 2024/Learn2Reg meetings.

The WBIR 2024 proceedings, published in Lecture Notes in Computer Science, were established through a rigorous peer-review process using OpenReview. We accepted only full papers. The full papers were reviewed in a double-blind fashion, with each submission evaluated by at least three members of the Program Committee. The international Program Committee consisted of 26 researchers established in the field of medical image registration. All papers and reviews were assessed and discussed by the General Chairs to reach decisions. From a total of 32 full paper submissions, 28 were selected for oral and poster presentation and are included in these proceedings. Prominent topics included novel deep learning architectures, strategies to assess and/or improve registration robustness, atlas construction and multi-modal image registration-based fusion strategies, learning-based approaches to feature extraction and image similarity evaluation, and sequential pipelines aiming at efficiency of image registration.

Many contributed to the organization and success of the WBIR 2024 meeting. In particular, we would like to thank the members of the Program Committee for their work that assured the high quality of the workshop. We also thank the organizers of the Learn2Reg challenge for their important contributions that undoubtedly enriched the WBIR 2024 program. We would like to thank the MICCAI 2024 team for their organizational effort and kind support of WBIR 2024, and kindly acknowledge the support of MICCAI's Special Interest Group on Biomedical Image Registration (SIG-BIR). Finally, we would like to thank all authors and participants of WBIR 2024 for their contributions.

August 2024

Marc Modat
Ivor Simpson
Žiga Špiclin
Wietske Bastiaansen
Alessa Hering
Tony C. W. Mok

Organization

General Chairs

Marc Modat	King's College London, UK
Ivor Simpson	University of Sussex, UK
Žiga Špiclin	University of Ljubljana, Slovenia
Wietske Bastiaansen	Erasmus MC, The Netherlands
Alessa Hering	Radboud University Medical Center, The Netherlands
Tony C. W. Mok	Hong Kong University of Science and Technology, China

Program Committee

Hari Om Aggrawal	Technical University of Denmark, Denmark
John Ashburner	University College London, UK
Wietske Bastiaansen	Erasmus MC, Netherlands
Aaron Carass	Johns Hopkins University, USA
Gary Christensen	University of Iowa, USA
Lina Felsner	Technical University of Munich, Germany
Mattias Heinrich	University of Lübeck, Germany
Alessa Hering	Radboud University Medical Center, The Netherlands
Leo Joskowicz	Hebrew University of Jerusalem, Israel
Bernhard Kainz	Imperial College London, UK
Fryderyk Kögl	Technical University of Munich, Germany
Deepa Krishnaswamy	Brigham and Women's Hospital, USA
Yihao Liu	Johns Hopkins University, USA
Lucas Mansilla	Ciudad Universitaria UNL, Argentina
Marc Modat	King's College London, UK
Tony C. W. Mok	Hong Kong University of Science and Technology, China
Bartlomiej Papiez	University of Oxford, UK
Josien Pluim	Eindhoven University of Technology, The Netherlands
Ivor Simpson	University of Sussex, UK
Ayushi Sinha	Philips Research, The Netherlands

Contents

Architectures

Robustness

Atlas/Fusion

Architectures

MultiGradICON: A Foundation Model for Multimodal Medical Image Registration

Başar Demir[1(✉)], Lin Tian[1], Hastings Greer[1], Roland Kwitt[2], François-Xavier Vialard[3], Raúl San José Estépar[4], Sylvain Bouix[5], Richard Rushmore[6], Ebrahim Ebrahim[7], and Marc Niethammer[1]

[1] University of North Carolina at Chapel Hill, Chapel Hill, USA
bdemir@cs.unc.edu
[2] University of Salzburg, Salzburg, Austria
[3] Université Gustave Eiffel, LIGM, Champs-sur-Marne, France
[4] Brigham and Women's Hospital, Boston, USA
[5] ÉTS Montréal, Montreal, Canada
[6] Boston University, Boston, USA
[7] Kitware, Inc., Carrboro, USA

Abstract. Modern medical image registration approaches predict deformations using deep networks. These approaches achieve state-of-the-art (SOTA) registration accuracy and are generally fast. However, deep learning (DL) approaches are, in contrast to conventional non-deep-learning-based approaches, anatomy-specific. Recently, a universal deep registration approach, uniGradICON, has been proposed. However, uniGradICON focuses on monomodal image registration. In this work, we therefore develop multiGradICON as a first step towards universal *multimodal* medical image registration. Specifically, we show that 1) we can train a DL registration model that is suitable for monomodal *and* multimodal registration; 2) loss function randomization can increase multimodal registration accuracy; and 3) training a model with multimodal data helps multimodal generalization. Our code and the multiGradICON model are available at https://github.com/uncbiag/uniGradICON.

Keywords: Medical image registration · Deep learning · Multimodal

1 Introduction

Learning-based medical image registration [5,9,53,59] has significantly improved over the last decade. Current SOTA learning-based deep registration networks [35,36,47,48] are faster and more accurate than conventional registration methods using numerical optimization. However, learning-based approaches

Supplementary Information The online version contains supplementary material available at https://doi.org/10.1007/978-3-031-73480-9_1.

M. Modat et al. (Eds.): WBIR 2024, LNCS 15249, pp. 3–18, 2024.
https://doi.org/10.1007/978-3-031-73480-9_1

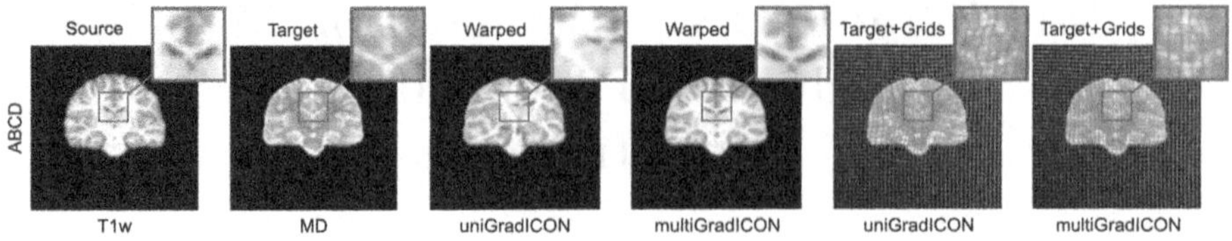

Fig. 1. Comparison of uni- and multiGradICON on T1w MRI-mean diffusivity (MD) registration from ABCD. Note the much better matching of the ventricles for multiGradICON.

generally 1) focus on monomodal image registration and 2) are trained for a specific anatomical region (often the brain), making them much less generically applicable than conventional approaches. While many monomodal learning-based approaches have been developed, uniGradICON [47] is currently the only learning-based approach that is designed to be a *universal* method supporting registrations for different anatomies in *one* model working directly with images and SynthMorph [23] may generalize to other anatomies by training on synthetic shapes. Further, uniGradICON focuses on monomodal registration and multimodal generalization is primarily achieved by instance optimization (IO). Hence, our goal is to further close the gap between conventional and learning-based registration approaches by developing multiGradICON, a multimodal generalization of uniGradICON. multiGradICON extends uniGradICON by 1) choosing a multimodal similarity measure, 2) incorporating multimodal registration tasks into training, and 3) exploring monomodal, multimodal, and randomized strategies for image similarity loss based on multimodal registration network inputs.

Our specific contributions are to show that

1) while uniGradICON is a strong baseline for monomodal registration, even for unseen modalities, it does not generalize well to multimodal registration when modalities are drastically different (see Fig. 1 for an example);
2) training a monomodal model with a squared local normalized cross correlation image similarity loss ($1 - \text{LNCC}^2$) does not lead to multimodal generalization in the absence of multimodal registration tasks during training;
3) multiGradICON allows for multimodal generalization while retaining good monomodal registration accuracy;
4) similarity loss randomization (i.e., randomly picking which modalities should be compared in the loss) improves registration accuracy for datasets containing multi-parametric data, even when scalar images are used for inference.

2 Related Work

Following uniGradICON [47] (which builds on ICON [15] and GradICON [48]), we focus on non-parametric image registration [34] where a displacement field is predicted. Such image registrations are generally formulated as a balance between an image (dis)similarity term and a regularizer encouraging spatial smoothness [34]. By extending uniGradICON, we retain its gradient inverse consistent regularizer (see Sect. 3) and focus on multimodal alternatives to the negative local normalized cross correlation (1 − LNCC) loss used in uniGradICON.

Multimodal Similarity Measures. Multimodal registration requires similarity measures that extend beyond the direct intensity similarities measured by mean squared or absolute error (MSE/MAE) losses. The goal is to assess whether an image pair is spatially aligned even though image appearance may be quite different across modalities. Conventional approaches maximize measures of statistical dependence such as normalized cross correlation (NCC) or squared NCC (if the sign of the correlation is unknown). Even more general statistical dependencies can be captured by maximizing mutual information (MI) between image pairs [51]. Allowing for more local control, these measures have been extended to local NCC (LNCC; as in uniGradICON) or its squared variant as well as local MI; see [22,34] for an overview of these conventional similarity measures. Such conventional similarity measures have been used to train multimodal deep registration networks [16,42,48]. More modern approaches target similarity measures depending on image self-similarity (such as the modality independent neighborhood descriptor (MIND) [20] and its improved version with self-similarity context (MIND-SSC) [21]) or local entropy images [52]. The general idea of these more modern similarity measures is to sidestep image differences by using a representation that remains similar even if the underlying image pairs look different. On the extreme end of the similarity measure spectrum, one can then use image segmentations via a Dice loss or try to transform one modality into the other via image synthesis. As such segmentations are generally not directly available and the intensity relationship between image pairs is not a-priori known learning-based approaches are commonly employed.

Learning-based Approaches for Multimodal Similarity Quantification. Learning better multimodal similarity measures has been explored via non-deep-learning [30] and deep learning approaches [10,31,37,43,44,49,57]. These multimodal similarity measures can then be used for conventional optimization-based image registration or to train a multimodal deep registration network. In general, multimodal registration can be simplified by 1) converting the multimodal problem into a monomodal one via image synthesis [39,41,54]; or 2) if already aligned images of the same subject are available (e.g., multi-parametric sequences in magnetic resonance imaging (MRI)) by only using monomodal pairings for the similarity loss between subjects but multimodal pairings as the input to the learning formulation [6,58]; or by 3) training a segmentation model on both modalities and then using the segmentations for the similarity loss [13]. Several works use the Dice loss computed from image segmentations to train a multimodal registration network [23,26,29,45]. Here, segmentations are either obtained manually [26,45], via automatic segmentation algorithms [29], or via label-image pair synthesis [23]. See the recent review articles on deep-learning-based image registration for further details on multimodal formulations [9,53].

multiGradICON can use any multimodal similarity measure. However, as a first step, we focus on $1-\text{LNCC}^2$ *as the similarity training loss as it is closely related to uniGradICON's* $1-\text{LNCC}$ *loss. We use* $1-\text{LNCC}^2$ *as well as MIND-SSC* [21] *for instance optimization. This allows us to keep experiments simple.*

3 Methodology

We follow the uniGradICON approach with *three key differences*: 1) we adjust the image similarity measure so that it is appropriate for multimodal image registration, 2) we use a larger training dataset that contains multimodal registration tasks, and 3) we explore image similarity loss randomization. We introduce the uniGradICON methodology below and highlight our modifications.

Notation. We denote our source and target images by (I^A, I^B), and consider them to be functions from an image domain to the real numbers. We denote a registration network by Φ_θ. This network Φ_θ operates on a pair of images to yield a function $\Phi_\theta[I^A, I^B] : \mathbb{R}^N \to \mathbb{R}^N$ which, when precomposed with the source image, is intended to align the images: $I^A \circ \Phi_\theta[I^A, I^B] \sim I^B$.

3.1 Architecture

Following the principles outlined in uniGradICON [47], we use the registration network from GradICON [48], i.e., we create a multi-step, multi-resolution network using the TwoStep (TS) and DownSample (DS) operators from [48]:

$$\texttt{TS}\{\Psi_\theta^1, \Psi_\theta^2\}[I^A, I^B] := \Psi_\theta^1[I^A, I^B] \circ \Psi_\theta^2[I^A \circ \Psi_\theta^1[I^A, I^B], I^B], \tag{1}$$

$$\texttt{DS}\{\Psi_\theta\}[I^A, I^B] := \Psi_\theta[\texttt{averagePool}(I^A, 2), \texttt{averagePool}(I^B, 2)]. \tag{2}$$

Specifically, we use U-Nets [11] (Ψ_θ^i) that predict displacement fields and construct the registration network as $\Phi_\theta := \texttt{TS}\{\texttt{TS}\{\texttt{DS}\{\texttt{TS}\{\texttt{DS}\{\Psi_\theta^1\}, \Psi_\theta^2\}\}, \Psi_\theta^3\}, \Psi_\theta^4\}$.

3.2 Training Losses

Baseline. The training loss proposed in GradICON [48] is

$$\begin{aligned}\mathcal{L} = &\mathcal{L}_{\text{sim}}\left(I^A \circ \Phi_\theta\left[I^A, I^B\right], I^B\right) + \mathcal{L}_{\text{sim}}\left(I^B \circ \Phi_\theta\left[I^B, I^A\right], I^A\right) \\ &+ \lambda \left\|\nabla\left(\Phi_\theta\left[I^A, I^B\right] \circ \Phi_\theta\left[I^B, I^A\right]\right) - \mathbf{I}\right\|_F^2.\end{aligned} \tag{3}$$

We use the negative squared localized normalized cross correlation ($1-\text{LNCC}^2$) as image similarity measure, $\mathcal{L}_{\text{sim}}(\cdot,\cdot)$, in contrast to uniGradICON's ($1-\text{LNCC}$) loss which assumes locally positive correlations. In contrast, ($1-\text{LNCC}^2$) is agnostic to the signs of the correlations. Therefore, it is more appropriate for general multimodal image registration where it is unclear if image intensity pairs are positively or negatively correlated locally.

Image Similarity Loss Randomization. Some multimodal medical datasets provide different modalities for the same patient and anatomical region, e.g., for multi-parametric MRI where images from multiple MR sequences are available (this is, for example, the case for brain MRIs of the Human Connectome Project (HCP) or from BratsReg; see Table 1 for details)[1]. These within-patient images are already aligned since they are derived from the same acquisition. We propose a training strategy to further benefit from these paired multimodal images.

Most deep registration approaches first predict the transformation map to warp the source image to the space of the target image. Image similarity is then calculated between the warped source image and the target image. If the source and target image come from a different modality, this will require a multimodal image similarity measure. If each patient has multiple paired images of different modalities, then the two simplest possible extensions of this approach are to 1) pick one specific modality per patient and proceed with a multimodal image similarity loss or 2) if all patients have the same set of modalities to simply use vector-valued images. The latter approach complicates training a universal model as it would require all datasets to share the same set of modalities which is unrealistic. Instead, we propose extending the former approach. However, we do not pick a specific modality per patient but rather pick a random modality per patient as the input to the deep registration network and another random one to compute the multimodal similarity loss. In expectation, this strategy will train the network with all possible input and loss combinations.

Formally, we assume that our dataset $D = \{P_i = \{I_i^1, I_i^2, ..., I_i^m\}\}_{i\in[n]}$ consists of m scans in different modalities for each patient P_i. We first sample a patient pair (P_A, P_B). We then uniformly sample a source and target image pair (I^A, I^B) with $I^A \in P_A, I^B \in P_B$ *and* a source and target image pair (I_L^A, I_L^B) with $I_L^A \in P_A, I_L^B \in P_B$ for similarity loss computations. The pair (I^A, I^B) is used for transformation map prediction, and (I_L^A, I_L^B) is used for similarity loss calculation. The resulting training loss is

$$\begin{aligned}\mathcal{L} = \mathcal{L}_{\text{sim}}\left(I_L^A \circ \Phi_\theta\left[I^A, I^B\right], I_L^B\right) + \mathcal{L}_{\text{sim}}\left(I_L^B \circ \Phi_\theta\left[I^B, I^A\right], I_L^A\right) + \\ + \lambda \left\|\nabla\left(\Phi_\theta\left[I^A, I^B\right] \circ \Phi_\theta\left[I^B, I^A\right]\right) - \mathbf{I}\right\|_F^2 .\end{aligned} \tag{4}$$

To explore the effects of choosing the modalities for image similarity calculations we experiment with both sampling I_L^A and I_L^B randomly or restricting the random sampling to picking the same modality for I_L^A and I_L^B. Similar to our baseline approach, we use $(1 - \text{LNCC}^2)$ as our similarity measure. We train our model with the same hyperparameters as for uniGradICON. We set $\lambda = 1.5$.

3.3 Dataset

We created a comprehensive training dataset by combining monomodal and multimodal datasets across anatomical regions; see Table 1 for details. We extend

[1] For notational simplicity, we denote multiple MR sequences also as multimodal rather than multi-sequence data.

Table 1. Datasets used for training and testing.

Dataset	Anatom. region	# of patients	# of pairs	Type	Modality	Label Randomization	% Training Set	% Finetuning Set
COPDGene [40]	Lung	899	899	Intra-pat.	CT	✗	2.12	8.33
OAI [38]	Knee	2532	7,398,400	Inter-pat.	DESS/T2 MRI	✗	6.38	12.5
HCP [50]	Brain	1076	4,605,316	Inter-pat.	T1w/T2w MRI	✓	6.38	8.33
L2R-Abdomen [56]	Abdomen	30	450	Inter-pat.	CT	✓	6.38	6.25
BratsReg [4]	Brain	140	2,240	Intra-pat.	T1w/T1ce/T2w/FLAIR MRI	✓	21.27	8.33
ABCD [7]	Brain	302	364,816	Inter-pat.	FA/MD	✓	6.38	0
L2R-AbdomenMRCT [2,12,14,32]	Abdomen	97	11,025	Inter-pat.	CT/T1w MRI	✗	12.76	6.25
UK Biobank [46]	Neck-to-Knee	90	194,400	Inter-pat.	Fat/Water DIXON	✓	38.29	27.08
L2R-ThoraxCBCT-train [27,28]	Lung	14	1,764	Inter-pat.	CT/CBCT	✗	0	8.33
Pancreatic-CT-CBCT-SEG [24]	Pancreas	40	720	Intra-pat.	CT/CBCT	✗	0	6.25
ABCD [7]	Brain	307	1,483,524	Inter-pat.	T1w/T2w/FA/MD	✗	0	8.33
Dirlab-COPDGene [8]	Lung	10	10	Intra-pat.	CT	-	0	0
OAI-test [38]	Knee	301	301	Inter-pat.	DESS MRI	-	0	0
HCP-test [50]	Brain	32	100	Inter-pat.	T1w/T2w MRI	-	0	0
L2R-NLST-val [1,12]	Lung	10	10	Intra-pat.	CT	-	0	0
L2R-OASIS-val [25,33]	Brain	20	19	Inter-pat.	T1w MRI	-	0	0
IXI-test[9]	Brain	115	115	Atlas-pat.	T1w MRI	-	0	0
L2R-ThoraxCBCT-val [27,28]	Lung	3	6	Intra-pat.	CT/CBCT	-	0	0
L2R-AbdomenMRCT-val [2,12,14,32]	Abdomen	2	3	Intra-pat.	CT/T1w MRI	-	0	0
UK Biobank-test [46]	Neck-to-Knee	10	360	Inter-pat.	Fat/Water DIXON	-	0	0
Pancreatic-CT-CBCT-SEG [24]	Pancreas	40	80	Intra-pat.	CT/CBCT	-	0	0

[9] https://brain-development.org/ixi-dataset/

the uniGradICON [47] training dataset which contains only monomodal datasets. We also used additional modalities available in the uniGradICON datasets. Further, we added new datasets containing a wide range of brain MRI sequences (T1w, T1ce, T2w, FLAIR), contrasts derived from diffusion tensors (fractional anisotropy-FA, mean diffusivity-MD), CT-T1w abdomen MRIs, and DIXON MRIs for fat and water covering anatomical regions across the entire body from neck to knee. Our final corpus is composed of 16 different datasets, contains 5 different anatomical regions (lung, knee, brain, abdomen, pancreas) in addition to a whole body MR dataset, and 12 different image modalities (T1w, T1ce, T2w, T2, FLAIR, DESS, FA, MD, CT, CBCT, Fat/Water DIXON).

Data augmentation. We utilize affine data augmentation which randomly flips the input images and applies random affine transforms, as described in [48]. Further, inverting 0–1 normalized CT scans $(1-\text{CT})$ may enhance the segmentation accuracy of CT images when using a network trained on T1w MRIs [18]. This indicates that inverted CT scans may more strongly resemble T1w MRIs. Consequently, we integrate inverted CT scans into our training L2R-Abdomen and L2R-AbdomenMRCT datasets which already include CT images. During finetuning, we only apply random affine augmentation and do not use CT inversion to further fit our model on real image modalities.

Data balancing. The number of patients, scans, and provided modalities varies across datasets. To ensure a balanced dataset with respect to the number of possible modality-anatomical region combinations, we start by randomly selecting 4,000 image pairs from each dataset. These pairs are then assigned weights based on the dataset they belong to, ensuring an equal representation of observations from each (modality/region) combination during training. We then sample 4,000 3D image pairs per epoch using weighted sampling, consistent with the number of pairs per epoch used in uniGradICON [47]. For *finetuning*, we recompute our weights to account for the additional finetuning datasets; further, we use an anatomic-region-based weighting strategy that balances the number of seen anatomical regions by equally weighting anatomical regions for finetuning.

Please refer to Tab. 1 for the diversity of the datasets and the percentages of each in the training and finetuning sets.

Data preprocessing. We clip the Hounsfield Units (HU) to the range [−1000, 1000] for all CT images and then normalize them to [0, 1]. For all MR images (T1w, T1ce, T2w, T2, FLAIR, DESS, DIXON, FA, MD), we clip the maximum intensity at the 99th percentile and then normalize them to [0, 1]. For the pancreatic CT-CBCT dataset, we follow the preprocessing steps outlined in [17]. We resize all images to a shape of [175, 175, 175] using trilinear interpolation. The spacing across the datasets varies; however, the input pairs (within a dataset) have the same spacing. During inference, we always evaluate our model on the original images by interpolating the transformation maps.

4 Results

We conduct experiments to assess multiGradICON's performance and effectiveness compared to the existing monomodal foundational registration model, uniGradICON [47], and optimization-based registration method SyN [3], using default hyperparameters and MI as the similarity measure. We analyze monomodal and multimodal performance separately.

General hypothesis and questions. We hypothesize that multiGradICON can adapt to multimodal data while maintaining comparable monomodal performance to uniGradICON. However, our goal is for multiGradICON to be appropriate for multimodal *and* monomodal registration. Hence, for the monomodal setting, we seek answers to the questions: 1) How does the performance on monomodal datasets in the training set compare between our approach and uniGradICON?; 2) What is the performance difference in additional monomodal datasets that multiGradICON trained on compared to uniGradICON?; 3) Does multiGradICON generalize to unseen cases as well as uniGradICON?

Training design questions. We also investigate the impact of different factors such as training similarity loss selection ($1 - \text{LNCC}$ or $1 - \text{LNCC}^2$), instance optimization similarity loss selection ($1 - \text{LNCC}^2$ or MIND-SSC), and training loss calculation strategy (baseline or label randomization). For this, we first train uniGradICON with $1 - \text{LNCC}^2$ to investigate the effect of loss selection on generalization to multimodal pairs. Then, we introduce three variants of multiGradICON based on their image similarity loss calculation strategy: 1) the baseline multiGradICON-B approach which uses the same image pairing as input to the network and the loss; 2) multiGradICON-F which uses loss randomization but always samples from the same modality for the loss; 3) multiGradICON-R which also uses loss randomization but allows for sampling from different modalities. Finally, we obtain multiGradICON by further training our best-performing approach multiGradICON-R by including additional datasets to the training set (ThoraxCBCT, Pancreas, ACBD Diffusion (MD or FA)-Structure (T1w or T2w)). For this further training, we do not use $1 - \text{CT}$ for data augmentation, we use lung-masked images for the COPDGene dataset, and we recompute the

Table 2. Performance comparison on the monomodal datasets used for both uni- and multiGradICON training.

		Lung				Brain		Abdomen		Knee	
		COPDGene				HCP		Abdomen CTCT		OAI	
		CT/CT (masked)		CT/CT		T1w/T1w		CT/CT		DESS/DESS	
		mTRE	$\%\|J\|_{<0}$	mTRE	$\%\|J\|_{<0}$	DICE(%)	$\%\|J\|_{<0}$	DICE(%)	$\%\|J\|_{<0}$	DICE(%)	$\%\|J\|_{<0}$
w/o IO	SyN	8.20	0	15.18	0	75.8	0	25.2	0	65.7	0
	uniGradICON	2.26	9.3e-5	6.71	5.7e-3	76.2	6.4e-5	48.3	3.1e-1	68.9	6.9e-2
	uniGradICON-LNCC^2	2.62	9.5e-5	6.59	1.3e-2	76.6	5.9e-5	49.8	3.2e-1	69.5	9.0e-2
	multiGradICON - B	5.62	1.4e-3	6.34	3.3e-3	75.6	6.4e-5	39.2	2.2e-1	64.8	2.1e-2
	multiGradICON - F	5.12	2.6e-4	5.56	2.0e-3	76.8	3.9e-5	40.5	5.1e-2	66.0	2.0e-2
	multiGradICON - R	6.18	5.0e-4	6.60	3.6e-3	76.4	4.6e-5	40.2	2.7e-1	65.2	5.6e-2
	multiGradICON	3.14	7.2e-4	5.85	4.0e-3	76.4	3.7e-5	39.5	8.6e-1	65.4	4.3e-2
1-LNCC^2	uniGradICON	1.44	2.4e-4	2.80	1.3e-3	78.4	2.0e-4	52.9	9.4e-1	69.8	4.8e-2
	uniGradICON-LNCC^2	1.46	4.2e-4	2.97	1.7e-3	78.7	1.3e-4	53.4	8.9e-1	70.2	1.0e-2
	multiGradICON - B	1.75	5.4e-5	2.65	3.6e-4	78.1	9.3e-5	46.5	9.7e-1	68.4	3.9e-2
	multiGradICON - F	1.69	1.4e-4	2.48	5.3e-4	78.4	4.9e-5	48.1	6.2e-1	69.3	1.8e-2
	multiGradICON - R	1.78	5.9e-5	2.91	3.8e-4	78.1	6.7e-5	47.6	7.2e-1	68.4	3.6e-2
	multiGradICON	1.63	1.2e-4	2.92	5.2e-4	78.2	7.6e-5	46.9	6.5e-1	68.2	3.7e-2
MIND-SSC	uniGradICON	1.77	2.6e-5	3.99	4.4e-5	77.6	3.7e-7	50.8	4.1e-1	69.3	4.9e-7
	uniGradICON-LNCC^2	1.80	6.7e-5	4.30	1.9e-4	77.7	1.6e-6	51.4	3.8e-1	69.7	9.8e-5
	multiGradICON - B	2.22	0	3.79	7.4e-6	76.8	0	42.7	3.1e-1	66.6	0
	multiGradICON - F	2.13	1.3e-5	3.39	5.5e-6	77.4	1.8e-7	44.5	8.6e-3	67.4	0
	multiGradICON - R	2.25	0	3.92	7.4e-6	76.9	1.8e-7	44.4	3.6e-2	66.4	0
	multiGradICON	2.03	1.3e-5	3.99	1.86e-6	77.1	7.4e-7	43.5	3.0e-2	66.4	0

dataset weights (see Table 1). We report results for all our methods without instance optimization (w/o IO) or with 50 steps of IO using either $1 - \text{LNCC}^2$ or MIND-SSC as similarity measures. Note that we perform all of the instance optimization operations on a given pair without any image similarity loss randomization, and we optimize the network parameters for the displacement field using an Adam optimizer with a learning rate of 2×10^{-5}.

4.1 Performance on Monomodal Registration

Here, we discuss the performance of multiGradICON on monomodal datasets compared to uniGradICON. We split our evaluations into three categories based on the evaluation datasets: 1) Datasets that exist in both uni- and multiGradICON training sets; 2) Datasets that exist only in the multiGradICON training set; 3) Unseen datasets during training for uniGradICON and multiGradICON.

Datasets used for both uni- and multiGradICON training. The uni- and multiGradICON training datasets both contain lung (COPDGene CT), brain (HCP T1w MRI), abdomen (L2R Abdomen CT), and knee (OAI MRI) images. Table 2 shows that uniGradICON outperforms multiGradICON-B,F,R on the lung, abdomen, and knee datasets based on the initial prediction without instance optimization. This performance difference is around ~3 mm for COPDGene, ~8.5% Dice score for the abdomen, and ~3% Dice score for the knee dataset. This result is expected, as uniGradICON is exclusively trained on these datasets and thus has better expertise in these areas. Conversely, multiGradICON is trained on diverse datasets where these specific tasks have lower weight during training. This is further supported by the brain registration results, where multiGradICON-B,F,R show similar performance, with multiGradICON-F even outperforming uniGradICON-LNCC^2 by a ~0.2% Dice score. Since brain datasets are more prevalent in the training sets (e.g., ABCD and BratsReg), multiGradICON performs similarly to uniGradICON on brain registration. After finetuning with anatomical region-based sampling, we observe a performance improvement on the COPDGene dataset, which forms 2.1% of the training set but is sampled at 8.3% during finetuning. The performance on the remaining datasets remains similar since their percentages do not change drastically. We observe similar performance improvement on the unseen NLST lung dataset (see Sect. 4.1).

Instance optimization with $1 - \text{LNCC}^2$ narrows the performance gap between uniGradICON and multiGradICON. The difference decreases to ~0.19 mm for COPDGene, ~6% Dice score for the abdomen, and ~1.6% Dice score for the knee dataset. Instance optimization with MIND-SSC always under-performs instance optimization with $1 - \text{LNCC}^2$ across all these datasets.

Different multimodal loss strategies perform differently for monomodal registration. Using the same modalities multiGradICON-F improves performance on the datasets to which this strategy is applied, compared to the baseline multiGradICON-B and random modality sampling in multiGradICON-R.

Lung masking also affects registration performance. We train our multiGradICON variants using full lung CT images, whereas uniGradICON uses masked images that are zeroed out outside the lung. We observe that a registration model trained without lung masking cannot generalize to register fine details of the lung even if we provide masked lungs during inference. Therefore, during the finetuning process, we further train our model with region of interest (ROI)-masked lung images. After that, we achieve approximately 3 mm improvement in mTRE on masked lung registration without IO. Additionally, we observe approximately 0.75 mm performance improvement on non-masked lung images, even though we do not incorporate any non-masked lung images in the finetuning process. We hypothesize that both the sampling amount in diverse datasets and ROI-masked training are crucial for achieving good registration performance.

Monomodal datasets that only exist in multiGradICON training. We additionally introduce new monomodal datasets to the multiGradICON training while retaining the existing uniGradICON training datasets. These new monomodal datasets comprise a wide variety of image modalities, such as T2w MRI, FLAIR, and DIXON, which have not been previously seen by uniGradICON. Table 3 shows the results across these datasets. Overall, the multiGradICON variants perform slightly better than uniGradICON on the Brats-Reg and UK Biobank datasets, since multiGradICON is trained on these domains. However, both approaches converge to similar performance after 50 steps of IO. These results demonstrate that even on previously unseen monomodal domains, uniGradICON remains a strong baseline, while multiGradICON performs better in initial predictions on the modalities it has seen during training.

Table 3. Comparison on monomodal datasets seen by multiGradICON but not by uniGradICON during training.

		Brain										Neck to Knee			
		HCP		Brats-Reg								UK Biobank			
		T2w/T2w		T1w/T1w		T2w/T2w		T1ce/T1ce		FLAIR/FLAIR		WDIXON/WDIXON		FDIXON/FDIXON	
		DICE(%)	$\%\|J\|_{<0}$	mTRE	$\%\|J\|_{<0}$	mTRE	$\%\|J\|_{<0}$	mTRE	$\%\|J\|_{<0}$	mTRE	$\%\|J\|_{<0}$	DICE(%)	$\%\|J\|_{<0}$	DICE(%)	$\%\|J\|_{<0}$
w/o IO	SyN	75.6	0	3.50	0	3.39	0	3.42	0	3.73	0	47.7	0	43.7	0
	uniGradICON	76.9	5.6e-4	3.27	1.0e-3	3.31	1.2e-3	3.24	1.4e-3	3.83	1.9e-3	42.2	8.1e-3	40.0	1.6e-2
	uniGradICON-LNCC²	77.3	5.0e-4	3.22	0	3.21	0	3.13	0	3.79	0	42.4	1.6e-2	40.5	3.6e-2
	multiGradICON - B	76.3	1.0e-4	3.10	6.1e-4	3.04	1.3e-3	2.91	6.9e-4	3.35	1.1e-3	43.6	3.9e-2	42.1	1.6e-2
	multiGradICON - F	77.2	6.8e-5	2.94	7.4e-4	2.95	1.6e-3	2.73	9.2e-4	3.14	1.3e-3	45.5	1.8e-2	44.5	1.2e-2
	multiGradICON - R	76.6	6.5e-5	3.07	1.2e-4	3.04	2.1e-3	2.87	1.3e-3	3.33	1.9e-3	43.6	4.4e-2	41.9	3.7e-2
	multiGradICON	76.5	1.4e-4	3.06	8.8e-4	3.04	1.6e-3	2.90	1.2e-3	3.38	1.6e-3	43.8	2.3e-2	42.1	2.0e-2
1-LNCC²	uniGradICON	77.5	6.1e-4	2.93	7.7e-4	2.84	1.1e-3	2.48	9.4e-4	3.02	1.9e-3	47.0	8.5e-3	45.2	6.7e-3
	uniGradICON-LNCC²	77.9	6.5e-4	2.92	8.8e-4	2.81	1.1e-3	2.45	9.1e-4	2.97	1.8e-3	46.8	1.3e-2	45.0	1.9e-2
	multiGradICON - B	77.2	1.0e-4	2.94	7.3e-4	2.79	1.0e-3	2.50	7.7e-4	2.99	1.3e-3	47.9	5.7e-3	46.2	5.6e-3
	multiGradICON - F	77.6	1.3e-4	2.92	8.1e-4	2.80	1.2e-3	2.49	9.9e-4	2.95	1.6e-3	48.6	5.2e-3	47.6	6.7e-3
	multiGradICON - R	77.2	1.6e-4	2.92	8.8e-4	2.79	1.1e-3	2.47	9.7e-4	2.99	1.6e-3	47.1	6.0e-3	45.4	8.1e-3
	multiGradICON	77.2	2.6e-4	2.92	9.7e-4	2.81	1.2e-3	2.48	1.0e-3	2.99	1.9e-3	47.3	6.9e-3	45.8	1.0e-2
MIND-SSC	uniGradICON	77.2	7.8e-6	2.70	2.1e-6	2.54	0	2.20	0	2.62	0	45.0	0	43.1	1.0e-7
	uniGradICON-LNCC²	77.5	2.0e-5	2.66	3.5e-5	2.48	0	2.14	0	2.55	2.6e-7	44.7	5.2e-8	43.1	0
	multiGradICON - B	76.7	0	2.72	0	2.53	0	2.23	9.3e-7	2.62	0	45.6	0	44.1	6.7e-7
	multiGradICON - F	77.2	7.4e-7	2.72	0	2.52	0	2.23	0	2.63	3.9e-7	46.8	0	46.0	0
	multiGradICON - R	76.6	0	2.69	0	2.53	1.3e-7	2.23	1.3e-7	2.64	6.2e-6	44.8	2.1e-7	43.5	0
	multiGradICON	76.6	7.4e-7	2.69	0	2.53	0	2.22	2.6e-7	2.62	1.3e-7	45.1	1.6e-7	43.7	0

Table 4. Performance comparison on unseen mono- and multimodal datasets for both uni- and multiGradICON.

		Lung				Brain				Pancreas	
		NLST		ThoraxCBCT		IXI		OASIS		Pancreatic-CT-CBCT-SEG	
		CT/CT		CT/CBCT		T1w/T1w		T1w/T1w		CT/CBCT	
		mTRE	$\%\|J\|_{<0}$	mTRE	$\%\|J\|_{<0}$	DICE(%)	$\%\|J\|_{<0}$	DICE(%)	$\%\|J\|_{<0}$	DICE(%)	$\%\|J\|_{<0}$
w/o IO	SyN	3.04	9.8-e1	57.4	0	64.5	1.0e-4	75.6	1.5e-2	78.2	0
	uniGradICON	2.07	4.7e-4	57.0	4.7e-4	70.6	7.4e-3	79.0	8.9e-4	81.1	6.9e-2
	uniGradICON-LNCC²	2.00	0	61.0	2.8e-3	69.7	2.1e-3	79.6	2.8e-3	81.0	8.1e-2
	multiGradICON - B	2.74	0	58.1	3.6e-3	69.9	2.2e-4	78.6	1.7e-3	80.9	4.1e-2
	multiGradICON - F	2.42	0	59.9	8.4e-3	71.6	2.9e-4	79.2	1.7e-3	80.5	2.3e-2
	multiGradICON - R	2.66	0	58.9	3.8e-2	70.7	3.4e-4	78.5	2.3e-3	80.6	5.6e-2
	multiGradICON	2.27	0	58.7	3.2e-3	71.0	1.8e-3	78.7	2.0e-3	81.8	2.1e-2
1-LNCC²	uniGradICON	1.77	8.7e-5	60.9	2.3e-1	70.4	1.5e-3	79.7	6.5e-3	82.2	2.4e-2
	uniGradICON-LNCC²	1.76	4.8e-5	62.1	2.5e-2	70.8	1.6e-3	80.1	9.1e-3	82.0	4.2e-2
	multiGradICON - B	1.84	3.1e-4	60.1	2.0e-2	70.8	1.0e-3	79.5	5.6e-3	82.0	1.9e-2
	multiGradICON - F	1.82	2.1e-4	61.4	4.3e-1	70.7	1.1e-3	79.9	6.9e-3	82.3	7.9e-3
	multiGradICON - R	1.86	2.3e-4	60.2	2.7e-1	70.7	1.4e-3	79.6	7.1e-3	82.1	8.8e-3
	multiGradICON	1.79	2.7e-5	60.6	2.8e-1	71.0	1.8e-3	79.6	6.3e-3	82.2	8.3e-3
MIND-SSC	uniGradICON	1.87	0	57.9	2.3e-2	71.7	1.6e-6	78.9	3.4e-5	82.0	1.1e-6
	uniGradICON-LNCC²	1.84	0	58.3	1.8e-2	72.2	8.9e-7	79.3	0	81.8	1.5e-5
	multiGradICON - B	1.99	0	59.4	9.1e-1	72.0	2.5e-7	78.6	3.1e-6	81.5	3.4e-6
	multiGradICON - F	1.95	0	63.5	1.4e-3	71.7	1.0e-6	79.1	0	81.8	0
	multiGradICON - R	2.03	0	63.0	6.6e-6	71.6	3.2e-6	78.7	2.3e-6	81.7	2.3e-7
	multiGradICON	1.94	0	63.4	0	71.8	4.3e-6	78.2	0	81.9	0

Unseen monomodal datasets. We also test on datasets that are never seen during training. Table 4 shows performance metrics for lung CT-CT NLST and brain T1w MRI registrations from the IXI and OASIS datasets. For NLST, uniGradICON achieves an mTRE of 2.07 mm, whereas the multiGradICON-B,F,R variants show a range between 2.42 and 2.74 mm. We observe a performance improvement on the NLST dataset of approximately 0.4 mm after finetuning.

Instance optimization with $1 - \text{LNCC}^2$ decreases the performance gap to ∼0.02 mm on the NLST dataset. However, for brain registration,

Table 5. Performance comparison on the multimodal datasets used for multiGradICON training.

		Brain														Abdomen		Neck to Knee	
		HCP		Brats-Reg												Abdomen MRCT		UK Biobank	
		T1w/T2w		T1w/T2w		T1w/T1ce		T1w/FLAIR		T2w/T1ce		T2w/FLAIR		T1ce/FLAIR		MR/CT		FDIXON/WDIXON	
		DICE(%)	$\%\|J\|_{<0}$	mTRE	$\%\|J\|_{<0}$	mTRE	$\%\|J\|_{<0}$	mTRE	$\%\|J\|_{<0}$	mTRE	$\%\|J\|_{<0}$	mTRE	$\%\|J\|_{<0}$	mTRE	$\%\|J\|_{<0}$	DICE(%)	$\%\|J\|_{<0}$	DICE(%)	$\%\|J\|_{<0}$
w/o IO	SyN	71.9	0	3.74	0	3.62	0	3.91	0	3.74	0	3.96	0	3.88	0	45.0	0	33.9	0
	uniGradICON	10.0	2.6e-3	13.11	3.7e-3	4.16	2.1e-3	6.38	2.3e-3	12.30	4.6e-3	8.12	5.0e-3	6.92	3.9e-3	50.0	4.0e-2	17.5	2.4e-2
	uniGradICON-LNCC²	14.5	8.6e-4	13.10	4.9e-3	4.08	4.2e-3	6.33	3.2e-3	12.24	6.8e-3	7.95	6.0e-3	6.83	6.3e-3	49.7	2.6e-2	18.1	4.4e-2
	multiGradICON - B	69.2	8.0e-4	3.60	1.7e-3	3.84	8.1e-4	4.60	1.2e-3	3.73	2.4e-3	4.95	2.9e-3	4.95	2.3e-3	58.1	3.4e-2	19.7	1.2e-2
	multiGradICON - F	40.1	1.0e-3	9.10	2.2e-3	3.66	1.2e-3	5.06	1.4e-3	8.56	2.8e-3	6.50	3.2e-3	5.60	2.3e-3	58.2	6.0e-2	20.1	1.2e-2
	multiGradICON - R	68.2	9.0e-4	3.58	2.3e-3	3.78	1.5e-3	4.48	2.1e-3	3.70	2.9e-3	4.90	3.8e-3	4.76	3.1e-3	61.1	8.4e-2	19.8	2.9e-2
	multiGradICON	69.0	7.9e-4	3.64	2.5e-3	3.83	1.6e-3	4.68	1.9e-3	3.78	3.4e-3	5.05	4.1e-3	5.04	3.3e-3	61.8	4.0e-2	19.7	1.3e-2
1-LNCC²	uniGradICON	7.8	3.3e-3	12.06	3.2e-3	3.63	1.6e-3	5.13	2.0e-3	11.69	3.8e-3	6.81	4.8e-3	5.71	3.3e-3	75.1	6.9e-1	17.0	6.9e-3
	uniGradICON-LNCC²	8.3	2.4e-3	11.94	2.7e-3	3.59	1.8e-3	5.05	1.9e-3	11.48	3.2e-3	6.60	4.5e-3	5.63	3.2e-3	80.3	5.8e-1	17.4	8.6e-3
	multiGradICON - B	72.8	6.3e-4	3.27	7.4e-4	3.56	1.3e-3	4.28	1.1e-3	3.34	1.1e-3	4.58	2.5e-3	4.76	2.1e-3	73.3	4.4e-1	18.7	5.5e-3
	multiGradICON - F	71.9	4.9e-4	3.34	9.5e-4	3.55	1.5e-3	4.29	1.4e-3	3.39	1.5e-3	4.57	3.1e-3	4.78	2.6e-3	74.8	3.0e-1	18.9	6.9e-3
	multiGradICON - R	72.8	5.7e-4	3.30	9.4e-4	3.55	1.5e-3	4.24	1.4e-3	3.37	1.4e-3	4.55	2.8e-3	4.70	2.5e-3	74.8	1.6e-1	18.6	6.6e-3
	multiGradICON	73.1	9.3e-4	3.32	1.1e-3	3.56	1.8e-3	4.27	1.6e-3	3.38	1.8e-3	4.60	3.2e-3	4.72	3.0e-3	73.3	2.0e-1	18.6	8.0e-3
MIND-SSC	uniGradICON	39.8	1.4e-5	2.66	0	2.61	8.7e-7	2.81	1.9e-6	2.55	6.7e-8	2.88	6.7e-8	2.75	7.3e-7	75.4	1.9e-4	20.2	0
	uniGradICON-LNCC²	53.0	9.5e-6	2.58	6.7e-8	2.55	6.7e-8	2.74	2.7e-7	2.47	6.7e-8	2.80	0	2.68	6.7e-8	77.4	1.7e-3	20.8	3.1e-7
	multiGradICON - B	74.5	0	2.70	0	2.61	8.0e-7	2.83	6.7e-8	2.62	0	2.94	0	2.79	3.3e-7	68.1	5.0e-3	23.3	0
	multiGradICON - F	73.0	9.3e-7	2.72	0	2.62	0	2.87	6.7e-8	2.65	2.7e-7	2.94	2.0e-7	2.82	0	70.9	3.2e-4	25.2	0
	multiGradICON - R	74.3	0	2.73	0	2.62	6.7e-8	2.88	6.7e-7	2.66	0	2.97	0	2.83	6.7e-8	70.3	1.0e-3	23.7	0
	multiGradICON	74.4	1.4e-6	2.70	1.3e-7	2.62	1.2e-6	2.85	1.6e-6	2.62	0	2.94	1.3e-7	2.80	4.7e-7	70.8	2.7e-4	23.3	0

multiGradICON shows similar performance to uniGradICON, outperforming it by a ~0.6% Dice score on the IXI dataset and underperforming it by a ~0.1% Dice score on the OASIS dataset. These results show that multiGradICON scales well to monomodal tasks, providing performance close to that of uniGradICON, which *specializes* in monomodal registration.

4.2 Performance on Multimodal Registration

In this section, we evaluate the multimodal registration performance of multiGradICON on several datasets including a wide variety of anatomical structures and modalities. We again investigate its performance on both seen and unseen datasets. We use uniGradICON as our main comparison model.

Multimodal training datasets. Our training dataset consists of several anatomical regions and modalities such as T1w, T1ce, T2w, FLAIR brain MRIs, CT abdominal scans, and fat and water-weighted DIXON images (see Table 1 for details). Table 5 shows registration performances for the HCP, Brats-Reg, Abdomen MR/CT, and UK Biobank datasets. In both the HCP and Brats-Reg datasets, we observe that uniGradICON fails to register pairs that contain images with large appearance differences. For instance, uniGradICON cannot register pairs containing T2w images, even with instance optimization. This is one of the key problems that we aim to solve with multiGradICON. We observe a significant performance improvement on the HCP (~59% Dice score improvement), Brats-Reg, and MR-CT (~11% Dice score improvement) datasets without instance optimization with our multiGradICON approach.

On the other hand, the UK Biobank fat-water weighted DIXON registration is challenging due to the different underlying information across modalities. Although this is a clinically irrelevant scenario since they are acquired together, this result demonstrates the limitation of our approach. We will address this type of registration, where the pairs share the same anatomies but capture entirely different properties, in future work by incorporating semantic information. Additionally, we note that we obtain UK Biobank segmentations using MRSegmentator [19] and evaluate our models directly based on MRSegmentator predictions. Therefore, we can only provide a silver-standard performance metric that may also include possible segmentation errors arising from the MRSegmentator.

We observe that multiGradICON-F cannot generalize to multimodal pairs compared to the other multiGradICON variants. It shows poor performance on several registration tasks such as T1w-T2w, T1ce-T2w, and T1ce-FLAIR brain registration. Although multiGradICON-F performs better than uniGradICON, these results suggest that sampling the same modality for similarity loss calculation causes a performance drop on multimodal datasets, while it improves the performance on monomodal pairs, as we discussed in Sect. 4.1. Additionally, we do not observe any significant multimodal performance improvement in uniGradICON when $1 - \text{LNCC}^2$ is used as similarity loss during training. This indicates that using appropriate losses for multimodal registration may not lead to multimodal generalization if multimodal datasets were not used during training. We hypothesize that the diversity of the training dataset is more important for generalization.

Moreover, we observe improved registration results with instance optimization (IO) using the MIND-SSC loss in the Brats-Reg dataset compared to IO with $1 - \text{LNCC}^2$. The Brats-Reg dataset consists of pre-operative and follow-up brain scans that include tumors with varying shapes and resections where each modality captures different tumor properties. As MIND-SSC provides better alignment for these inconsistent structures than $1 - \text{LNCC}^2$, we conclude that task-specific similarity loss selection can be important for good multimodal registration.

Unseen multimodal datasets. During the training of multiGradICON, we initially did not include the ThoraxCBCT and pancreatic CT-CBCT datasets in our training set. Although the performances of uniGradICON and multiGradICON-B,F,R on these datasets are close to each other, uniGradICON underperforms by $\sim$2% Dice scores on the ThoraxCBCT dataset and outperforms by $\sim$0.5% Dice score on the pancreatic CT-CBCT dataset. However, finetuning that incorporates the pancreatic dataset leads to a better Dice score for multiGradICON which outperforms uniGradICON by $\sim$0.7% Dice score. The performance on the ThoraxCBCT dataset remains similar even when it is included in the training set during finetuning. We hypothesize that our model already converges on CT images during the initial training and that additional CBCT images do not affect the performance since they look sufficiently similar to CT images.

5 Conclusion

We developed multiGradICON, a *universal* deep network for *mono- and multimodal* registration. multiGradICON extends uniGradICON, the first deep medical registration network for *monomodal* registration across different anatomies.

We observed that uniGradICON remains a strong baseline for monomodal registration but multiGradICON shows improved performance for multimodal registration, in particular, when modalities look so different that not even instance optimization recovers good registrations for uniGradICON. We also demonstrated that similarity loss randomization can bring multimodal registration benefits. Although multiGradICON showed encouraging performance, there are many different avenues for future improvements. For example, we only investigated using $1 - \text{LNCC}^2$ as a training loss, while many other multimodal similarity measures exist. Further, multiGradICON cannot reliably register between DIXON fat and water images. Although this is somewhat expected, it points to the existence of multimodal datasets that share underlying anatomy but have such large appearance differences that a similarity measure such as $1 - \text{LNCC}^2$ is insufficient. Using segmentations as part of the image similarity loss may show benefits in such cases. In general, training with segmentations [55] or simulated data for the network input and the similarity loss could be a fruitful avenue for future work. Lastly, just like uniGradICON, multiGradICON is built on top of the exact same network architecture as GradICON. Exploring networks with increased capacity and further increasing training dataset sizes would be desirable.

Acknowledgements. This work was supported by NIH grants 1R01AR072013, 1R01AR082684, 1R01EB028283, 1R21MH132982, RF1MH126732, 1R01HL149877, 5R21LM013670, and R01NS125307. The work expresses the views of the authors, not of NIH. RK was supported in part by projects EXDIGIT 20204-WISS/263/6-6022, 0102-F1901166-KZP, and 20204-WISS/225/197-2019; SB by NSERC grants RGPIN-2023-05443 and CRC-2022-00183. Data are from the following sources. Knee: Osteoarthritis Initiative: NIMH Data Archive Collection ID: 2343; Brain: Human Connectome Project (1U54MH091657), IXI (EPSRC GR/S21533/02), the Brain Tumor Sequence Registration challenge, and the Adolescent Brain Cognitive Development^{SM} (ABCD) Study. ABCD consortium investigators did not participate in the analysis or writing of this article. The ABCD data used for this work can be found at https://doi.org/10.15154/8v6w-yr62. Lung: COPDGene. Abdomen: The Cancer Imaging Archive (https://doi.org/10.7937/TCIA.ESHQ-4D90). Neck to Knee: UK Biobank under Application Number 22783. We also used the Learn2Reg challenge data across various anatomies.

References

1. Aberle, D.R., et al.: Reduced lung-cancer mortality with low-dose computed tomographic screening. New Engl. J. Med. **365**(5), 395–409 (2011)
2. Akin, O., et al.: Radiology data from the cancer genome atlas kidney renal clear cell carcinoma [TCGA-KIRC] collection. Cancer Imaging Arch. **1310** (2016)
3. Avants B.B., et al.: Symmetric diffeomorphic image registration with cross-correlation: evaluating automated labeling of elderly and neurodegenerative brain. Med. Image Analysis **12**(1), 26–41 (2008)
4. Baheti, B., et al.: The brain tumor sequence registration challenge: establishing correspondence between pre-operative and follow-up MRI scans of diffuse glioma patients. arXiv:2112.06979 (2021)

5. Balakrishnan, G., et al.: VoxelMorph: a learning framework for deformable medical image registration. TMI **38**(8), 1788–1800 (2019)
6. Cao, X., et al.: Deep learning based inter-modality image registration supervised by intra-modality similarity. In: MLMI/MICCAI, pp. 55–63 (2018)
7. Casey, B.J., et al.: The adolescent brain cognitive development study: imaging acquisition across 21 sites. Dev. Cogn. Neurosci. **32**, 43–54 (2018)
8. Castillo, R., et al.: A reference dataset for deformable image registration spatial accuracy evaluation using the COPDgene study archive. Phys. Med. Biology **58**(9), 2861 (2013)
9. Chen, J., et al.: A survey on deep learning in medical image registration: New technologies, uncertainty, evaluation metrics, and beyond. arXiv:2307.15615 (2023)
10. Cheng, X., et al.: Deep similarity learning for multimodal medical images. Comput. Methods Biomech. Biomed. Eng. Imaging Vis. **6**(3), 248–252 (2018)
11. Çiçek, Ö., et al.: 3D U-Net: learning dense volumetric segmentation from sparse annotation. In: MICCAI, pp. 424–432 (2016)
12. Clark, K., et al.: The cancer imaging archive (TCIA): maintaining and operating a public information repository. J. Digital Imaging **26**, 1045–1057 (2013)
13. Demir, B., et al.: Multimodal image registration guided by few segmentations from one modality. In: MIDL (2024)
14. Erickson, B.J., et al.: The cancer genome atlas liver hepatocellular carcinoma collection (TCGA-LIHC) (2016)
15. Greer, H., et al.: ICON: learning regular maps through inverse consistency. In: ICCV (2021)
16. Guo, C.K.: Multi-modal image registration with unsupervised deep learning. Ph.D. thesis, Massachusetts Institute of Technology (2019)
17. Han, X., et al.: Deep-learning-based image registration and automatic segmentation of organs-at-risk in cone-beam CT scans from high-dose radiation treatment of pancreatic cancer. Med. Phys. **48**(6), 3084–3095 (2021)
18. Häntze, H., et al.: Improve cross-modality segmentation by treating MRI images as inverted CT scans. arXiv:2405.03713 (2024)
19. Häntze, H., et al.: Mrsegmentator: Robust multi-modality segmentation of 40 classes in MRI and CT sequences. arXiv:2405.06463 (2024)
20. Heinrich, M.P., et al.: MIND: modality independent neighbourhood descriptor for multi-modal deformable registration. Med. Image Analysis **16**(7), 1423–1435 (2012)
21. Heinrich, M.P., et al.: Towards realtime multimodal fusion for image-guided interventions using self-similarities. In: MICCAI, pp. 187–194. Springer (2013)
22. Hermosillo, G., et al.: Variational methods for multimodal image matching. IJCV **50**(3), 329–343 (2002)
23. Hoffmann, M., et al.: SynthMorph: learning contrast-invariant registration without acquired images. TMI **41**(3), 543–558 (2021)
24. Hong, J., et al.: Breath-hold CT and cone-beam CT images with expert manual organ-at-risk segmentations from radiation treatments of locally advanced pancreatic cancer [data set]. TCIA https://doi.org/10.7937/TCIA. ESHQ-4D90 (2021)
25. Hoopes, A., et al.: Hypermorph: amortized hyperparameter learning for image registration. In: IPMI (2021)
26. Hu, Y., et al.: Weakly-supervised convolutional neural networks for multimodal image registration. MedIA **49**, 1–13 (2018)
27. Hugo, G.D., et al.: Data from 4D lung imaging of NSCLC patients. Cancer Imaging Arch. **10**, K9 (2016)

28. Hugo, G.D., et al.: A longitudinal four-dimensional computed tomography and cone beam computed tomography dataset for image-guided radiation therapy research in lung cancer. Med. Phys. **44**(2), 762–771 (2017)
29. J. E. Iglesias: A ready-to-use machine learning tool for symmetric multi-modality registration of brain MRI. Sci. Rep. **13**(1), 6657 (2023)
30. Lee, D., et al.: Learning similarity measure for multi-modal 3D image registration. In: CVPR, pp. 186–193 (2009)
31. Li, Z., et al.: Samconvex: fast discrete optimization for CT registration using self-supervised anatomical embedding and correlation pyramid. In: MICCAI, pp. 559–569 (2023)
32. Linehan, M., et al.: The cancer genome atlas cervical kidney renal papillary cell carcinoma collection (TCGA-KIRP) (2016)
33. Marcus, D.S., et al.: Open access series of imaging studies (OASIS): cross-sectional MRI data in young, middle aged, nondemented, and demented older adults. J. Cognitive Neurosci.**19**(9), 1498–1507 (2007)
34. Modersitzki, J.: Numerical methods for image registration (2003)
35. Mok, T.C., et al.: Fast symmetric diffeomorphic image registration with convolutional neural networks. In: CVPR (2020)
36. Mok, T.C., et al.: Large deformation diffeomorphic image registration with Laplacian pyramid networks. In: MICCAI (2020)
37. Mok, T.C., et al.: Modality-agnostic structural image representation learning for deformable multi-modality medical image registration. arXiv:2402.18933 (2024)
38. Nevitt, M., et al.: The osteoarthritis initiative. Protoc. Cohort study **1**, 737 (2006)
39. Qin, C., et al.: Unsupervised deformable registration for multi-modal images via disentangled representations. In: IPMI, pp. 249–261 (2019)
40. Regan, E.A., et al.: Genetic epidemiology of COPD (COPDGene) study design. COPD: J. Chronic Obstructive Pulm. Dis. **7**(1), 32–43 (2011)
41. Roy, S., et al.: Magnetic resonance image example-based contrast synthesis. TMI **32**(12), 2348–2363 (2013)
42. Shen, Z., et al.: Networks for joint affine and non-parametric image registration. In: CVPR (2019)
43. Siebert, H., et al.: Learning a metric for multimodal medical image registration without supervision based on cycle constraints. Sensors **22**(3), 1107 (2022)
44. Simonovsky, M., et al.: A deep metric for multimodal registration. In: MICCAI, pp. 10–18 (2016)
45. Song, X., et al.: Cross-modal attention for multi-modal image registration. MedIA **82**, 102612 (2022)
46. Sudlow, C., et al.: UK biobank: an open access resource for identifying the causes of a wide range of complex diseases of middle and old age. PLoS Med. **12**(3), e1001779 (2015)
47. Tian, L., et al.: uniGradICON: A foundation model for medical image registration. arXiv:2403.05780 (2024)
48. Tian, L., et al.: GradICON: approximate diffeomorphisms via gradient inverse consistency. In: CVPR (2023)
49. Tian, L., et al.: SAME++: A self-supervised anatomical embeddings enhanced medical image registration framework using stable sampling and regularized transformation. arXiv:2311.14986 (2024)
50. Van Essen, D.C., et al.: The Human Connectome Project: a data acquisition perspective. Neuroimage **62**(4), 2222–2231 (2012)
51. Viola, P., et al.: Alignment by maximization of mutual information. IJCV **24**(2), 137–154 (1997)

52. Wachinger, C., et al.: Entropy and Laplacian images: structural representations for multi-modal registration. MedIA **16**(1), 1–17 (2012)
53. Xiao, H., et al.: A review of deep learning-based three-dimensional medical image registration methods. Quant. Imaging Med. Surg. **11**(12), 4895 (2021)
54. Xu, Z., et al.: Adversarial uni-and multi-modal stream networks for multimodal image registration. In: MICCAI, pp. 222–232 (2020)
55. Xu, Z., et al.: DeepAtlas: joint semi-supervised learning of image registration and segmentation. In: MICCAI, pp. 420–429 (2019)
56. Xu, Z., et al.: Evaluation of six registration methods for the human abdomen on clinically acquired CT. TBE **63**(8), 1563–1572 (2016)
57. Yan, P., et al.: Adversarial image registration with application for MR and TRUS image fusion. In: MLMI/MICCAI, pp. 197–204 (2018)
58. Yang, X., et al.: Fast predictive multimodal image registration. In: ISBI, pp. 858–862 (2017)
59. Yang, X., et al.: Quicksilver: fast predictive image registration–a deep learning approach. NeuroImage **158**, 378–396 (2017)

XSynthMorph: Generative-Guided Deformation for Unsupervised Ill-Posed Volumetric Recovery

Alexandre Cafaro[1,2,3](✉), Amaury Leroy[1,2,3], Guillaume Beldjoudi[3], Pauline Maury[2], Charlotte Robert[2], Eric Deutsch[2], Vincent Grégoire[3], Vincent Lepetit[4], and Nikos Paragios[1]

[1] TheraPanacea, Paris, France
a.cafaro@therapanacea.eu
[2] Gustave Roussy, Inserm 1030, Paris-Saclay University, Villejuif, France
[3] Department of Radiation Oncology, Centre Léon Bérard, Lyon, France
[4] LIGM, Ecole des Ponts, Univ Gustave Eiffel, CNRS, Champs-sur-Marne, France

Abstract. We introduce a novel unsupervised approach to recovering and registering a 3D volume from only two planar projections that exploits a previously-captured 3D volume of the patient. Such precapturing volume is readily available in many important medical procedures and previous methods already used such a volume. Earlier methods that work by deforming this volume to match the projections can fail when the number of projections is very low as the alignment becomes underconstrained. We show how to use a generative model of the volume structures to constrain the deformation and obtain a correct estimate. Moreover, our method is independant of the number, calibration and geometry of projections and could be adapted to new configurations without retraining. We evaluate our approach on a challenging dataset and show it outperforms state-of-the-art methods. As a result, our method could be used in treatment scenarios such as surgery and radiotherapy while drastically reducing patient radiation exposure.

1 Introduction

In medical treatments, the reconstruction and registration of a partial 3D volume of the patient's body is essential for both initial treatment planning and ongoing monitoring of the patient's progress. This process, crucial for accurately targeting areas affected by tumors in interventions like radiotherapy or surgery, traditionally requires numerous X-ray projections, leading to significant radiation exposure and extended time commitments. In practice, however, only two orthogonal planar projections are captured to reduce radiation exposure; they are used for rigid registration, a basic form of alignment.

Supplementary Information The online version contains supplementary material available at https://doi.org/10.1007/978-3-031-73480-9_2.

M. Modat et al. (Eds.): WBIR 2024, LNCS 15249, pp. 19–33, 2024.
https://doi.org/10.1007/978-3-031-73480-9_2

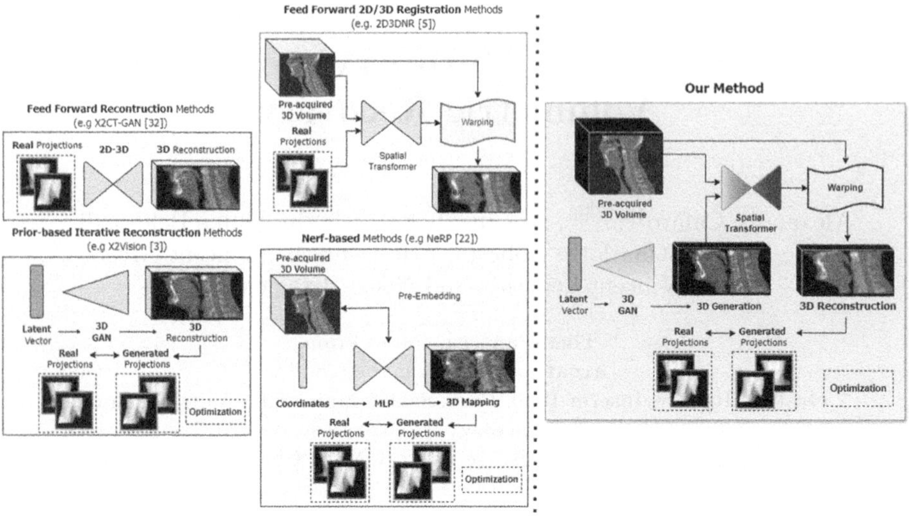

Fig. 1. Our method can recover an accurate 3D volume given two projections of the patient and a volume acquired at the beginning of the therapy, to unlock better therapy procedures. Combining correctly these sources of information is however challenging: **(1)** Feed-forward reconstruction [11,13,18,23–26,31,32,34] directly predict a volume from a set of projections. Their performance significantly degrades when the number of projections is very low. **(2)** Methods such as 2D3DNR [5] deform a pre-acquired volume of the patient based on input projections. However, the predicted deformation can become under-constrained when the number of projections is very low as two. **(3)** Methods such as X2Vision [3] first learn a volume generative model and optimize the parameters of this model to match the projections. However, they are not able to exploit the pre-acquired volume. **(4)** NeRF-based methods [22] can take the pre-acquired volume as input and optimize on the volume to match the projections—however, they do not exploit any anatomical knowledge besides the pre-acquired volume. **(5)** To avoid predicting incorrect deformations when the number of projections gets too low as one or two, we propose to guide the deformations using a volume generative model. Note that this is different from X2Vision, which relies on a generative model to directly create the predicting volume. Instead, our approach deforms the pre-acquired volume under the guarantee that the resulting volume is anatomically plausible.

However, using only two projections currently falls short of providing the full details of 3D reconstructions. Despite attempts to accurately reconstruct anatomy with few projections, existing methods with just two projections lack the necessary detail for radiotherapy. We aim to develop a method that delivers precise reconstructions with only two projections, balancing the need for accuracy in treatment with reduced radiation exposure and shorter procedure times.

Figure 1 illustrates existing approaches to decreasing the number of projections when reconstructing a volume. One approach is to train a deep model to regress the volume from a number of projections in a supervised way [11,13,18,23–27,31,32,34,35], but such direct inference often results in poor reconstruction.

Instead, X2Vision [3] learns an anatomical prior and optimizes its parameters to match the projections. Such optimization approaches (see also [22]) tend to generalize much better.

In modern medical practice, CT and MRI scans are now widely used for treatment planning and diagnostics. This provides a volume captured under a different patient pose and different from the volume to reconstruct by medically-relevant changes such as weight loss or tumor transformation. X2Vision ignores this pre-captured volume but 2D3DNR [5] deforms it given several projections.

How can we exploit both sources of information, anatomy knowledge from a generative model and the pre-acquired volume, to improve the reconstruction quality to the point it can actually be used for medical applications? This is the goal of our method.

Getting the best of these two worlds is however not straightforward as methods like X2Vision and 2D3DNR work in very different ways. We start by observing that when 2D3DNR deforms the pre-acquired volume given very few projections, unrealistic deformations may occur. We therefore propose to guide the deformations using a volume generative model. Note that this is different from X2Vision, which directly optimizes the generative model parameters to match the projections: Our approach allows to deform the pre-acquired volume under the guarantee that the resulting volume is anatomically possible.

To do so, we optimize over the generative model parameters so that the pre-acquired model matches well the projections after being mapped to the generated volume. This is illustrated in Fig. 1(5). Compared to 2D3DNR for example, we guarantee that the deformed pre-acquired volume is anatomically possible, since it is constrained to be close to a generated (thus anatomically correct) volume. Moreover, we also have the guarantee that the deformed volume match well the projections. Compared to X2Vision, because our approach predicts the pre-acquired volume after deformation, it captures the patient's unique anatomy or abnormalities accurately. This is by contrast with a generated volume, as done by X2Vision, which often lacks critical details.

In our evaluations, we focus on head-and-neck CT scans from cancer patients undergoing regular radiotherapy from cohorts at two different medical centers. The head and neck region is a very challenging part of the human anatomy, maybe the most challenging one, as it is very heterogeneous with complex shapes (larynx, jaw, teeth, etc.) and its deformations are complex—combination of twist of the neck, articulation of the jaw, and compression and extension due to patient weight and tumor variations.

We compare our method against top-tier techniques, including X2Vision [3], 2D3DNR [5], and the NeRF-based approach [22]. The results demonstrate our method's superiority and ability to capture important anatomical details.

Our method demonstrates high-quality deformable and rigid registration, indicating a move towards more precise biplanar systems over traditional 3D visualization. Unlike typical 2D/3D registrations focused on bones, our method enables finer adjustments due to our detailed 3D reconstructions. Our approach closely matches patient anatomy, potentially enhancing daily treatment precision and enabling daily adaptation, like adaptive radiotherapy, while significantly reducing irradiation. It delivers volumes accurate enough for procedures, showing potential to transform medical practices.

A portion of the data used in our study, specifically for training our generative model, is publicly available. We plan to request the release of the longitudinal radiotherapy dataset for training our deformation model and testing.

2 Related Work

2.1 3D Reconstruction from Few X-Ray Projections

Many methods have already been proposed to reduce the number of X-ray projections when reconstructing a 3D volume [3,11,13,18,22–26,31–34]. (1) Training a deep learning model for volume estimation from multiple projections in a supervised manner, as in [11,13,18,23–26,31,32,34] requires unchanging sensor calibrations from training to deployment, and this direct volume estimation often yields poor reconstructions. Most current methods employ feature embedding from 2D projections [11,24] with a fusion mechanism [13,18,25,32,34], often using an adversarial loss [13,31,32]. To reduce the space of potential solutions, some methods introduce geometric constraints [23] with refinement of a first estimate with a 3D network like U-Net [18,23]. Self-attention mechanisms [25] and transformer architectures [31,34] have also been introduced. [26] relies on semi-supervised learning within a teacher-student framework to address the limited availability of paired volume-projection data. Our method advances beyond these limitations by being fully unsupervised, thereby eliminating the need for paired datasets.

Methods [3,22] iteratively optimizing on the reconstructed volume at inference given projections typically perform better. X2Vision [3] relies on a learned 3D manifold to find a realistic and matching solution, while others rely on NeRF [22,33]. These methods ensure projection consistency at inference and offer better generalization. [22] already exploits the pre-acquired volume but simply by using the pre-acquired volume to initialize the NeRF. When using very few projections, this fails as shown in [3]. Our approach efficiently combines such a pre-acquired volume thanks to a prior on its possible deformation.

2.2 2D/3D Deformable Registration

Our approach is directly related to 2D/3D deformable registration problems, as we deform the pre-acquired 3D volume by comparing its projections to the captured X-ray projections. This usually involves solving an optimization problem to find the best transformation parameters that explain the deformation between a volume and a set of 2D projections, as discussed in [7,20,28,36]. This can be

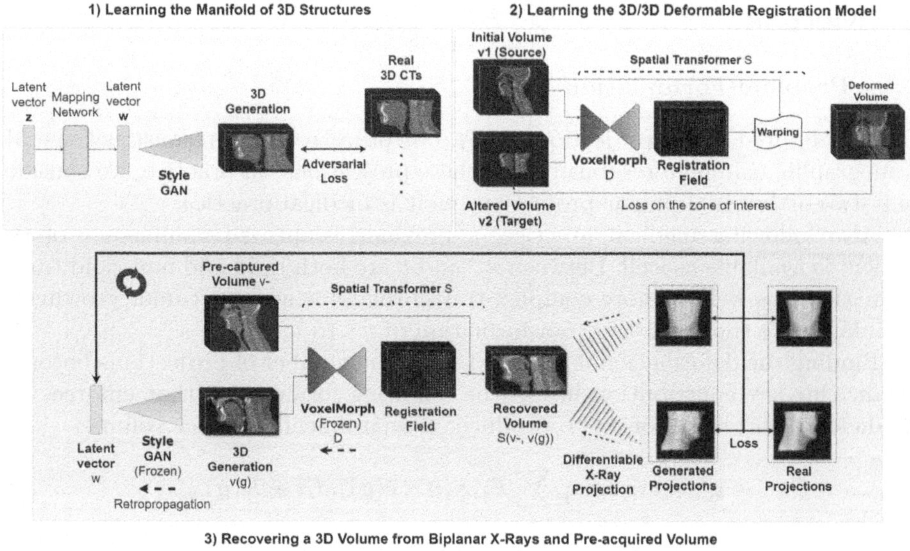

Fig. 2. Our pipeline. We first train a generative model to generate 3D volumes in a low-dimensional manifold and a 3D/3D deformable registration model between two volumes. Now, given two projections and a pre-captured volume of the patient, we recover a 3D volume corresponding to the two projections by finding the latent vectors that generate the best 3D volume so that the pre-captured volume well deforms on it to match the projections. We iteratively refine the generation and deformations based on the discrepancy between the generated and actual projections.

done by comparing the actual CT projections with their corresponding simulated CT projections. Nonetheless, when only a limited number of projections is used, traditional approaches struggle because they rely on many projections for accurate alignment. Fewer projections lead to underconstrained alignment, resulting in unrealistic or poorly matched deformations. Some learning-based methods use feedfoward predictions for faster registration [8,17,19,35], but this still struggles with ambiguity in the 2D projections used as input. [17,19] attempt to overcome this by using prior data to constrain deformations within a realistic range using PCA, yet this does not fully eliminate the spatial ambiguity of 2D measurements.

To solve the ambiguities, deep learning can be a solution as it can learn a reliable prior. Building on this, [5] introduced a method we reference by the name "2D3DNR" that transitions from 2D biplanar projections to 3D space, by estimating a 3D feature map from the projections followed by a 3D-3D deformation learning process using a U-Net-based model [4]. Our method exploits two priors: a prior on the predicted volume and a prior on the deformation of the pre-acquired volume. We learn both in an unsupervised way. As our results show, this is key to perform better than the previous state-of-the-art methods.

3 Method

3.1 Problem Formulation

Given a limited set of projections $\{I_i\}_i$, our objective is to reconstruct the 3D tomographic volume v responsible for these projections. In practice, we consider only two orthogonal planar projections as it is medical practice.

Also following medical practice, a previously-captured volume v^- of the patient is available as well. Between v^- and v are both rigid and non-rigid transformations, as well as more complex transformations such as tumor growing or shrinking. We thus seek the transformation of v^- to v.

Finding the deformation is ill-posed when the number of projections becomes small. Our key contribution lies in the following formulation that enforces the predicted deformation of v^- to produce an anatomically correct volume:

$$\mathbf{g}^* = \arg\min_{\mathbf{g}} \sum_i \mathcal{L}_i(S(v^-, v(\mathbf{g})), I_i) + \mathcal{R}(\mathbf{g})\,. \tag{1}$$

We briefly describe below each component of this formulation, then describe them in more details in the rest of the section:

- $v(.)$ is a generative model of volumes of parameters $\mathbf{g}$, i.e., $v(\mathbf{g})$ is a generated volume. In practice, we use a model similar to the one in X2Vision. However, X2Vision uses this model to directly predict the final volume v. The key difference in our method is that we use it here to guide the deformation of v^-.
- $S(v^-, v(\mathbf{g}))$ is a spatial transformer [12][1] trained to predict directly the transformation between two volumes. Here, $S(v^-, v(\mathbf{g}))$ returns volume v^- after deformation to align on the generated volume with the deformation consistent with training data. We use a spatial transformer very close to the one proposed by VoxelMorph [1].
- $\mathcal{L}_i$ is a loss term that compares the projections of deformed volume $S(v^-, v(\mathbf{g}))$ with the input projections I_i.

Intuitively, the optimization on $\mathbf{g}$ generates a volume $v(\mathbf{g})$ that guides the deformation of v^- thanks to the first term and exploits prior knowledge on volumes to recover. After optimization, our method returns $S(v^-, v(\mathbf{g}))$, the pre-acquired volume after deformation. In practice, we used the same optimization parameters as X2Vision.

Figure 2 illustrates our pipeline that implements our approach: We first train generative model $v(\mathbf{g})$ as well as spatial transformer S, both in an unsupervised way. Given two input projections, we then optimize the parameters $\mathbf{g}$ of the generative model, which gives us deformed volume $S(v^-, v(\mathbf{g}))$. We describe the generative model $v(\mathbf{g})$, spatial transformer S, and loss term $\mathcal{L}_i$ below.

[1] Not to be confused with Transformers [30].

3.2 Generative Model $v(.)$

Like X2Vision [3], we learn generative model $v(.)$ using GANs. We decompose parameters $\mathbf{g}$ into a latent vector $\mathbf{w}$ and Gaussian noise vectors $\mathbf{n} = \{\mathbf{n}_j\}_j$: $\mathbf{g} = [\mathbf{w}, \mathbf{n}]$. Latent vector $\mathbf{w} \in \mathcal{N}(\mathbf{w}|\mu, \sigma)$ is computed from an initial latent vector $\mathbf{z} \in \mathcal{N}(0, I)$ mapped using a learned network m: $\mathbf{w} = m(\mathbf{z})$. $\mathbf{w}$ controls the global structure of the predicted volumes at different scales by its components $\mathbf{w}_i$, while the noise vectors $\mathbf{n}$ allow more fine-grained details. Mean μ and standard deviation σ of the mapped latent space can be computed by mapping over initial latent space $\mathcal{N}(0, I)$ after training. Like [3], we optimize on the noise vectors $\mathbf{n}$ as well as they are useful to generate high-resolution details.

To ensure that the predicted volume remains in the manifold of possible volumes, $\mathcal{R}(\mathbf{g})$ is defined as a sum of regularization terms on $\mathbf{w}$ and $\mathbf{n}$:

$$\mathcal{R}(\mathbf{g}) = \mathcal{R}(\mathbf{w}, \mathbf{n}) = \lambda_w \mathcal{L}_w(\mathbf{w}) + \lambda_c \mathcal{L}_c(\mathbf{w}) + \lambda_n \mathcal{L}_n(\mathbf{n}) \,. \tag{2}$$

Terms $\mathcal{L}_w(\mathbf{w}) = -\sum_k \log \mathcal{N}(\mathbf{w}_k|\mu, \sigma)$ and $\mathcal{L}_n(\mathbf{n}) = -\sum_j \log \mathcal{N}(\mathbf{n}_j|\mathbf{0}, I)$ ensure that $\mathbf{w}$ and $\mathbf{n}$ remain on their respective distributions learned during training. Term $\mathcal{L}_c(\mathbf{w}) = -\sum_{i,j} \log \mathcal{M}(\theta_{i,j}|0, \kappa)$ encourages the $\mathbf{w}_i$ vectors to be collinear to keep the generation of coarse-to-fine structures coherent. $\mathcal{M}(\cdot; \mu, \kappa)$ is the density of the Von Mises distribution of mean μ and scale κ, which we take fixed, and $\theta_{i,j} = \arccos(\frac{\mathbf{w}_i \cdot \mathbf{w}_j}{\|\mathbf{w}_i\|\|\mathbf{w}_j\|})$ is the angle between vectors $\mathbf{w}_i$ and $\mathbf{w}_j$. The λ_* are fixed weights.

3.3 Spatial Transformer S

We use a spatial transformer S very close to the one introduced by VoxelMorph [1]. It can be decomposed into:

$$S(v_1, v_2) = \mathcal{W}(v_1, D(v_1, v_2)) \,, \tag{3}$$

where $D(v_1, v_2)$ is a deep network predicting a deformation field from v_1 to v_2; $\mathcal{W}(v_1, D(v_1, v_2))$ deforms volume v_1 according to the deformation field predicted by D. Model D is trained to predict deformation $\mathcal{W}$ between two volumes v_1 and v_2 by minimizing $\lambda_s \|v_2 - S(v_1, D(v_1, v_2))\|^2 + \lambda_D \|\nabla D(v_1, v_2))[x]\|^2$, over a training set of corresponding volumes $\{(v_1, v_2)\}$. The second term is a smoothing loss that mitigates sharp local fluctuations and promote smoothness of the predicted field. λ_s and λ_D are balancing weights that adjust the emphasis between similarity and regularization during training.

Maintaining 1-to-1 mapping in medical image registration is crucial to prevent tearing or overlapping. Our model, inspired by the VoxelMorph approach, predicts a velocity field. By integrating this velocity field over time, we obtain smooth, invertible transformations that naturally avoid singularities. This approach ensures the deformation remains diffeomorphic.

3.4 Loss Term $\mathcal{L}_i$

As in X2Vision, we take term $\mathcal{L}_i(v, I_i)$ as the weighted sum of the Euclidean distance and the perceptual loss [14] as we observed that this combination results in the best results. To generate projections, we used a realistic differentiable cone-beam projector as in X2Vision [3].

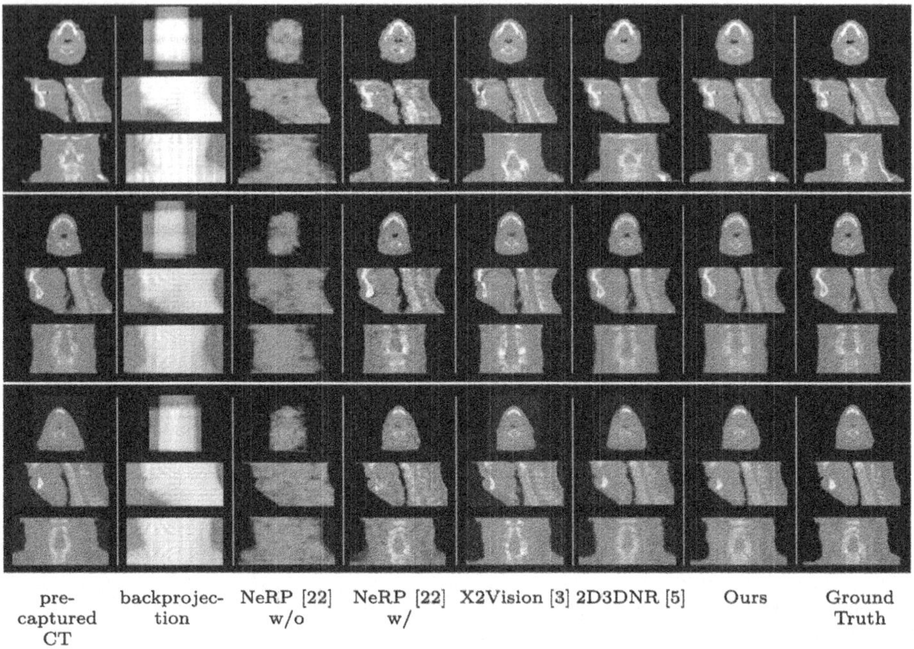

Fig. 3. Visual analysis of recovered volumes from two projections by previous methods and our approach. In the absence of a pre-captured CT volume, NeRP struggles due to lack of constraints. When exploiting the pre-captured CT volume, NeRP still tends to introduce artifacts in an attempt to align with the projections and alters the anatomy without ensuring anatomical accuracy. In contrast to X2Vision, our method predicts a reconstruction that captures patient-specific details and nuances. 2D3DNR results in deformations that do not adequately match the anatomy.

4 Experiments

We evaluate our method for our target application, head-and-neck cancer radiotherapy. Head-and-neck imagery exhibits many fine details and complex deformations and is representative of many of the different challenges of volume recovery. In this section, we introduce our dataset, models for learning key priors, and present both quantitative and qualitative comparisons with state-of-the-art methods. We also include an ablation study to evaluate the contribution of our priors. Our method recovers high-quality volumes in only 1 min. While some other methods are faster, the trade-off fidelity/runtime is well acceptable as clinical CBCT acquisition and FDK reconstruction [6] currently require more than 2–4 minutes.

4.1 Datasets and Metrics

Volume Generator Learning. We trained our GAN model for $v(\mathbf{g})$ on a large dataset of 3500 CTs of patients with head-and-neck cancer, more exactly 2297 patients from the publicly available TCIA [2,10,15,16,29,37] and 1203 from private internal data, after obtaining ethical approbation. We split this data into 3000 cases for training, 250 for validation, and 250 for testing.

We targeted CT scans of the head-and-neck region with a resolution of $80 \times 96 \times 112$ ($1.3 \times 2.4 \times 1.9\ mm^3$). Using an automatic segmentation of the mouth with a pre-trained U-Net [21], we centered these scans around it. We applied min-max normalization and post-clipping values between -1024 and 2000 Hounsfield Units (HU) to the CT scans as pre-processing.

Longitudinal Radiotherapy Data. In radiotherapy, a "planning CT" capture is used to design the therapy plan, while daily 3D Cone-Beam CTs (CBCTs) are captured during treatment to ensure the patient's positioning matches this initial CT. With patient consent, we compiled planning CT scans and subsequent CBCT scans from 242 patients across two medical centers, one contributing 177 and the other 65 cases. These datasets, distinct in protocols and scanning equipment, offer a diverse basis for evaluation.

As depicted in Fig. 3, notable differences emerge between the initial CT scans and subsequent CBCT scans, because of both treatment-induced alterations and patient pose variations. By aligning planning CTs with CBCTs using MRF minimization [9], we derived 3D volumes as CTs. We used some of these volumes for training our 3D/3D deformable registration model and the rest for evaluating the predicted reconstructions by the different methods.

3D/3D Deformable Registration Training. More precisely, to train our 3D/3D deformable registration model, we randomly selected 146 patients for training, 16 for validation, and 10 for testing. We paired each initial CT with any subsequent CT from the same patient to obtain a large training set.

Volume Recovery. The second part of volumes, used for evaluation, includes 70 patients with the most marked longitudinal alterations. We selected them by comparing their CBCTs with planning CTs. We paired the planning CT with each patient's final CT—which underscores the utmost discrepancies. We used the planning CT as pre-captured volume. Biplanar projections were synthesized from the last 3D volumes, focusing on the reconstruction area.

Table 1. Metrics on volumes from two projections by previous methods and our approach. Standard deviations are provided in parentheses. (w/) and (w/o) stand for the use or not of the pre-captured volume respectively.

Method	PSNR (dB) ↑	SSIM ↑
Backprojection	10.29 (±0.5)	0.23 (±0.01)
NeRP (w/o) [22]	19.81 (±1.7)	0.21 (±0.03)
NeRP (w/) [22]	25.32 (±1.6)	0.34 (±0.02)
X2Vision [3]	27.80 (±1.4)	0.89 (±0.03)
2D3DNR [5]	29.07 (±1.6)	0.92 (±0.02)
Ours	**33.23** (**±0.62**)	**0.96** (**±0.01**)

Method	Dice ↑	
	Mouth	Larynx
2D3DNR [5]	0.91 (±0.03)	0.80 (±0.07)
Ours	**0.95** (**±0.01**)	**0.91** (**±0.02**)
Method	Rigid Registration Error (6 DoF)	
	Rotation (°) ↓	Translation (mm) ↓
2D3DNR [5]	0.52 (±0.29)	0.88 (±0.45)
X2Vision [3]	0.45 (±0.31)	0.50 (±0.26)
Ours	**0.16** (**±0.15**)	**0.20** (**±0.07**)

Metrics. We assessed the reconstruction performance using two quantitative metrics: PSNR, which quantifies reconstruction error, and SSIM, which measures the perceptual quality of the images. We also evaluated the accuracy of the deformation between the pre-acquired volume and the recovered volume for the two methods that estimate this deformation: 2D3DNR and ours. To this end, we consider the Dice score for the mouth and the larynx, two structures that are likely to deform significantly. To compute it, we segmented these structures on the groundtruth and recovered volumes using a trained U-Net model using about 1000 head-and-neck CTs.

Additionnaly, we compared 3D (6DoF) rigid registration differences, between initial full CT scans and our reconstructions against the ground truth. This analysis highlights the precision of our method in capturing and reconstructing the nuanced deformations of critical anatomical features.

4.2 Results and Analysis

Table 1 reports the quantitative results. We detail below the methods we compare to and discuss their results after they were retrained on our data. Figure 3 compares visually our reconstruction to these methods on several examples. Additional results and reconstructions are provided in the supplementary material, but we summarize below our visual analysis of the results.

The **backprojection method** is a very simple baseline inspired from [6]. It estimates the value of each voxel as the average of the values at the projected voxel locations in the input X-ray projections. When enough input projections are available, this method can provide satisfying results. However, it fails when only two projections are used.

The **NeRP** method [22] optimizes the 3D volume to match the projections. It also struggles when very few projections are given since they lack prior anatomical knowledge. Even when conditioned on the pre-captured volume, it is often not able to eliminate the many artefacts.

We also considered the recent **X2Vision** [3] method to highlight the advantages of exploiting the pre-captured volume as we do—which X2Vision does not. It provides a reasonable reconstruction but still misses important details.

2D3DNR [5] predicts in a feedforward way the deformation between the pre-captured 3D volume and the new one given the precaptured volume itself and the available biplanar projections. Since the original code was unavailable, we used the same VoxelMorph backbone as ours to reimplement the 3D/3D registration method. Further details can be found in the supplementary material. The volumes predicted by 2D3DNR do not reproject well on the input projections in general. Because it is a feedforward method, it also tends to generalize poorly. Our method recovers better the deformation of the tissues.

Like X2Vision and NeRP, our method optimizes on the volume during inference for consistency with the input projections, which helps generalization. It also introduces a prior on the anatomical volume thanks to its GAN, in a way related to X2Vision. Our method has however an original way to exploit the pre-acquired volume by controling its deformations. This contrast with 2D3DNR, which takes this volume as input to a feedforward process, and with NeRP, which uses this volume only as conditioning. Our approach appears to be more powerful as it yields the best results.

Although X2Vision provides a capable reconstruction, it lacks in detailing critical aspects. Our method integrates patient-specific details to surpass the constraints of the generative model manifold, which might not capture the patient's unique anatomy or abnormalities accurately. By leveraging the pre-acquired volume, our method obtains a more accurate depiction of the patient's real anatomy rather than depending on a generalized learned manifold. This focus on patient specificity is crucial for achieving detailed and realistic anatomy reconstructions.

In contrast to 2D3DNR, we adopt an optimization strategy to inform the deformation prior with the capabilities of the generative model. This model lays down a realistic 3D support for the optimization process, ensuring the deformations are not just plausible but supported by the anatomical description of the generative model. This leads to more precise reconstructions, showcasing a significant step forward in anatomical fidelity.

4.3 Validation for Medical Applications

As shown on the right of Table 1, our method demonstrates superior deformation accuracy, evidenced by high dice coefficients indicating precise structure deformation. Moreover, we attain impressive rigid registration with an average precision way below 1mm. This level of detail and accuracy underscores the effectiveness of our approach in capturing the complex nuances of patient anatomy. Such high-quality volume reconstruction from only biplanar projections opens the way for accurate 3D rigid registration and daily assessment of the evolution of the patient, without needing 3D acquisition.

Table 2. Ablation Study. This table shows the contribution of our two priors. More details can be found in in Sect. 4.4.

Metric	deformation of the pre-acquired volume without any prior	generative model only followed by 1 deformation	generative model	deformation of the pre-acquired volume only	full method
PSNR (dB) ↑	27.04 (±1.9)	27.80 (±1.4)	29.24 (±1.8)	30.75 (±1.19)	**33.23 (±0.62)**
SSIM ↑	0.88 (±0.03)	0.88 (±0.03)	0.92 (±0.02)	0.93 (±0.01)	**0.96 (±0.01)**

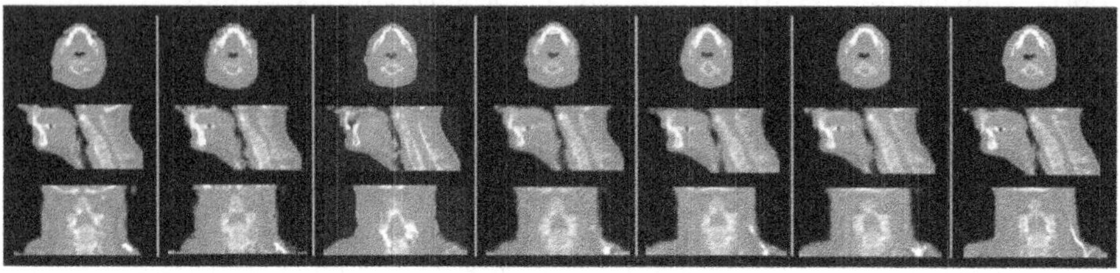

pre-captured CT	deform. of the pre-acquired volume w/o prior	generative model only	generative model followed by 1 deformation	deformation of the pre-acquired volume only	full method	Ground Truth

Fig. 4. Visual Analysis of the Ablation Study. Deforming the pre-acquired volume without any prior results in erratic and anatomically inconsistent changes. Reconstruction solely with the generative model may overlook details and lead to mismatches. Deforming the pre-acquired volume on it introduces patient-specific features but may retain initial misalignments. While introducing prior on deformation aids guiding the direction process, it leads to unnatural distortions of body contour and bone structures. Our method, by leveraging both anatomical and deformation priors, yields more realistic and anatomically preserving results.

4.4 Ablation Study

Table 2 presents an ablation study highlighting the benefits of our loss function in Eq. (1) by comparing it to different possible variants. Figure 4 presents a visual comparison of the results obtained with these variants. We considered four variants. The reader should compare the loss functions for these variants to the loss function we introduced in Eq. (1):

- 'deformation of the pre-acquired volume without any prior': This variant returns volume $\mathcal{W}(v^-, \phi^*)$ with $\phi^* = \arg\min_\phi \sum_i \mathcal{L}_i(\mathcal{W}(v^-, \phi), I_i)$, where $\mathcal{W}(v^-, \phi)$ applies a deformation field ϕ to the pre-acquired volume v^-. This approach retrieves the deformation parameters iteratively without any prior learning. The retrieved deformations tend to be erratic, blending structures and leading to artifacts that compromise anatomical accuracy.

- ‘generative model only’: This variant returns $v(\mathbf{g}^G)$ with $\mathbf{g}^G = \arg\min_{\mathbf{g}} \sum_i \mathcal{L}_i(v(\mathbf{g}), I_i) + \mathcal{R}(\mathbf{g})$. It uses only the generative model to predict the volume and corresponds to the X2Vision method.
- ‘generative model followed by 1 deformation’: This variant returns volume $S(v^-, v(\mathbf{g}^G))$. This approach deforms the pre-acquired volume to fit the generative model's reconstruction, introducing patient-specific details but potentially retaining initial mismatches. This shows the advantage of combining volume v^- and the generative model $v(\mathbf{g})$ during optimization.
- ‘deformation of the pre-acquired volume only’: This variant returns volume $S(v^-, \mathbf{v}^*)$ with $\mathbf{v}^* = \arg\min_{\mathbf{v}} \sum_i \mathcal{L}_i(S(v^-, \mathbf{v}), I_i)$, where $\mathbf{v}$ is a volume represented by a voxel grid, with each voxel encompassing an intensity to optimize. This approach uses only the pre-acquired volume and the spatial transformer, but not the generative model. This results in local deformations that are not anatomically realistic, such as bone extensions or body contour distortions, stemming from its lack of anatomical prior.

The quantitative results in Table 2 clearly show that our loss function exploits both priors well.

5 Conclusion

By managing to exploit both patient-specific data and anatomical constraints, we achieve unmatched accuracy in anatomical reconstructions, avoiding the need for intensive 3D scans. Our method promises improved patient care with daily adjustable treatments like adaptive radiotherapy, enhancing precision and outcomes while reducing treatment times and radiation exposure.

Acknowledgement. This study was funded in part by the grant N°ANR-21-RHU5-0005 within the FRANCE2030 investment plan.

References

1. Balakrishnan, G., Zhao, A., Sabuncu, M.R., Guttag, J., Dalca, A.V.: VoxelMorph: a learning framework for deformable medical image registration. IEEE Trans. Med. Imaging **38**(8) (2019)
2. Beichel, R.R., et al.: Data from QIN-HEADNECK (2015)
3. Cafaro, A., et al.: X2Vision: 3D CT reconstruction from biplanar X-Rays with deep structure prior. In: International Conference on Medical Image Computing and Computer-Assisted Intervention, pp. 699–709 (2023)
4. ÇIçek, Ö., Abdulkadir, A., Lienkamp, S.S., Brox, T., Ronneberger, O.: 3D U-Net: learning dense volumetric segmentation from sparse annotation. In: Conference on Medical Image Computing and Computer Assisted Intervention, pp. 424–432 (2016)
5. Dong, G., et al.: 2D/3D non-rigid image registration via two orthogonal X-Ray projection images for lung tumor tracking. Bioengineering **10**(2) (2023)
6. Feldkamp, L.A., Davis, L.C., Kress, J.W.: Practical Cone-Beam Algorithm. Science (1984)

7. Flach, B., Brehm, M., Sawall, S., Kachelrieß, M.: Deformable 3D–2D registration for CT and its application to low dose tomographic fluoroscopy. Phys. Med. Biol. **59**(24) (2014)
8. Foote, M.D., Zimmerman, B.E., Sawant, A., Joshi, S.C.: Real-time 2D-3D deformable registration with deep learning and application to lung radiotherapy targeting. In: Information Processing in Medical Imaging, pp. 265–276 (2019)
9. Glocker, B., Komodakis, N., Tziritas, G., Navab, N., Paragios, N.: Dense image registration through MRFs and efficient linear programming. Med. Image Anal. **12**(6), 731–741 (2008)
10. Grossberg, A., et al.: Anderson Cancer Center Head and Neck Quantitative Imaging Working Group. HNSCC (2020)
11. Henzler, P., Rasche, V., Ropinski, T., Ritschel, T.: Single-image tomography: 3D volumes from 2D cranial X-Rays. In: Computer Graphics Forum, vol. 37, pp. 377–388 (2018)
12. Jaderberg, M., Simonyan, K., Zisserman, A., Kavukcuoglu, K.: Spatial Transformer Networks. In: Advances in Neural Information Processing Systems, vol. 28 (2015)
13. Jiang, Y.: MFCT-GAN: multi-information network to reconstruct CT volumes for security screening. J. Intell. Manufact. Spec. Equipment **3**(1), 17–30 (2022)
14. Johnson, J., Alahi, A., Fei-Fei, L.: Perceptual losses for real-time style transfer and super-resolution. In: European Conference on Computer Vision, pp. 694–711 (2016)
15. Kinahan, P., Muzi, M., Bialecki, B., Coombs, L.: Data from the ACRIN 6685 Trial HNSCC-FDG-PET/CT (2020)
16. Kwan, J.Y.Y., et al.: Data from radiomic biomarkers to refine risk models for distant metastasis in oropharyngeal carcinoma (2019)
17. Li, P., Pei, Y., Guo, Y., Ma, G., Xu, T., Zha, H.: Non-Rigid 2D-3D Registration Using Convolutional Autoencoders. In: International Symposium on Biomedical Imaging, pp. 700–704 (2020)
18. Lu, S., Li, S., Wang, Y., Zhang, L., Hu, Y., Li, B.: Prior information-based high-resolution tomography image reconstruction from a single digitally reconstructed radiograph. Phys. Med. Biology **67**(8) (2022)
19. Pei, Y., et al.: Non-rigid craniofacial 2D-3D registration using CNN-based regression. In: Deep Learning in Medical Image Analysis and Multimodal Learning for Clinical Decision Support: Third International Workshop, DLMIA 2017, and 7th International Workshop, ML-CDS 2017, Held in Conjunction with MICCAI (2017)
20. Prümmer, M., Hornegger, J., Pfister, M., Dörfler, A.: Multi-modal 2D-3D non-rigid registration. In: Medical Imaging 2006: Image Processing, vol. 6144, pp. 297–308 (2006)
21. Ronneberger, O., Fischer, P., Brox, T.: U-net: convolutional networks for biomedical image segmentation. In: Conference on Medical Image Computing and Computer Assisted Intervention, pp. 234–241 (2015)
22. Shen, L., Pauly, J., Xing, L.: NeRP: implicit neural representation learning with prior embedding for sparsely sampled image reconstruction. IEEE Trans. Neural Netw. **35**(1), 770–782 (2022)
23. Shen, L., Zhao, W., Capaldi, D., Pauly, J., Xing, L.: A geometry-informed deep learning framework for ultra-sparse 3D tomographic image reconstruction. Comput. Biology Med. **148** (2022)
24. Shen, L., Zhao, W., Xing, L.: Patient-specific reconstruction of volumetric computed tomography images from a single projection view via deep learning. Nature **3**(11), 880–888 (2019)

25. Tan, Z., Li, J., Tao, H., Li, S., Hu, Y.: XctNet: reconstruction network of volumetric images from a single X-ray image. Comput. Med. Imaging Graph. **98** (2022)
26. Tan, Z., Li, S., Hu, Y., Tao, H., Zhang, L.: Semi-XctNet: volumetric images reconstruction network from a single projection image via semi-supervised learning. Comput. Biology Med. **155** (2023)
27. Tian, L., Lee, Y.Z., San José Estépar, R., Niethammer, M.: LiftReg: limited angle 2D/3D deformable registration. In: International Conference on Medical Image Computing and Computer-Assisted Intervention, pp. 207–216 (2022)
28. Tian, L., et al.: Fluid registration between lung CT and stationary chest tomosynthesis images. In: Conference on Medical Image Computing and Computer Assisted Intervention, pp. 307–317 (2020)
29. Vallières, M., et al.: Data from Head-Neck-PET-CT (2020)
30. Vaswani, A., et al.: Attention Is All You Need. In: Adv. Neural Inf. Process. Syst.(2017)
31. Wang, Y., Xia, Q.: TPG-rayGAN: CT reconstruction based on transformer and generative adversarial networks. In: Third International Conference on Intelligent Computing and Human-Computer Interaction (ICHCI 2022), vol. 12509, pp. 347–354 (2023)
32. Ying, X., Guo, H., Ma, K., Wu, J., Weng, Z., Zheng, Y.: X2CT-GAN: reconstructing CT from biplanar X-Rays with generative adversarial networks. In: Conference on Computer Vision and Pattern Recognition, pp. 10619–10628 (2019)
33. Zha, R., Zhang, Y., Li, H.: NAF: neural attenuation fields for sparse-view CBCT reconstruction. In: Conference on Medical Image Computing and Computer Assisted Intervention, pp. 442–452 (2022)
34. Zhang, C., et al.: XTransCT: Ultra-Fast Volumetric CT Reconstruction Using Two Orthogonal X-Ray Projections via a Transformer Network. In: arXiv Preprint arXiv:2305.19621 (2023)
35. Zhang, Y.: An unsupervised 2D–3D deformable registration network (2D3D-RegNet) for cone-beam CT estimation. Phys. Med. Biology **66**(7) (2021)
36. Zikic, D., Groher, M., Khamene, A., Navab, N.: Deformable registration of 3D vessel structures to a single projection image. In: Medical imaging 2008: image processing, vol. 6914, pp. 348–359 (2008)
37. Zuley, M.L., et al.: The cancer genome atlas head-neck squamous cell carcinoma collection (TCGA-HNSC) (2015)

Large Deformation Registration with A Confidence-Guided Network

Xiaolong Wu[1,2], Bo Hu[1], Ruikang Xu[1], Shenglong Zhou[1], Ruisheng Gao[1], and Zhiwei Xiong[1,2](✉)

[1] MoE Key Laboratory of Brain-inspired Intelligent Perception and Cognition, University of Science and Technology of China, Hefei, China
wuxiaolong@mail.ustc.edu.cn, zwxiong@ustc.edu.cn

[2] Anhui Province Key Laboratory of Biomedical Imaging and Intelligent Processing, Institute of Artificial Intelligence, Hefei Comprehensive National Science Center, Hefei, China

Abstract. Large deformation is a challenging optimization problem in medical image registration. To deal with it, recent deep learning-based registration methods employ multi-step network structures to break the large deformation into smaller ones and register them separately. However, these methods have two problems. First, they cannot effectively discriminate between hard-to-optimize large deformations and easy-to-optimize small deformations. This indistinctive registration manner may result in inaccurate registration results. Second, since the deformation range to register is wide but local, existing convolution-based and transformer-based methods cannot have a proper receptive field. In this paper, we propose a novel unsupervised registration network, with a confidence-guided part to focus on different scale deformations with different degrees. Furthermore, we propose a recurrent large kernel attention registration part to register the image with the proper receptive field. Comprehensive experimental results on the brain and liver datasets show that our proposed confidence-guided network significantly improves registration accuracy over existing methods for large deformation registration. Code is available at https://github.com/aicaomei/CGRNet.

Keywords: Large Range Deformation · Confidence Guidance · Deformable Image Registration

1 Introduction

Medical image registration aims to learn a deformation field to align an image pair and is crucial in medical image analysis tasks [1,2]. In recent years, deep learning methods based on convolutional neural networks (CNN) [3–5] dominate in the registration area [6–14] with high efficiency and accuracy. Since manual annotations are hard to obtain, unsupervised methods adopting similarity measures have become the common practice.

M. Modat et al. (Eds.): WBIR 2024, LNCS 15249, pp. 34–44, 2024.
https://doi.org/10.1007/978-3-031-73480-9_3

The pioneering unsupervised registration network VoxelMorph [15] and many following works [4,16] take the U-Net structure as their backbone, and estimate the deformation field in a one-step manner. Since these one-step methods cannot deal with large deformation well, recent studies have shifted towards the development of multi-step networks with both affine and deformable registration. They initially estimate linear transformations with an affine registration network, followed by the deformable registration using a multi-step estimation network. Cascade and pyramid methods are representative ones. Cascade methods [3,17,18], such as RCN [3], propose to integrate several identical networks to facilitate the network learning, which breaks a large deformation into several small ones. Pyramid methods [5,19,20], such as Dual-PRNet [20], propose to estimate a deformation field in a coarse-to-fine manner with multi-level features. Although the multi-step estimation can solve the large deformation problem to a certain extent, it lacks the discrimination to hard-to-optimize large deformations and easy-to-optimize small deformations. Only a few areas in an image pair have large deformation and need the multi-step registration. Taking preoperative and postoperative tumor images as an example, large deformations only exist in the tumor area, while the remaining areas are almost undeformed. This indistinctive registration manner may result in inaccurate registration results. Furthermore, existing methods based on CNNs all adopt small kernels. Although stacking them can realize a large receptive field, each convolution only has a limited local receptive field, which constrains the estimation ability of the network.

With the introduction of Transformer architectures [21,22], registration networks [23,24] can exploit the long-range feature dependencies, which means the receptive field is global to the whole image. However, the too-large receptive field may also be harmful to registration since the real deformation only exists in local areas. Therefore, neither CNN nor Transformer can adaptively offer a proper receptive field for large deformation registration yet.

In this paper, we propose a confidence-guided registration network (CGRNet). CGRNet consists of a confidence-guided (CG) part and a recurrent large kernel attention registration part. The latter is realized by a large kernel attention block (LKAB). Specifically, CG generates confidence scores that feature the degree of the deformation with self and cross-correlations, to accurately offer the network different focuses on different scale deformations. LKAB is made up of existing CNNs, but its convolution kernel is much larger. Recent research proves that this manner benefits motion and displacement estimation [25], which is similar to deformation estimation. With these two components, CGRNet effectively solves the problems of CNN and Transformer based structures. Extensive experiments on the brain and liver datasets demonstrate the superiority of our proposed method over existing methods for large deformation registration.

2 Method

Our proposed CGRNet consists of two sequential parts: a confidence-guided (CG) part and a recurrent large kernel attention registration part. The overall pipeline of CGRNet is shown in Fig. 1.

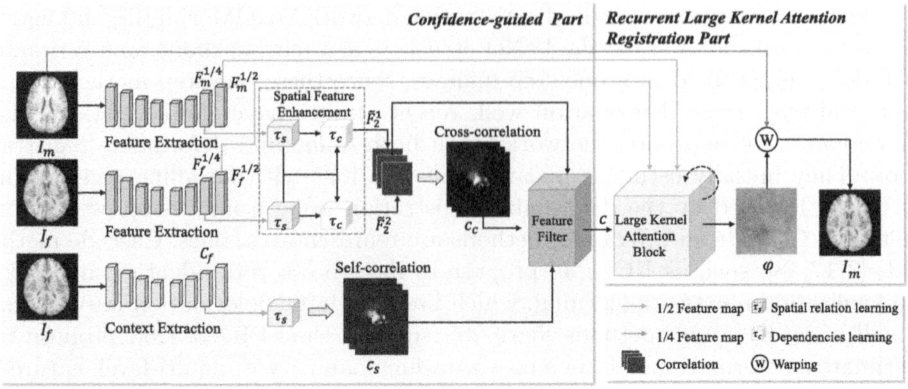

Fig. 1. The illustration of the proposed CGRNet. The confidence-guided part identifies and emphasizes large deformation areas using spatial feature enhancement and correlation maps. The recurrent large kernel attention registration part iteratively refines these features to produce the final deformation field, achieving accurate image registration.

First, in the CG part, there are both the feature extractors and the context extractor. The feature extractors capture the feature and deformation representations between a given fixed image I_f and moving image I_m, while the context extractors capture contextual information of the registration target. Specifically, the feature extractors produce the 1/2-resolution fixed and moving features $F_f^{1/2}$,$F_m^{1/2}$, as well as the 1/4-resolution features $F_f^{1/4}$, $F_m^{1/4}$. The context extractor produces a 1/4-resolution guidance map C_f. After feature extraction, the confidence-guided part computes the cross-correlation c_c and self-correlation c_s using the previously extracted features. These correlations are procedureed through the feature enhancement modules τ_s and τ_c, respectively. The resulting features are then fused using a feature filter to generate the confidence graph c. The cross-correlation and the self-correlation denote the confidence of the large deformation area. Next, the features $F_f^{1/2}$, $F_m^{1/2}$, and confidence map c are input into the recurrent large kernel attention registration part. After several iterations of the large kernel attention block, this procedure yields the final deformation field, denoted as φ. Finally, the deformation field φ is applied to warp I_m using a grid sampler, achieving spatial alignment with I_f.

2.1 Confidence-Guided Part

The confidence-guided part consists of four modules: feature extraction, spatial feature enhancement, correlation generation, and feature filter. Firstly, through the feature extraction, the images I_f and I_m can extract deeper information extraction, as introduced above. Following this, the remaining three modules are executed sequentially, as follows.

Spatial Feature Enhancement. The Spatial Feature Enhancement (SFE) module is utilized to obtain the correlation of image pairs. Specifically, we design an SFE module to capture long-range and cross-image relationships for representation enhancement. This module contains a spatial relation learning block, $\tau_s(\cdot)$, and a dependencies learning block, $\tau_c(\cdot)$. By employing $\tau_s(\cdot)$ and $\tau_c(\cdot)$, we can generate the enhanced features $\tilde{F}_f^1$ and $\tilde{F}_m^1$, denoted as:

$$\tilde{F}_2^1 = \tau_c\left(\tau_s\left(F_m^{1/4}\right), \tau_s\left(F_f^{1/4}\right)\right), \quad \tilde{F}_2^2 = \tau_c\left(\tau_s\left(F_f^{1/4}\right), \tau_s\left(F_m^{1/4}\right)\right). \tag{1}$$

$\tau_s(\cdot)$ is employed to enhance the spatial relations of features and mitigate ambiguity in constructing correlations between image pairs. $\tau_c(\cdot)$ denotes a cross-attention block, which is utilized to convey image-level mutual information. Both $\tau_s(\cdot)$ and $\tau_c(\cdot)$ are implemented using a large kernel attention block (see Fig. 2).

Correlation Generation. After obtaining the enhanced features $\tilde{F}_1^1$ and $\tilde{F}_1^2$, we can calculate the correlation across the image. We can use softmax-based feature matching to obtain the cross-correlation map. Specifically, given the features $\tilde{F}_1^1$, $\tilde{F}_1^2 \in \mathbb{R}^{h \times w \times t \times d_m}$, we generate a cross-correlation map $c_c \in \mathbb{R}^{h \times w \times t}$, denoted as:

$$c_c = \max_{h,w,t}\left(\text{softmax}\left(\frac{\tilde{F}_2^1 \times \tilde{F}_2^{2\top}}{\sqrt{d_m}}\right)\right), \tag{2}$$

where h, w, t and d_m represent the height, width, depth, and channel of the features $\tilde{F}_1^1$ and $\tilde{F}_1^2$, respectively. We use $\times$ and $\top$ to denote the matrix multiplication and the matrix transpose, respectively. softmax$(\cdot)$ is the result of applying the softmax function to the elements of a matrix.

Feature Filter. After obtaining the cross-correlation graph, we can further calculate the confidence graph. First, we need to get the cross-correlation score f_s, which is obtained by multiplying the correlation graph c_c with the pre-produced coordinate grid. Meanwhile, using the spatial relation learning block τ_s, the contextual guidance C_f provides the self-correlation map c_s within the fixed image I_f. The self-correlation map c_s is denoted as:

$$c_s = \tau_s\left(C_f\right). \tag{3}$$

Additionally, to avoid using low-confidence points, we mask these points with a mask m. The mask m consists of values 0 and -1, and its size is the same as c_s. Specifically, where the value of c_s is greater than threshold α, the corresponding value in the mask is set to 0; otherwise, it is set to -1. Then given c_s and $m \in \mathbb{R}^{h \times w \times t}$, we can refine the cross-correlation score f_s to obtain a correlation score $\hat{\mathrm{f}}_s$, denoted as:

$$\hat{\mathrm{f}}_s = \text{softmax}\left(\frac{c_s \times c_s^\top}{\sqrt{d_c}} + m\right) \odot \mathrm{f}_s, \tag{4}$$

where d_c refers to the number of channels for c_s, we use $\odot$ to denote the matrix dot product. We find that regions with high correlations are highly consistent with regions exhibiting large deformations. Therefore, we use a threshold value α to represent the correlation map. When the cross-correlation value exceeds α, we consider that part of the region as a large deformation region, and the network needs to focus on this region. The correlation map after applying the threshold α is expressed as c_r. Specifically, c_r is set to 1 when the value of c_c exceeds α; otherwise, the original value is retained. After that, the final confidence graph c can be expressed as:

$$c = c_r \times \mathrm{f}_s + (1 - c_r) \times \hat{\mathrm{f}}_s. \tag{5}$$

The specific procedure of the feature filtering module can be seen in Fig. 1 (a).

2.2 Recurrent Large Kernel Attention Registration Part

To obtain accurate registration results, we design a recurrent large kernel attention registration part. It involves estimating a precise deformation field through an iterative large kernel attention block. Specifically, in the first iteration, we input the features $F_m^{1/2}$, $F_f^{1/2}$, and the confidence graph c_c, and output a residual field $\Delta\varphi$ to warp $F_m^{1/2}$, resulting in the warped feature $F_{m_1}^{1/2}$. In subsequent iterations, the input is the warped feature from the previous iteration and $F_f^{1/2}$, with the output again being the residual field $\Delta\varphi$. In each iteration, we update the current deformation field as $\varphi_{k+1} = \Delta\varphi + \varphi_k$, where $k \in [0, n]$. After obtaining the final deformation field φ, we warp the image I_m to achieve the final registration result.

Large Kernel Attention Block (LKAB). Although the current CNN-based method improves the receptive field of networks by stacking networks, in practice, the effective receptive field of convolutional networks is much smaller than that in theory. Therefore, we use LKAB to increase the effective receptive field. LKAB is composed of a series of convolutional layers and large kernel attention units. As shown in Fig. 2, in each large kernel attention unit, there are four parallel sub-layers, including a separable large kernel (SLK) convolution layer of size $K \times K \times K$, a convolution layer of $1 \times 1 \times 1$, a convolution layer of $3 \times 3 \times 3$, and an identity shortcut. The outputs of these sub-layers are then element-wise added to generate the final output of a large kernel attention unit. The parallel paths within each large kernel attention unit not only address distant spatial information but also capture and fuse spatial details at a finer scale.

The SLK convolution plays an important role here. It can not only increase the range of the receptive field but also overcome the computational and parametric burdens of self-attention and large kernel convolutions. Specifically, we can decompose a convolution with the kernel size of $K \times K \times K$ into a $\lceil \frac{K}{d} \rceil \times \lceil \frac{K}{d} \rceil \times \lceil \frac{K}{d} \rceil$ depth-wise dilation convolution with dilation rate d, a $(2d-1) \times (2d-1) \times (2d-1)$ depth-wise convolution, and a $1 \times 1 \times 1$ convolution. Through this decomposition, we can capture long-range relationships with

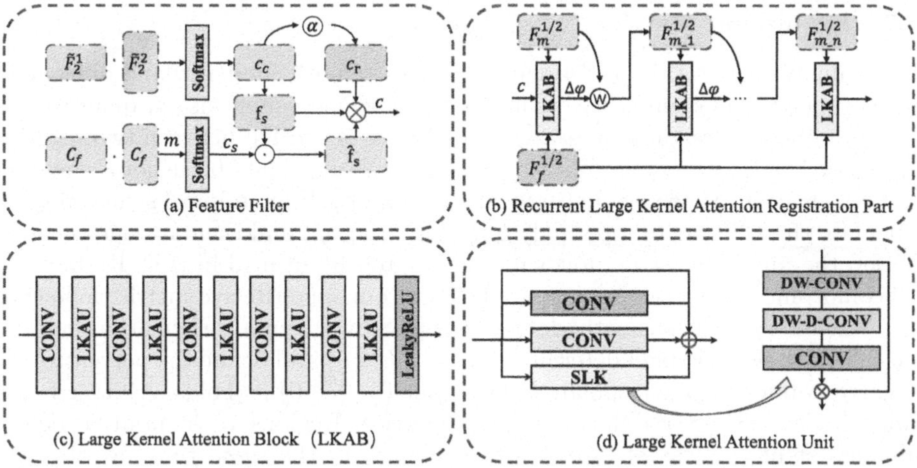

Fig. 2. (a) Feature Filter, (b) Recurrent Large Kernel Attention Registration Part, (c) Large Kernel Attention Block, and (d) Large Kernel Attention Unit. Illustrations (b), (c), and (d) are presented progressively. Where the blue CONV block is a convolution with a kernel size of $3 \times 3 \times 3$, and the green CONV block is a convolution with a kernel size of $1 \times 1 \times 1$. (Color figure online)

a slight computational cost and fewer parameters. After obtaining the long-range relationships, we can estimate the importance of the relevant voxels and generate an attention map. The SLK convolution can be formulated as

$$Attention = Conv_{1\times1\times1}\left(DW{-}D{-}Conv\left(DW{-}Conv\left(F\right)\right)\right), \tag{6}$$

where $DW{-}D{-}Conv\left(\cdot\right)$ is depth-wise dilation convolution, $DW{-}Conv\left(\cdot\right)$ is depth-wise convolution and $Conv_{1\times1\times1}\left(\cdot\right)$ is $1 \times 1 \times 1$ convolution.

3 Experiments

3.1 Datasets and Registration Strategy

We conduct comprehensive experiments on four datasets: LPBA, IXI, Mindboogle, and SLIVER. We divide the LPBA into 30 training scans, 2 validation scans, and 8 tests using a subject-to-subject registration strategy with a resolution of $160 \times 196 \times 160$. The IXI dataset consists of 576 T1-weighted brain MRI images. The data sets were divided into a training set (403 volumes), a validation set (58 volumes), and a test set (115 volumes), using an atlas-to-subject registration strategy. The images are preprocedureed using FreeSurfer [26] and cropped to $160 \times 192 \times 224$. The Mindboggle dataset consists of five subsets, with the training set composed of three subsets: MMR-21, NKI-RS-22, and NKI-TRT-20. The validation set and test set are OASIS and Extra, respectively. The resolution of the images is uniformly cropped to $192 \times 192 \times 192$. For the SLIVER dataset, using subject-to-subject registration, we divide it into 16 training scans (16×15 pairs) and 4 test scans (4×3 pairs) with a resolution of $128 \times 128 \times 128$.

3.2 Evaluation Metrics

We evaluate registration performance in terms of both similarity and diffeomorphic properties. For similarity, we utilize the DSC percentage as our quantitative metric, which is designed to measure the degree of overlap between two regions. A higher DSC percentage indicates better similarity, and 100 denotes the perfect registration. When evaluating scans with multiple anatomical structures, we calculate the DSC score separately for each structure and then average them to obtain the final DSC score, following the approach outlined in [15]. To evaluate diffeomorphic properties, we employ Folds as our quantitative metric, following the Voxelmorph-diff [4] methodology. Specifically, given a deformation field φ, we calculate the Jacobian determinant of φ $(|J(\varphi)|)$ and define the percentage of negative values in the Jacobian determinant $(|J_{\varphi}| < 0)$ as Folds. A lower Folds value indicates better diffeomorphic properties. Regions with negative determinants signify a loss of one-to-one mapping. At the same time, we use the 95% percentile of Hausdorff distance (HD95), which was used to evaluate the difference between the segmentation result and the real annotation. HD95 is particularly useful for dealing with large deformations or misalignments because it can effectively reflect the performance of segmentation methods in dealing with shape changes and boundary errors. In experiments, the smaller the HD95 measurement, the better the registration performance.

3.3 Implementation Details

We implement our CGRNet using PyTorch on an NVIDIA GeForce RTX 3090 with 24 GB memory. We use the ADAM [27] optimizer with a learning rate of 0.0001 to train CGRNet for 26,000 iterations. We set the kernel size to 7 and the number of iterations of the recurrent sub-net to 5. Our method is to model the deformation field, which is very easy to change to the diffeomorphism version.

4 Results

Table 1 presents the registration performance of our method compared to the baseline method. Our method achieves significantly better performance in terms of DSC on the LPBA, Mindboggle, and SLIVER datasets compared to the baseline method. Notably, on the LPBA dataset, our model's DSC is 3.2 points higher than the second-best Dual-PRNet index. This advantage is particularly pronounced in the SLIVER dataset, where our method achieves a DSC of 92.8. While TransMorph performs significantly better than CNN-based methods on datasets with large displacements, such as Mindboggle and IXI, our method surpasses TransMorph on both indicators. Overall, our method outperforms the other registration methods compared in DSC across multiple datasets, demonstrating its effectiveness and generalization capabilities. Furthermore, the folding consistency of our method across all datasets further affirms the comprehensive performance of our proposed approach. For an intuitive comparison, we also provide qualitative results in Fig. 3.

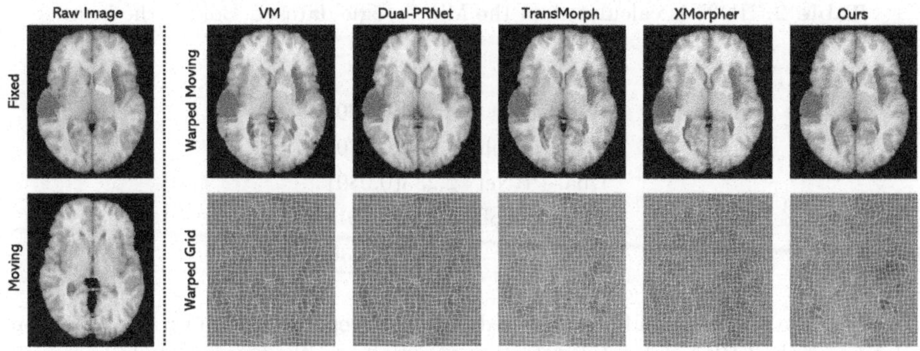

Fig. 3. Visualization of an example in the brain dataset LPBA. Left: Fixed image and moving image. Right: the first row shows the corresponding warped moving images, and the second row shows the warped grids to visualize diffeomorphic properties. Pink shows the overlap and blue shows the rest. Yellow arrows indicate the location to distinguish performance. (Color figure online)

Table 1. Comparison results on the LPBA, IXI, SLIVER, and Mindboggle datasets. **Bold** indicates the best DSC performance. The values in parentheses represent standard deviations.

	LPBA		IXI		SLIVER		Mindboggle	
Methods	DSC (%)	Folds (%)	DSC (%)	Folds (%)	DSC (%)	Folds (%)	DSC (%)	Folds (%)
Raw	52.5 (4.1)	-	40.9 (3.6)	–	80.4 (3.3)	–	33.5 (2.0)	-
VoxelMorph	62.3 (2.6)	0.7 (0.25)	71.6 (2.6)	0.6 (0.1)	87.3 (3.4)	2.6 (0.2)	51.4 (2.1)	0.9 (0.11)
Dual-PRNet	66.7 (1.1)	0.1 (0.03)	71.3 (2.7)	0.2 (0.1)	89.4 (2.3)	0.3 (0.1)	42.6 (2.4)	0.1 (0.02)
TransMorph	65.9 (2.2)	0.8 (0.18)	73.8 (2.7)	1.2 (0.1)	88.7 (3.6)	5.6 (0.4)	58.0 (2.2)	2.2 (0.18)
XMorpher	66.3 (1.4)	0.2 (0.05)	–	–	87.3 (2.9)	0.2 (0.1)	-	-
Ours	**70.9** (1.2)	0.4 (0.07)	**74.2** (2.6)	0.6 (0.1)	**92.8** (2.1)	0.8 (0.3)	**60.6** (2.3)	0.6 (0.08)

At the same time, in order to demonstrate the advantages of our model in dealing with large deformation problems, we select the Mindboogle dataset with the lowest DSC score of the original image to calculate the HD95 score. On the one hand, the low DSC score can be understood as a result of the non-coincidence of some regions caused by large deformation. On the other hand, the distance score HD95 is introduced to better capture the overall shape difference between images, which can especially reflect the problem of large deformation. It can be seen from Table 2 that our method achieves the optimal HD95 score with the more effective receptive field, which is 60% lower than the HD95 score of the original image. Our approach is also superior to TransMorph, which has a global receptive field.

In Table 3, we further explore the impact of confidence-guided (CG) and SLK convolution on our approach. The CG part and SLK we proposed have significantly enhanced the accuracy of registration. The SLK increases the receptive field while reducing the number of parameters in the model.

Table 2. HD95 is calculated on the Mindboogle dataset. Bold is the best.

Methods	HD95
Raw	3.356(0.270)
VoxelMorph	1.691(0.170)
Dual-PRNet	2.283(0.089)
TransMorph	1.414(0.101)
Ours	**1.368**(0.130)

Table 3. The effects of confidence-guided (CG) and separable large kernel (SLK) convolution on the LPBA dataset registration accuracy and the number of model parameters are compared.

Our		LPBA		Params (K)
CG	SLK	DSC (%)	Folds (%)	
		67.8 (1.7)	0.45 (0.06)	11,857
	✓	69.5 (1.8)	0.39 (0.07)	**10,385**
✓		69.7 (1.4)	0.43 (0.07)	14,352
✓	✓	**70.9** (1.2)	**0.37** (0.07)	12,880

5 Conclusion

In this paper, we propose a confidence-guided registration network (CGRNet). CGRNet consists of two main components: the confidence-guided(CG) part and the recurrent large kernel attention registration part. The CG part generates confidence scores for the degree of deformation using self-correlation and cross-correlation characteristics, thereby providing the network with precise attention to deformations at different scales. Additionally, we introduce a recurrent large kernel attention registration part with an extensive receptive field to better manage regions with significant deformations. Comprehensive experiments on brain and liver datasets demonstrate the superiority of our network.

Acknowledgement. This work was supported in part by the National Natural Science Foundation of China under Grant 62021001.

Disclosure of Interests. The authors have no competing interests to declare that are relevant to the content of this article.

References

1. Xu, Z., Niethammer, M.: DeepAtlas: Joint semi-supervised learning of image registration and segmentation. In Medical Image Computing and Computer Assisted Intervention–MICCAI 2019: 22nd International Conference, Shenzhen, China, October 13–17, 2019, Proceedings, Part II 22, pages 420–429. Springer, 2019.
2. Du, J., Li, W., Lu, K., Xiao, B.: An overview of multi-modal medical image fusion. Neurocomputing **215**, 3–20 (2016)
3. Zhao, S., et al.: Recursive cascaded networks for unsupervised medical image registration. In: Proceedings of the IEEE/CVF International Conference on Computer Vision, pp.10600–10610 (2019)
4. Balakrishnan, G., Zhao, A., Sabuncu, M.R., Guttag, J., Dalca, A.V.: Voxelmorph: a learning framework for deformable medical image registration. IEEE Trans. Med. Imaging **38**(8), 1788–1800 (2019)
5. Kang, M., Hu, X., Huang, W., Scott, M.R., Reyes, M.: Dual-stream pyramid registration network. Med. Image Analysis **78**,102379 (2022)
6. Ashburner, J.: A fast diffeomorphic image registration algorithm. Neuroimage **38**(1), 95–113 (2007)
7. Avants, B.B., Epstein, C.L., Grossman, M., Gee, J.C.: Symmetric diffeomorphic image registration with cross-correlation: evaluating automated labeling of elderly and neurodegenerative brain. Med. Image Anal. **12**(1), 26–41 (2008)
8. Bajcsy, R., Kovačič, S.: Multiresolution elastic matching. Comput. Vision Graph. Image Process. **46**(1), 1–21 (1989)
9. Shen, Z., Han, X., Xu, Z., Niethammer, M.: Networks for joint affine and non-parametric image registration. In: Proceedings of the IEEE/CVF Conference on Computer Vision and Pattern Recognition, pp. 4224–4233 (2019)
10. Hu, B., Zhou, S., Xiong, Z., Wu, F.: Self-recursive contextual network for unsupervised 3d medical image registration. In: Machine Learning in Medical Imaging: 11th International Workshop, MLMI 2020, Held in Conjunction with MICCAI 2020, Lima, Peru, October 4, 2020, Proceedings 11,pp. 60–69,springer (2020). https://doi.org/10.1007/978-3-030-59861-7_7
11. Zhou, S., et al.: Fast and accurate electron microscopy image registration with 3D convolution. In: Medical Image Computing and Computer Assisted Intervention–MICCAI 2019: 22nd International Conference, Shenzhen, China, October 13–17, 2019, Proceedings, Part I 22, pp. 478–486. Springer (2019). https://doi.org/10.1007/978-3-030-32239-7_53
12. Hu, B., Zhou, S., Xiong, Z., Wu, F.: Cross-resolution distillation for efficient 3D medical image registration. IEEE Trans. Circuits Syst. Video Technol. **32**(10), 7269–7283 (2022)
13. Zhou, S., Hu, B., Xiong, Z., Wu, F.: Self-distilled hierarchical network for unsupervised deformable image registration. IEEE Trans. Med. Imaging (2023)
14. Hu, B., Zhou, S., Xiong, Z., Wu, F.: Recursive decomposition network for deformable image registration. IEEE J. Biomed. Health Inform. **26**(10), 5130–5141 (2022)
15. Balakrishnan, G., Zhao, A., Sabuncu, M.R., Guttag, J., Dalca, A.V.: An unsupervised learning model for deformable medical image registration. In: Proceedings of the IEEE Conference on Computer Vision and Pattern Recognition, pp. 9252–9260 (2018)

16. Dalca, A.V., Balakrishnan, G., Guttag, J., Sabuncu, M.R.: Unsupervised learning for fast probabilistic diffeomorphic registration. In: Medical Image Computing and Computer Assisted Intervention–MICCAI 2018: 21st International Conference, Granada, Spain, September 16-20, 2018, Proceedings, Part I, pp. 729–738. Springer (2018)
17. Zhao, S., Lau, T., Luo, J., Eric, I., Chang, C., Xu, Y.: Unsupervised 3D end-to-end medical image registration with volume tweening network. IEEE J. Biomed. Health Inform. **24**(5), 1394–1404 (2019)
18. Yu, H., Chen, X., Shi, H., Chen, T., Huang, T.S., Sun, S.: Motion pyramid networks for accurate and efficient cardiac motion estimation. In: Medical Image Computing and Computer Assisted Intervention–MICCAI 2020: 23rd International Conference, Lima, Peru, October 4–8, 2020, Proceedings, Part VI 23
19. Lv, J., Wang, Z., Shi, H., Zhang, H., Wang, S., Wang, Y., Li, Q.: Joint progressive and coarse-to-fine registration of brain MRI via deformation field integration and non-rigid feature fusion. IEEE Trans. Med. Imaging, **41**, 1–1 (2022)
20. Xiaojun Hu, Miao Kang, Weilin Huang, Matthew R. Scott, Roland Wiest, and Mauricio Reyes. Dual-stream pyramid registration network. In Dinggang Shen, Tianming Liu, Terry M. Peters, Lawrence H. Staib, Caroline Essert, Sean Zhou, Pew-Thian Yap, and Ali Khan, editors, *Medical Image Computing and Computer Assisted Intervention – MICCAI 2019*, pages 382–390, Cham, 2019. Springer International Publishing.
21. Vaswani, A., et al.: Attention is all you need. Adv. neural inf. process. syst. **30** (2017)
22. Liu, Z., et al.: Swin transformer: hierarchical vision transformer using shifted windows. In Proceedings of the IEEE/CVF International Conference on Computer Vision, pp. 10012–10022 (2021)
23. Shi, J., et al.: XMorpher: full transformer for deformable medical image registration via cross attention. In: International Conference on Medical Image Computing and Computer-Assisted Intervention, pp. 217–226. Springer (2022). https://doi.org/10.1007/978-3-031-16446-0_21
24. Chen, J., Frey, E.C., He, Y., Segars, W.P., Li, Y., Du, Y.: TransMorph: transformer for unsupervised medical image registration. Med. Image Anal. **82**, 102615 (2022)
25. Jia, X., Bartlett, J., Zhang, T., Lu, W., Qiu, Z., Duan, J.: U-net vs transformer: is u-net outdated in medical image registration? In: International Workshop on Machine Learning in Medical Imaging, pp. 151–160. Springer (2022). https://doi.org/10.1007/978-3-031-21014-3_16
26. Fischl, B.: Freesurfer. Neuroimage **62**(2), 774–781 (2012)
27. Kingma, D.P., Ba, J.: Adam: a method for stochastic optimization. arXiv preprint arXiv:1412.6980 (2014)

Unleashing Registration: Diffusion Models for Synthetic Paired 3D Training Data

Fenja Falta(✉), Wiebke Heyer, Christoph Großbröhmer, and Mattias P. Heinrich

University of Lübeck, Lübeck, Germany
f.falta@uni-luebeck.de

Abstract. Deep learning-based medical image analysis often comes with the drawback of a very limited amount of publicly available data due to privacy concerns. For the task of image registration, this is especially relevant, since networks rely on image pairs for training. Augmenting the dataset with synthetic deformations does not suffice for anonymous publication, as patient-specific topologies are kept. To handle this, we propose to leverage vector quantisation- and diffusion-based image generation to synthesise paired data for training of registration networks. Our method introduces a new concept that directly synthesises paired data. We make this method applicable to both image data in the form of 2D brain MRI slices and sparse point cloud data in the form of 3D vessel representations of intra-patient lung scan pairs by volumetric rasterisation. With paired synthetic data, only minimal fine-tuning on privacy-sensitive real data is needed to achieve comparable or even better results to training on a large real dataset. The code is publicly available at https://github.com/multimodallearning/UnleashingRegistration.

Keywords: image generation · image registration · pulmonary vessels

1 Introduction

Deep-learning based medical image analysis – and image registration in particular – presents a difficult task in the realm of computer vision, since access to large amounts of data is limited due to patient privacy and tight regulations [20]. For chest X-ray 2D image classification, those restrictions have led to the emergence of new methods for synthetic data generation that reduce re-identification risks [22]. Image registration builds upon paired images, limiting the amount of accessible training data even more. In addition, since expert annotations are only scarcely available, supervision for image registration is often infeasible and solving the task relies strongly upon good coverage of the distribution of the data domain.

Since acquisition of medical data is not standardised, domain-gaps between data from different centres are prevalent. In order to employ methods into clinical practice, they need to either be domain-agnostic enough to cover this variety

M. Modat et al. (Eds.): WBIR 2024, LNCS 15249, pp. 45–59, 2024.
https://doi.org/10.1007/978-3-031-73480-9_4

or need to be trained to specifically work on a clinic's in-house data. Being able to make data publicly available that directly represents the distribution and domain of a private dataset would thus be advantageous for developmental purposes. The Learn2Reg OncoReg challenge[1] has demonstrated the difficulties associated with private in-house data. The challenge winner [26] used a generic encoder pre-trained on non-medical data in addition to modality-independent feature extraction to handle this.

While classic augmentation is a typical way of generating additional training data, it is limited by the applied transformations, and only generates samples that are identical to the existing training cases in terms of topology. It also offers limited benefit in concerns to privacy-preservation and only small advantages when the available data is limited. For registration, augmentation may include using arbitrary deformations to create new training pairs. This has the additional disadvantage, that these pairs are perfect deformations of each other, which is not properly reflective of the real task.

In the last years, image generation has become a very prominent field of research in the deep learning community, and the quality of generated images has made enormous progress. Denoising diffusion probabilistic models (DDPM) [14] re-formulate the process of image generation as a sequential denoising. While they are able to achieve great performance and create high-quality images, they come with a high computational effort and long training times. Vector-quantised image generation [4,9,30] presents a method that can generate high-resolution images with significantly less computational effort than competing diffusion probabilistic models. Because of this, they pose as a promising alternative for medical tasks. However, while some common anatomical shapes in 3D CT such as the rib cage can be synthesised [6] – to the best of our knowledge – no currently available approach is able to create highly complex 3D vessel trees that are unique in each person. Inner-lung anatomy would be highly relevant for training intra-patient registration models (cf. [10]).

In our work, we will explore the use of synthetically generated training data, especially synthetically generated lung vessels, to train a registration network.

1.1 Related Work

Synthetic Training Data. Previous work has shown that artificially enlarging a dataset by creating additional synthetic data can enhance the performance of a network for various computer vision tasks. Tremblay et al. [28] use domain randomisation to create synthetic data for object detection and show that training a network with this synthetic data and then fine-tuning on real data yields better performance for object detection than solely relying on real data. The benefits of synthetic training data have also been explored in the context of medical images, where small dataset sizes are very prevalent. Zhang et al. [33] showed that conditional data generation as a form of data augmentation surpasses classic data augmentation for classification of thyroid function. SinGAN [27]

[1] https://learn2reg.grand-challenge.org/oncoreg/.

generates synthetic endoscopic images with corresponding segmentation masks. They showed that training only on this synthetic data can improve the network's performance when the real training dataset is small. Shin et al. [25] followed a similar approach by generating synthetic brain MRI scans with corresponding segmentations. SynthMorph [15] showed that contrast-agnostic registration can be learned by training on synthetic geometric data, even if topologically it does not represent a realistic medical image domain.

Synthetic Deformations. Outside of the medical domain, synthetic data has been shown to be practical for scene flow estimation, where synthetic data can be created using arbitrary deformations and bears the possibility of full supervision during training. Many scene flow datasets like FlyingChairs [7] consist of synthetic data that come with ground truth flow fields. This concept can also be applied to lung registration, where synthetical deformations can be applied to lung images to create augmented data and provide a way to supervise the training [8]. Depending on the applied deformations, however, this may create a domain shift between training and test data [3] and an ample amount of real data is still needed.

Diffusion Models for Medical Applications. With the recent progress on diffusion models, there has consequently also been a lot of research concerning the applicability of them to medical tasks. A major challenge that has to be tackled are memory- and cost-efficient solutions for volumetric images. Hence, there is so far very limited work on 3D generative diffusion models. PatchDDM [2] gains its memory-efficiency by training only on coordinate-encoded patches, while sampling can be performed on a whole 3D volume. Make-A-Volume [35] fine-tunes a diffusion model with a 2D backbone on 3D data to synthesise 3D brain MRI. Friedrich et al. [11] addressed these challenges by abstracting the 3D shape synthesis process as a sparse point cloud generation and designed a point cloud diffusion network for 3D skull shape reconstruction. This topic is also actively pursued in 3D computer vision [29]. In Ying et al. [32] the VQ-GAN is used in a transformer-based diffusion model to reconstruct approximate CT renderings from multiple 2D X-ray views. Note, that the latter only employed a 2D token space for the diffusion model and hence lacks the ability to create arbitrarily complex 3D volumes – an important shortcoming we address in this work.

1.2 Our Contribution

We propose a new approach to generate synthetic paired volumetric data by training an unconditional generative diffusion model with multi-channel images. By creating a joint vector-quantised latent space that represents both fixed and moving images, the transformer-based sampler generates realistic paired scans with small computation effort. By employing volumetric representations of relevant vessels, we make this approach applicable to deformable lung registration. For registration, we make use of multi-step registration and inverse consistency. Furthermore, we evaluate the effect of employing test-time adaptation.

2 Methods

2.1 Synthetic Image Generation

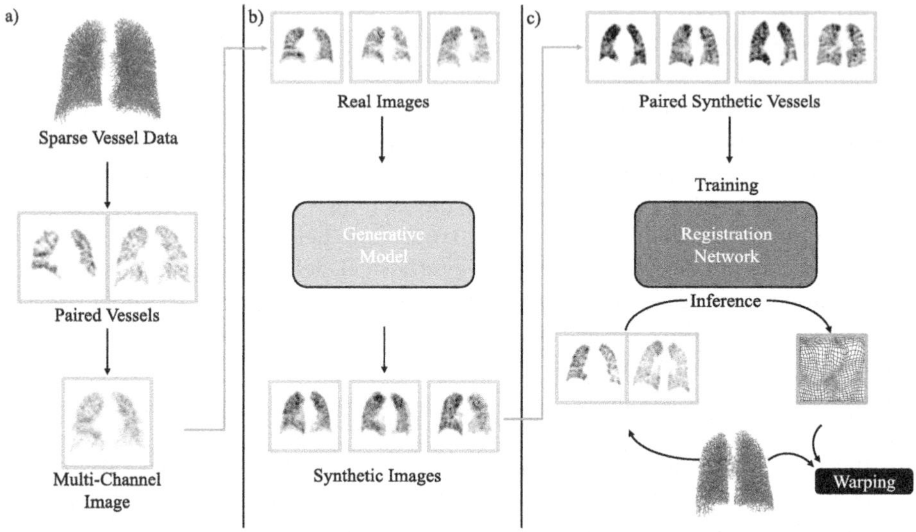

Fig. 1. Overview of our proposed method, exemplified using the PVT1010 dataset. a) Point cloud rasterisation and combination of fixed and moving images to a multi-channel image. b) Training of a generative model to synthesise images that represent paired data. c) Training of the registration network with paired synthetic data.

When training a neural network, the training data should reflect the distribution and variability of the test data. Keeping particularly image registration in mind, there are some characteristics that we try to address within the synthetic dataset generation.

1. **Similarity of the images**: The domain gap between the synthetically generated training images and the real test images should be minimal.
2. **Variability of the images**: The synthetic data should capture the range of images present in the real data. For medical images, this can be e.g. anatomical varieties or pathologies.
3. **Similarity and variability of the deformations**: In addition to the images, the deformations present between the synthetic images should be similar to those between real images. When training on synthetic data, the network should learn to predict deformations of the same scope as present in the test data.
4. **Relationship between paired images**: For intra-patient registration tasks, images within one pair stem from the same patient and correspond to different respiratory stages. Accordingly, images within one synthetic pair should share

those same similarities, while still being different to images from another image pair. Furthermore, within one pair, the images often correspond to specific time points or modalities. In lung registration, we expect the same vessel topology within one pair, while the deformation between them reflects their movement during respiration.

Conventional generative models try to address the first two points. As basis of our image generation, we use a vector-quantised variational autoencoder (VQ-VAE) in combination with an absorbing diffusion sampler as proposed by Bond-Taylor et al. [4]. The VQ-VAEs encodes an image into discrete latent representations using a finite set of learnable embeddings – so-called codebooks – and reconstructs the image from these discrete codes. This model is trained with reconstruction loss and a codebook loss. The absorbing diffusion model is then trained to predict the vector-quantised image embedding through masking based on the evidence lower bound (ELBO) loss. This combination provides a scalable, and computationally highly efficient method for image generation. Simply generating additional synthetic data with such a generative model, however, would not take into account the specifics of image registration data. Therefore, we propose to generate paired image samples as multi-channel data: One channel corresponds to fixed and moving image respectively and one channel to a combination of both. This is analogous to creating an overlaid RGB image of the grey-scale images. We can then re-extract grey-scale images from the generated multi-channel images and use them to train a registration network. (Fig. 1b)

By generating multi-channel data, the relationship between the channels – and thus between paired images – can be learned. This is crucial for intra-patient registration, but has another benefit: Instead of generating two typical images from the dataset, which could be very similar and thus encompass easier deformations on average, we generate a typical pair of images, which comes along with the differences of these images and thus difficulty of registering them. Compared to more complex alternatives – e.g. conditional image generation – our approach is independent of the employed network and even data domain.

2.2 Registration

In general, the previously described method can be applied independently of the registration network f itself.

For 2D registration, we employ a simple feature difference network that predicts a displacement field for fixed and moving image F and M either directly from $f(F) - f(M)$ or as $f_{\mathcal{D}}(f_{\mathcal{E}}(F) - f_{\mathcal{E}}(M))$ for an Encoder $f_{\mathcal{E}}$ and a Decoder $f_{\mathcal{D}}$.

For 3D registration, inspired by Mok et al. [18] and Zhao et al. [34], we employ a 2-step registration approach that predicts a deformation field φ^* for a pair of fixed and moving images F and M as the composition of two deformation fields φ_1 and φ_2. Additionally, we follow the very recent strategy of establishing symmetry and inverse consistency of the registration by construction following Greer et al. [12]. For this, instead of only predicting φ_i, we predict a backward-

and forward-deformation. The first deformation fields are obtained as

$$\varphi_1^M = \mathtt{SAS}(f_1(M,F) - f_1(F,M)), \qquad \varphi_1^F = \mathtt{SAS}(f_1(F,M) - f_1(M,F)), \quad (1)$$

where SAS represents the downsampling and exponential step of the scaling-and-squaring method [1]. The network f_1 is optimized to minimize the difference between F and $M \circ \varphi_1^M$ as well as M and $F \circ \varphi_1^F$. Based on the first deformation, mid-way warps

$$\hat{\varphi_1}^M = \mathtt{SAS}(\frac{f_1(M,F) - f_1(F,M)}{2}), \qquad \hat{\varphi_1}^F = \mathtt{SAS}(\frac{f_1(F,M) - f_1(M,F)}{2}) \quad (2)$$

are generated and used to warp M and F to

$$\hat{M} = M \circ \hat{\varphi_1}^M, \qquad \hat{F} = F \circ \hat{\varphi_1}^F, \quad (3)$$

similar to the ANTs SyN approach [1]. Equivalent to Eq. 1, φ_2^M and φ_2^F are obtained using the mid-way warps $\hat{M}$ and $\hat{F}$ as input to the network f_2. The final deformation field φ^* is then obtained as the composition

$$\varphi^* = \hat{\varphi_1}^M \circ \varphi_2^M \circ \hat{\varphi_1}^M. \quad (4)$$

For regularisation, we employ an implicit regularisation through B-spline transformation as well as an explicit one through a diffusion regularisation loss. Since we only use an image similarity loss and do not require annotations for supervision, we can use test-time adaptation to fine-tune the networks' weights.

3 Experiments and Results

3.1 Datasets

We test our method with three publicly available datasets, one dataset for 2D inter-patient registration and two datasets for 3D intra-patient vessel registration. All datasets include a comparably high number of samples, such that the training on real data should not be limited by too few data, and we can gain a fair comparison.

The **2D OASIS dataset** [16,17] consists of 414 skull-stripped intermediate brain MRI slices from different patients with segmentations of 24 structures. We randomly pair a subset of these slices for training and use fixed pairs of the remaining slices for testing. The registration accuracy is evaluated based on Dice overlap averaged over the 24 annotated structures. Note, that the expected accuracy will be limited due to unresolvable through-plane-deformations not captured when only registering mid-brain slices.

The **Lung250M-4B dataset** [10] comprises 248 3D chest CT scans of 124 patients from a diverse set of sources that show large motion between these two paired time points. The authors provided U-Net-based lung vessel segmentation and multiple thousand keypoints for evaluating performance very densely on 97 training pairs. We evaluate the registration performance using the target registration error (TRE) for these keypoints. It was shown that algorithms using exclusively vessel information and no intensity images can reach competitive performance without the need to release sensitive patient information encoded in CT scans.

The **PVT1010 dataset** [24] consists of 1000 pairs of sparse point clouds, automatically extracted using geometric (Hessian-like) filters of the pulmonary vessels during ex- and inspiratory phases. The test data consists of vessel point clouds for the 10 cases of the DIR-LAB COPDgene [5] dataset, annotated with manual landmarks for evaluation. We evaluate the registration accuracy using the TRE for these landmarks. To remove the increased complexity of creating synthetic point cloud representations (e.g. [29]) and directly training and evaluating graph-based registration algorithms such as PointPWC-Net [31] (adapted to lung registration in [24]), we propose to convert point clouds into rasterised vessel volumes.

Point Cloud Rasterisation. To make our method applicable to point cloud data, we employ the DiVRoC [13] approach, a differentiable rasterisation of point clouds to create vessel-like 3D volumes. When applying DiVRoC to point cloud data, the volumetric rasterisation converts point clouds onto a regular 3D grid and smooths the rasterisation with a specified Gaussian kernel. This leaves us with a smooth intensity image reflecting the positions and density of the pulmonary vessels (cf. Fig 1a). To ease the objective of the generative models, we use a rather large Gaussian kernel for rasterisation of $\sigma = 1.75$ for PVT1010 and dilated (binary) vessel segmentations for Lung250M-4B.

3.2 Image Generation

For the 2D task, we generate 8192 synthetic brain MRI pairs and obtain a total of 16 384 image samples that can be used to train the inter-patient registration network. For the 3D tasks, we generated 1280 synthetic vessel pairs from PVT1010 and 256 synthetic pairs from Lung250M-4B. Examples of real and generated synthetic images can be seen in Fig. 2 and Fig. 3. To train the VQ-AE for vessel reconstruction, we employ a cross-entropy loss. As can be seen in Fig. 3, the network seems to struggle with reconstruction of fine vessels. It achieves a Dice Score of $57.19 \pm 2.09\%$. Analogously, the synthetic vessel trees also consist of mostly larger vessels. The exact radius of the vessels is dependent on a chosen threshold, since the intensities of the synthetic images are continuous rather than binary (as can be seen from the smoother projections in Fig. 3). To omit this additional hyperparameter, we use the continuous vessel representation for registration.

Implementation Details. For the 2D generative diffusion model, we closely follow the publicly available implementation provided by Bond-Tayler et al. [4] and adapted the batch size (12) and iteration numbers (6000) to the capabilities of a single 48 GB GPU. The VQ-AutoEncoder is based on Esser et al. [9] and

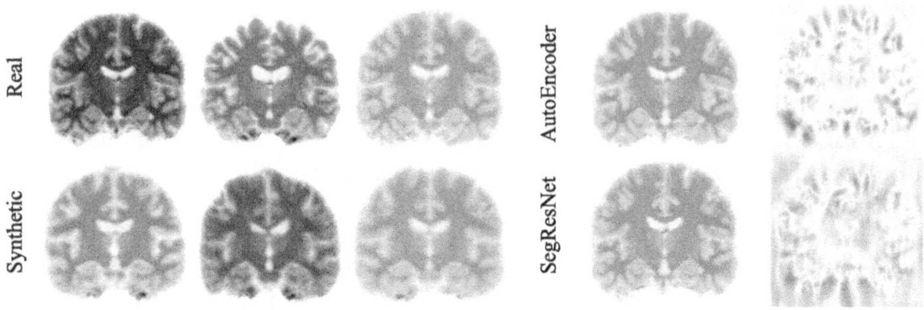

Fig. 2. Real and synthetically created data and registration results for 2D OASIS. On the top row (left) fixed and moving real images and the multi-component image as used to train the generative model are shown. Orange and blue hues correspond to differences in the images, while grey parts show similarities. The bottom row (left) shows the synthetically generated multi-component image and the two extracted greyscale images. On the right, registration results as overlay of fixed and transformed moving images and flow-fields are shown for the AutoEncoder (top) and the SegResNet (bottom). (Color figure online)

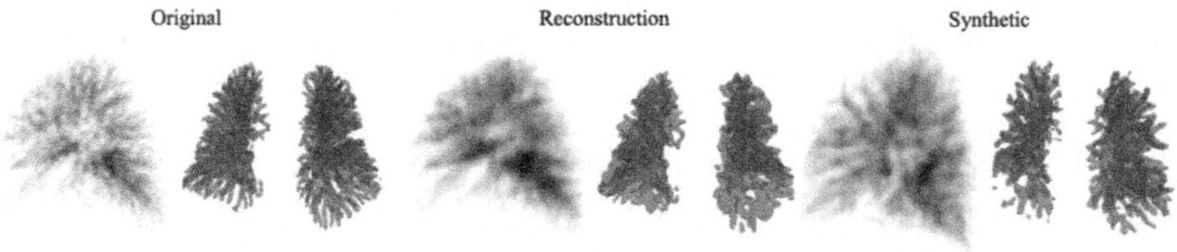

Fig. 3. Examples of a real vessel pair, the corresponding reconstructions with VQ-AE and a synthetic vessels pair. Top: Projections of fixed and moving vessels overlaid. Bottom: Polygon renderings of the inspiratory case.

encodes the whole 3-channel image of size 192×160 with a sequence of convolutional Res- and Attention-Blocks to a latent space of shape 12×10 with feature length of 512. The codebook size was left unaltered at 1024, but we reduced the initial feature channel width from 128 to 64 and included a perceptual VGG loss in addition to the L1 reconstruction and codebook losses. The 2D model can be trained in less than one hour. Subsequently, we generated the quantised latents for all training pairs and trained the absorbing transformer with 22 stacked (vanilla) transformer encoder layers for parallel token prediction with the weighted ELBO loss for 17500 iterations with a batch size of 16. The learning rate was warmed up within 500 steps to 0.0002.

For the 3D datasets, we used a patch-size of 128×128×128 and designed two own versions of the 3D VQ-AE implementation. We attempted to use the 3D VQ-VAE proposed by Ying et al. [32] for CT reconstruction, but found that it could not successfully reconstruct diverse 3D vessel geometries and resulted in a mode collapse (decoding always the same mean vessel tree). Therefore, a completely new 3D model was implemented based on the MONAI 3D AutoEncoder implementation with a channel size increasing from 48 up to 384 with eight 3D convolutional blocks in encoder and decoder each, with a total stride of 16 yielding a latent space of shape 8×8×8 and again a codebook of size 1024. No perceptual loss or discriminator was used (to avoid the mode collapse) and training for 42'000 iterations with a batch size of 8 lasted approx. 3 h. The 3D VQ-AE predicts four channels that are treated as two separate logits for fixed and moving vessels and supervised with a weighted cross-entropy loss. For the PVT1010 dataset all volumes were already rasterised to 128^3 whereas Lung250M-4B had varying dimensions. We padded/cropped the scans to 256×208×256 voxels (with a spacing of 1 mm). This yields a latent-space of 16×13×16 and a sequence length of 3'328, which was passed to a transformer with a width of 512, 16 heads, feed-forward dimension of 1024 and 12 blocks. The diffusion models were trained for 15'000 iterations with a batch size of 8.

3.3 Registration

Training Setup. We evaluate the registration accuracy with different combinations of the following datasets.

1. The complete real training dataset $\mathcal{R}_N$ with N data instances (315 cases for 2D OASIS, 988 pairs for PVT1010)
2. A small subset $\mathcal{R}_{10}$ of the real training dataset (10 cases)
3. Synthetically created data $\mathcal{S}_N$ with N synthetic data instances (up to 16'384 image samples for 2D OASIS, 1280 pairs for PVT1010)

2D Inter-Patient Registration. For 2D image registration, we employ a simple Mean Squared Error loss. As backbones, we use the AutoEncoder and SegResNet [21] architectures from the MONAI toolkit [19]. For regularisation, we employ B-spline smoothing and a diffusion regularisation loss. Furthermore, we limit the maximum displacements. Results are displayed in Table 1.

The best results overall are still achieved using the full real dataset as part of training data. The use of synthetic data in addition to the full real dataset as a form of augmentation (as proposed in e.g. [25,33]) results in comparable accuracy to only using the full real dataset. This shows that – as anticipated – our full real dataset appears to be large enough to not be limited by the number of samples. However, when limiting the amount of real training data the accuracy deteriorates by over 5% Dice Score. When then including our proposed synthetic data generation, this gap can be narrowed substantially.

When access to data is limited (i.e. only $\mathcal{R}_{10}$ may be used instead of $\mathcal{R}_{315}$), fine-tuning the network on the small real dataset gives the best results. How much the network can gain from synthetic data appears to be dependent on the network: Using the SegResNet backbone results in much smaller differences between real and synthetic training data than using the AutoEncoder backbone. The AutoEncoder model then profits greatly from fine-tuning. One reason could be a higher capacity of the SegResNet to capture information from a large training dataset due to a higher number of parameters and residual blocks.

Table 1. Quantitative evaluation of 2D inter-patient brain registration when training on different data. Accuracy is calculated as Dice overlap in percent after registration on 2D OASIS test cases. Results are averaged over 5 training runs. Standard deviations across runs are given in brackets. Improvements compared to $\mathcal{R}_{315}$ (top half of table) and to $\mathcal{R}_{10}$ (bottom) are shown. Best results overall are highlighted in orange, and best results without the full real dataset (bottom) are highlighted in bold.

Training data	AutoEncoder		SegResNet	
Initial	51.85		51.85	
$\mathcal{R}_{315}$	68.36 (0.19)		70.07 (0.13)	
$\mathcal{R}_{315} \cup \mathcal{S}_{315}$	68.44 (0.27)	+0.08	69.84 (0.21)	−0.23
$\mathcal{R}_{315} \cup \mathcal{S}_{16k}$	66.31 (0.23)	−2.05	68.34 (0.27)	−1.73
$\mathcal{R}_{10}$	64.95 (0.65)		64.79 (0.40)	
$\mathcal{S}_{16k}$	64.98 (0.27)	+0.03	67.64 (0.27)	+2.85
$\mathcal{R}_{10} \cup \mathcal{S}_{16k}$	65.00 (0.34)	+0.05	67.65 (0.27)	+2.86
$\mathcal{S}_{16k}$, fine-tuning on $\mathcal{R}_{10}$	**66.29 (0.28)**	**+1.34**	**67.74 (0.35)**	**+2.95**

3D Intra-patient Registration. For 3D intra-patient registration, we employ the registration method described in Sect. 2.2. As backbone, we use MONAI's U-Net [19,23]. For the dissimilarity loss, we choose the L_1-Norm. We optionally employ a test time adaptation step, that fine-tunes the network's parameters on the test data. During test time adaptation, we use the same unsupervised loss as during training and do not make use of the manual landmarks. The models are fine-tuned with exactly one real vessel pair, which is easily possible in clinical application, since this data is available.

Table 2. Quantitative results for intra-patient lung vessel registration. Target registration error (TRE) in mm after registration. Standard deviations across cases are given in brackets. Best results overall are highlighted in orange, and best results without training on the full real dataset are highlighted in bold.

	PVT1010		Lung250M-4B	
Training data	before TTA	after TTA	before TTA	after TTA
Initial	23.30		14.84	
$\mathcal{R}_{\text{all}}$	4.18 (2.06)	3.98 (1.47)	4.09 (3.80)	2.77 (2.33)
$\mathcal{R}_{10}$	8.03 (3.72)	4.38 (2.03)	10.99 (7.77)	3.78 (5.25)
$\mathcal{S}_{\text{all}}$	**6.31 (3.03)**	3.71 (1.17)	**8.50 (7.01)**	**3.10 (3.90)**

For the PVT1010 dataset, results are displayed in Table 2 and Fig. 4. Before test-time adaptation (TTA), there is a substantial performance difference (nearly 4 mm TRE) between using the large or small dataset for training. Additional synthetic training data helps to reduce this gap, but does not achieve a score

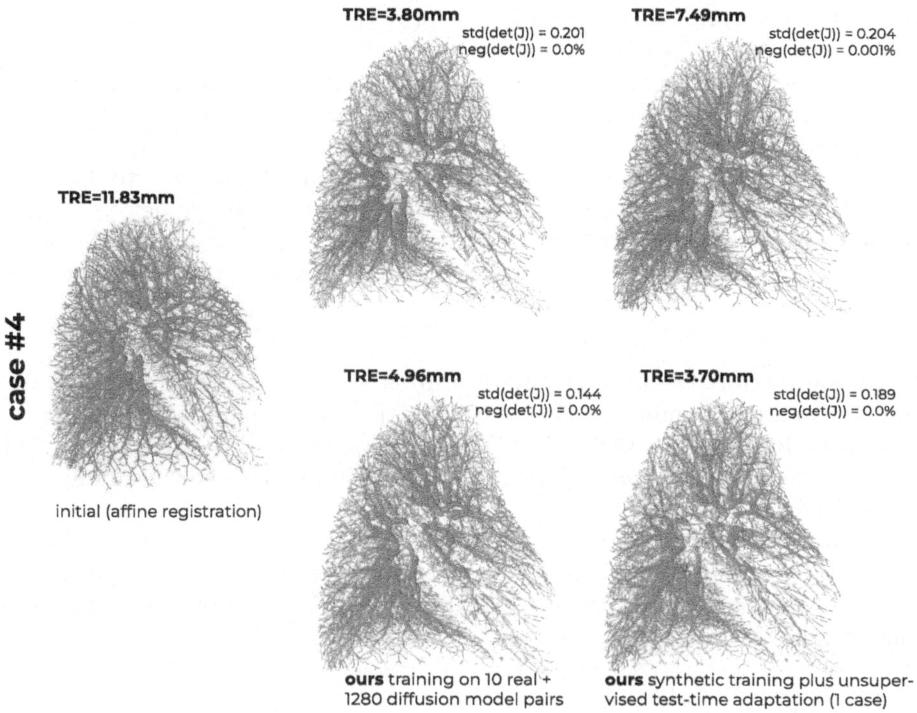

Fig. 4. Results after registration of a singular test case for intra-patient 3D vessel registration. Displayed are overlaid vessel trees (Red/Blue: Inspiratory/Expiratory phase) for different training setups with corresponding target registration error and statistics of the Jacobian determinant. (Color figure online)

comparable to training with the large real dataset. The improvement gained from TTA is only small (0.2 mm TRE) with the full real dataset, indicating that the training data is quite representative of the test data. The network trained on only very little data gets significantly better through TTA, but still hangs behind compared to the other. For the network trained on synthetic data, we can also see a strong improvement (2.6 mm TRE), that even results in outperforming the network trained on the large real dataset by about 0.3 mm TRE.

With Lung250M-4B, when training on $\mathcal{R}_{10}$, we achieve a TRE of 10.99 mm without TTA and 3.78 mm with 50 iterations of TTA. Training on the proposed synthetic dataset $\mathcal{S}_{256}$ yields a TRE of 8.50 mm without and 3.10 mm with TTA, outperforming $\mathcal{R}_{10}$ in both scenarios. Note that the reconstruction by the VQ-AE (as used as the basis for the generative model) already leads to a visual domain gap, but nevertheless a model entirely trained on this synthetic paired data performs much better (50% TRE reduction) compared to one that was trained on a smaller real dataset without domain gap. The advantage of our method remains at approx. 22% after adapting to each individual instance.

By design (cf. [12]) all our resulting deformations are inverse consistent and exhibit no negative Jacobian determinant values.

The best-performing learning-based method on this dataset benchmark, VoxelMorph++, uses the CT volumes and thus dense and patient-sensitive image information. In its basic form, it achieves a TRE of 5.50 mm. When trained with additional keypoint-based supervision, this can be improved to 4.20 mm. A subsequent instance optimisation that predicts the remaining deformation of the already deformed test images in a conventional optimisation-based manner results in the best reported performance within learning-based methods with 2.67 mm TRE.

Implementation Details and Run Times. Further implementation details can be found in the publicly released code: https://anonymous.4open.science/r/UnleashingRegistration-4374. The 2D registration network was trained for 1500 iterations with a batch size of 48 on an NVIDIA Titan RTX and runs for approx. 30–90 s depending on the backbone. For fine-tuning, it was trained for an additional 100 iterations. Inference takes a few milliseconds. The 3D registration networks can be trained within 45 min and inferred on new scan pairs in subseconds. When performing test time adaptation for 50 iterations on a single test pair, it takes about 15 s per case.

4 Discussion

We proposed an approach to generate synthetic paired data for the task of 2D and 3D deformable image registration. By creating data pairs as singular data instances made up of multiple components (fixed and moving image), we are able to gain synthetic pairs, that capture the differences and similarities of real data pairs. When using this synthetic data for training, only minimal fine-tuning on real data is necessary to achieve comparable performance – on the 2D task – to training with a large real dataset. With test-time adaptation, it is even possible to outperform a model trained on a large real dataset (3D lung vessel registration on PVT1010), showing the variability present in the synthetic dataset.

Different from data augmentation, synthetic data generation bears the possibility to make data publicly available where otherwise privacy-limitations would restrict a publication. Even with limited access to real training data, a model can be trained that achieves good registration performance. Its main advantage lies in the high variability, but a domain gap to real data is still present, as reflected in worse network performance when solely training on synthetic data. Fine-tuning and test-time adaptation can be performed even with very little computational effort and in a short time, so they present a viable approach for challenges as well as in clinical practice.

A main limitation in generating synthetic data lies in the type of data the generative network is able to robustly reconstruct and thus generate samples from. While we implemented a generic memory- and compute-efficient 3D generative network based on absorbing transformers in this work, a network that can robustly reconstruct even fine vessels topologically correctly could help to elevate the quality of the synthetic data and thus the trained network's performance even more.

Acknowledgments. This study was partly funded by the German Federal Ministry of Education and Research (BMBF) (grant number 16SV9254).

Disclosure of Interests. The authors have no competing interests to declare that are relevant to the content of this article.

References

1. Avants, B.B., Epstein, C.L., Grossman, M., Gee, J.C.: Symmetric diffeomorphic image registration with cross-correlation: evaluating automated labeling of elderly and neurodegenerative brain. Med. Image Anal. **12**(1), 26–41 (2008)
2. Bieder, F., Wolleb, J., Durrer, A., Sandkuehler, R., Cattin, P.C.: Denoising diffusion models for memory-efficient processing of 3D medical images. In: Medical Imaging with Deep Learning, pp. 552–567. PMLR (2024)
3. Bigalke, A., Hansen, L., Heinrich, M.P.: Adapting the mean teacher for keypoint-based lung registration under geometric domain shifts. In: International Conference on Medical Image Computing and Computer-Assisted Intervention, pp. 280–290. Springer (2022).
4. Bond-Taylor, S., Hessey, P., Sasaki, H., Breckon, T.P., Willcocks, C.G.: Unleashing transformers: parallel token prediction with discrete absorbing diffusion for fast high-resolution image generation from vector-quantized codes. In: European Conference on Computer Vision, pp. 170–188. Springer (2022).
5. Castillo, R., Castillo, E., Fuentes, D., Ahmad, M., Wood, A.M., Ludwig, M.S., Guerrero, T.: A reference dataset for deformable image registration spatial accuracy evaluation using the COPDgene study archive. Phys. Med. Biol. **58**(9), 2861 (2013)
6. Corona-Figueroa, A., et al.: Unaligned 2D to 3D translation with conditional vector-quantized code diffusion using transformers. In: Proceedings of the IEEE/CVF International Conference on Computer Vision, pp. 14585–14594 (2023)
7. Dosovitskiy, A., et al.: FlowNet: learning optical flow with convolutional networks. In: IEEE International Conference on Computer Vision (ICCV) (2015)
8. Eppenhof, K.A., Lafarge, M.W., Veta, M., Pluim, J.P.: Progressively trained convolutional neural networks for deformable image registration. IEEE Trans. Med. Imaging **39**(5), 1594–1604 (2019)
9. Esser, P., Rombach, R., Ommer, B.: Taming transformers for high-resolution image synthesis. In: Proceedings of the IEEE/CVF Conference on Computer Vision and Pattern Recognition, pp. 12873–12883 (2021)
10. Falta, F., Großbröhmer, C., Hering, A., Bigalke, A., Heinrich, M.: Lung250M-4B: a combined 3D dataset for CT-and point cloud-based intra-patient lung registration. Adv. Neural Inf. Process. Syst. **36** (2024)
11. Friedrich, P., Wolleb, J., Bieder, F., Thieringer, F.M., Cattin, P.C.: Point cloud diffusion models for automatic implant generation. In: International Conference on Medical Image Computing and Computer-Assisted Intervention, pp. 112–122. Springer (2023).

12. Greer, H., et al.: Inverse consistency by construction for multistep deep registration. In: International Conference on Medical Image Computing and Computer-Assisted Intervention, pp. 688–698. Springer (2023).
13. Heinrich, M.P., Bigalke, A., Großbröhmer, C., Hansen, L.: Chasing clouds: differentiable volumetric rasterisation of point clouds as a highly efficient and accurate loss for large-scale deformable 3D registration. In: Proceedings of the IEEE/CVF International Conference on Computer Vision, pp. 8026–8036 (2023)
14. Ho, J., Jain, A., Abbeel, P.: Denoising diffusion probabilistic models. Adv. Neural Inf. Process. Syst. **33**, 6840–6851 (2020)
15. Hoffmann, M., Billot, B., Greve, D.N., Iglesias, J.E., Fischl, B., Dalca, A.V.: SynthMorph: learning contrast-invariant registration without acquired images. IEEE Trans. Med. Imaging **41**(3), 543–558 (2021)
16. Hoopes, A., Hoffmann, M., Greve, D.N., Fischl, B., Guttag, J., Dalca, A.V.: Learning the effect of registration hyperparameters with hypermorph. J. Mach. Learn. Biomed. Imaging **1** (2022)
17. Marcus, D.S., Wang, T.H., Parker, J., Csernansky, J.G., Morris, J.C., Buckner, R.L.: Open access series of imaging studies (OASIS): cross-sectional MRI data in young, middle aged, nondemented, and demented older adults. J. cogn. neurosci. **19**(9), 1498–1507 (2007)
18. Mok, T.C., Chung, A.C.: Large deformation diffeomorphic image registration with laplacian pyramid networks. In: Medical Image Computing and Computer Assisted Intervention–MICCAI 2020: 23rd International Conference, Lima, Peru, October 4–8, 2020, Proceedings, Part III 23, pp. 211–221. Springer (2020).
19. MONAI Consortium: MONAI: Medical open network for AI. https://monai.io
20. Mostert, M., Bredenoord, A.L., Biesaart, M.C., Van Delden, J.J.: Big data in medical research and EU data protection law: challenges to the consent or anonymise approach. Eur. J. Hum. Genet. **24**(7), 956–960 (2016)
21. Myronenko, A.: 3D MRI brain tumor segmentation using autoencoder regularization. In: Brainlesion: Glioma, Multiple Sclerosis, Stroke and Traumatic Brain Injuries: 4th International Workshop, BrainLes 2018, Held in Conjunction with MICCAI 2018, Granada, Spain, September 16, 2018, Revised Selected Papers, Part II 4, pp. 311–320. Springer (2019).
22. Packhäuser, K., Folle, L., Nguyen, T.T., Thamm, F., Maier, A.: Privacy-enhancing image sampling for the synthesis of high-quality anonymous chest radiographs. In: BVM Workshop, pp. 27–32. Springer (2024).
23. Ronneberger, O., Fischer, P., Brox, T.: U-Net: Convolutional networks for biomedical image segmentation. In: Medical Image Computing and Computer-assisted Intervention–MICCAI 2015: 18th International Conference, Munich, Germany, October 5-9, 2015, proceedings, part III 18, pp. 234–241. Springer (2015).
24. Shen, Z.,et al.: Accurate point cloud registration with robust optimal transport. Adv. Neural Inf. Process. Syst. **34**, 5373–5389 (2021)
25. Shin, H.C., et al.: Medical image synthesis for data augmentation and anonymization using generative adversarial networks. In: Simulation and Synthesis in Medical Imaging: Third International Workshop, SASHIMI 2018, Held in Conjunction with MICCAI 2018, Granada, Spain, September 16, 2018, Proceedings 3, pp. 1–11. Springer (2018).
26. Song, X., Xu, X., Yan, P.: General purpose image encoder DINOv2 for medical image registration. arXiv preprint arXiv:2402.15687 (2024)
27. Thambawita, V., et al.: SinGAN-seg: synthetic training data generation for medical image segmentation. PloS one **17**(5), e0267976 (2022)

28. Tremblay, J., et al.: Training deep networks with synthetic data: bridging the reality gap by domain randomization. In: Proceedings of the IEEE Conference on Computer Vision and Pattern Recognition Workshops, pp. 969–977 (2018)
29. Vahdat, A., et al.: Lion: latent point diffusion models for 3D shape generation. Adv. Neural Inf. Process. Syst. **35**, 10021–10039 (2022)
30. Van Den Oord, A., Vinyals, O., et al.: Neural discrete representation learning. Adv. Neural Inf. Process. Syst. **30** (2017)
31. Wu, W., Wang, Z.Y., Li, Z., Liu, W., Fuxin, L.: PointPWC-Net: cost volume on point clouds for (self-) supervised scene flow estimation. In: Computer Vision–ECCV 2020: 16th European Conference, Glasgow, UK, August 23–28, 2020, Proceedings, Part V 16, pp. 88–107. Springer (2020).
32. Ying, X., Guo, H., Ma, K., Wu, J., Weng, Z., Zheng, Y.: X2CT-GAN: reconstructing CT from biplanar x-rays with generative adversarial networks. In: Proceedings of the IEEE/CVF Conference on Computer Vision and Pattern Recognition, pp. 10619–10628 (2019)
33. Zhang, H., Huang, Z., Lv, Z.: Medical image synthetic data augmentation using GAN. In: Proceedings of the 4th International Conference on Computer Science and Application Engineering, pp. 1–6 (2020)
34. Zhao, S., Dong, Y., Chang, E.I., Xu, Y., et al.: Recursive cascaded networks for unsupervised medical image registration. In: Proceedings of the IEEE/CVF international conference on computer vision, pp. 10600–10610 (2019)
35. Zhu, L., Xue, Z., Jin, Z., Liu, X., He, J., Liu, Z., Yu, L.: Make-a-volume: Leveraging latent diffusion models for cross-modality 3D brain MRI synthesis. In: International Conference on Medical Image Computing and Computer-Assisted Intervention. pp. 592–601. Springer (2023).

Feedback Attention for Unsupervised Cardiac Motion Estimation in 3D Echocardiography

Md. Kamrul Hasan(✉), Guang Yang, and Choon Hwai Yap

Department of Bioengineering, Imperial College London, London, UK
{k.hasan22,g.yang,c.yap}@imperial.ac.uk

Abstract. Echocardiography motion estimation is essential for clinical cardiac health assessments, such as myocardial strain and ejection fraction. Attaining reliable performance with deep learning image registration (DLIR) is traditionally challenging due to intrinsic noise and fuzzy anatomic boundaries in echocardiography. It is further advantageous to achieve DLIR in 3D, as the cardiac anatomy has complex 3D structures and motions that are difficult to understand in 2D, especially with disease and malformations. However, successful regularization strategies for 2D DLIR are often not as effective in 3D. To date, there have been few 3D DLIR implementations for adult echocardiography and none in fetal echocardiography, where small fetal cardiac structures pose additional challenges. Here, we propose a self-supervision module for an unsupervised DLIR network for 3D+time echocardiography and demonstrate its good performance in both adult and fetal echocardiography images. We introduce a novel feedback spatial transformer module where the registration outputs are used to generate a co-attention map that describes the remaining registration errors to guide the network's spatial emphasis during DLIR training. This effectively facilitates self-supervision. This feedback attention approach could be added to existing transformer-based approaches, including the co-attention spatial transformer with and without the spatial and channel attention in the DLIR backbone, bringing about non-overlapping benefits with existing approaches. Our results with 3D images here suggest that a focus on the resulting image after registration warping is key to good DLIR performance, and this is consistent with our earlier 2D DLIR investigations. Codes are available at https://github.com/kamruleee51/Feedback_DLIR.

Keywords: 4D (3D+time) echocardiography · Cardiac motion estimation · Echocardiography registration · Feedback attention · Self-supervised deep learning image registration

1 Introduction

Performing cardiac motion estimation on echocardiography is crucial for assessing cardiac function and identifying dysfunctions. Motion estimation can achieve

M. Modat et al. (Eds.): WBIR 2024, LNCS 15249, pp. 60–72, 2024.
https://doi.org/10.1007/978-3-031-73480-9_5

comparable precision as speckle tracking in deriving myocardial (MYO) strains [1], which are important indicators of cardiac contractility health. For example, it is an important criterion to assess chemotherapy-induced cardiotoxicity [2] to inform heart failure management [3]. In fetal echocardiography, this is useful for determining if the heart is sufficiently diseased to warrant interventions, such as placenta ablation for twin-to-twin transfusion syndrome [4] or catheter-based fetal heart intervention [5]. To date, cardiac deformation measurements via echocardiography have limited precision in both adult [6] and fetal [7] echocardiography, with high variability across different reports. This is especially poor in fetal echocardiography, where the right ventricle strains can vary by 2.5 times across different studies [7]. It is thus essential to develop effective deep learning (DL) approaches for resolving precision problems.

Although 3D+time echocardiography imaging is widely available, the clinical norm has remained with 2D due to the limited effectiveness of current 3D echocardiography visualization, where anatomic details are missed even by experts [8]. However, 2D imaging can have a misaligned imaging plane, causing normal hearts to be mistaken as defective and vice versa [9]. It is further difficult to mentally understand the complex 3D cardiac anatomy and motion in 2D, especially with complex diseases with malformations, and 3D images can achieve more accurate volumetric measurements. To date, however, no DLIR algorithm has been applied to 3D fetal echocardiography, which is one of the focuses here.

Various DLIR techniques have been proposed, and robust results have been obtained for 2D echocardiography. This includes the PWC-Net, which uses a feature pyramid extractor and an optical flow estimator [10] and a patch-based multilayer perceptron (MLP) and transformer network [11]. We have previously proposed a 2D DLIR incorporating an anatomic shape-encoder and an adversarial image texture constraint to a multi-scale DLIR [12]. However, techniques successful in 2D are often less effective in 3D due to the increased dimensionality and information in 3D images and the difficulty in scaling up training to 3D. To date, only one recent study demonstrated successful unsupervised DLIR in 3D adult echocardiography [13], where they proposed a co-attention [14] spatial transformer to provide this necessary regularization. However, there remains room for improvement in 3D echocardiography, and there are further challenges in fetal echocardiography with regard to resolution and image quality.

Here, we propose to add our novel feedback attention (FBA) module to the unsupervised DLIR transformer approaches. We use the warped image obtained from the registration output to compute a co-attention map that indicates locations where registration errors remain high and insert this co-attention map as feedback to the registration network to guide the training. This effectively facilitates the self-supervision of the unsupervised DLIR transformer approaches. This is a logical next step and is in line with our previous 2D DLIR studies, where we found that an additional focus on the warped image quality is capable of improving existing DLIR strategies and is sufficient to achieve robust results. We show that FBA can similarly enhance 3D DLIR and can produce robust results in 3D adult and fetal echocardiography. Here, we show that our FBA approach can provide non-overlapping benefits with existing transformers and can thus be useful add-ons for synergistic effects. Our specific contributions are:

- We propose a new feedback attention (FBA) module to guide the training of the unsupervised 3D DLIR, where the DLIR's outputs are used to compute a spatial co-attention map for feedback to the DLIR for self-supervision. We show that informing the remaining registration error map to the network via FBA improves 3D DLIR performance and show that this is due to an improved ability to correctly emphasize regions of interest.
- We couple our novel FBA to a spatial and channel attention (SCA) [15]-based backbone network to drive the network on what spatial locations and features to emphasize. We show that SCA has a non-overlapping advantage with FBA, and this coupling improves performance.
- We demonstrate that improved performance for 3D unsupervised DLIR with our novel FBA enables robust results in two different 3D echocardiography datasets. For the first time, we show that robust results in 3D fetal echocardiography are achievable despite poorer resolution and image quality than an adult echocardiography.

2 Methodology and Materials

2.1 Methods

Figure 1 depicts our proposed DLIR pipeline, consisting of a SCA-based backbone DLIR blocks (**A1** and **A2**) that predicts a deformation field from an image pair of fixed and moving images and two spatial attention blocks (**B** and **C**) that modify the input images to block **A** with a co-attention map between images in the pair. The first time point is considered a fixed image (here, it is the ED frame), while other time points are considered the moving image (here, it is the ES frame) for the cardiac motion estimation of the patient.

Blocks **A1** and **A2** are identical networks with the same shared weights. Block **B** generates the co-attention map between the original fixed and moving images to be applied to both images for the first registration block **A1** to highlight the spatial locale that needs emphasis during registration. Block **C** generates the co-attention map between the warped image (moving image deformed by initial registration output) and the fixed image, which is to be applied to the fixed image as feedback from the results for inputs into a second registration block **A2**, providing self-supervision. This co-attention map in block **C** plays a crucial role in highlighting the spatial locales of dissimilarities between the warped and fixed images and, thus, of registration errors. Since block **C** uses the outputs of registration to modify its inputs, it serves as a form of feedback mechanism, making the model adaptable and responsive. More details of blocks **A**, **B**, and **C** are provided below.

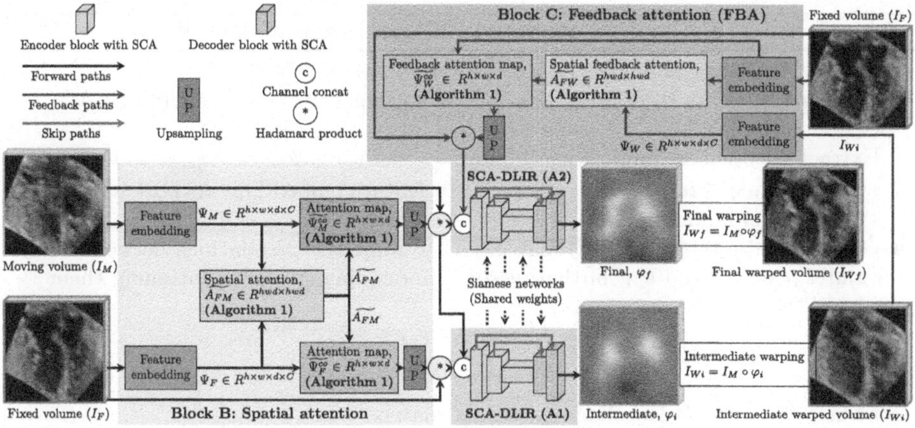

Fig. 1. The proposed self-supervised FBA-DLIR pipeline using our novel FBA (block **C**) module, incorporating Siamese SCA-DLIR networks (block **A1** and **A2**) and spatial co-attention (block **B**). All blocks **A**, **B**, and **C** are trainable. Our FBA module, block **C** (Algorithm 1), facilitates the SCA-DLIR's self-supervision.

The Unsupervised DLIR Networks (block A1 and A2). This backbone DLIRs consist of three encoding and decoding blocks in the UNet structure ($\mathcal{F}_{CNN}$), with 128, 256, and 512 filters in each block. $\mathcal{F}_{CNN}$ predict a deformation field $\varphi(x)$ from a pair of N-dimensional input images, $\varphi(x) = \mathcal{F}_{CNN}(I_F, I_M; \Theta)$; where Θ is training parameters and $\varphi(x) : \mathcal{R}^N \mapsto \mathcal{R}^N$. To highlight the essential anatomical regions and focus on relevant features across the channels, we propose to use SCA [15] in the encoder and decoder of Vanilla DLIR, named SCA-DLIR (blocks **A1** and **A2**). Blocks **A1** and **A2** are Siamese blocks using the same weights, where **A1** is used for the initial registration between the fixed and moving images, and **A2** is used for the subsequent registration after novel feedback is applied to the fixed image. To train $\mathcal{F}_{CNN}$ (Vanilla or SCA-DLIR), Eq. 1 is optimized.

$$\widehat{\Theta} = \arg\max_{\Theta} \left\{\mathcal{L}_{img}(I_F, I_M \circ \varphi) + \lambda_r \times \mathcal{L}_r(\varphi)\right\} \tag{1}$$

where φ is the deformation field and $\mathcal{L}_{img}$ is an image-based similarity metric such as normalized cross-correlation (NCC), mean square error (MSE), cross-entropy (CE), or mutual information (MI). Here, we choose to use MSE [16]. $\mathcal{L}_r$ is a smoothness regularization [16], and $\lambda_r \geq 0$ is the smoothness weight. However, this DLIR approach with only $\mathcal{L}_{img}$ does not produce optimal results, which could, in part, be due to echocardiography's challenges in resolution, image quality, noise, and fuzzy cardiac structures. Therefore, we propose to use our novel FBA as additional self-supervision (block **C**) and spatial co-attention (block **B**) to improve it.

Algorithm 1: Estimation of spatial FBA and spatial attention.

Input: An image pair $\{I_F, I_M\}$ or $\{I_F, I_W\} \in \mathbb{R}^{H \times W \times D}$
Output: Spatial attention ($\xi \in \mathbb{R}^{H \times W \times D}$) using $\{I_F, I_M\}$ or spatial FBA ($\vartheta \in \mathbb{R}^{H \times W \times D}$) using $\{I_F, I_W\}$

1 Extract encoded semantic feature maps $\Psi_F = \mathcal{E}(I_F)$, $\Psi_M = \mathcal{E}(I_M)$, and $\Psi_W = \mathcal{E}(I_W) \in \mathbb{R}^{C \times h \times w \times d}$, where $\mathcal{E}$ is a feature embedding network and C is the channel numbers

2 Compute affinity (similarity) matrix (A) to mine the correlations between Ψ_F and Ψ_M or Ψ_F and Ψ_W in their feature embedding space by flattening them ($\in \mathbb{R}^{C \times hwd}$),

$$A_{FM} = \Psi_F^T \cdot W_{fm} \cdot \Psi_M \in \mathbb{R}^{hwd \times hwd},$$

$$A_{FW} = \Psi_F^T \cdot W_{fw} \cdot \Psi_W \in \mathbb{R}^{hwd \times hwd},$$

where $W \in \mathbb{R}^{C \times C}$ is a diagonal weight matrix.

3 Normalize the affinity matrix using a softmax activation (σ_1) as $\widetilde{A_{FM}} = \sigma_1(A_{FM})$ or $\widetilde{A_{FW}} = \sigma_1(A_{FW})$, which reflects the relevance of each feature ($k \in \{1, 2, ..., hwd\}$) in Ψ_F to the kth feature in Ψ_M or Ψ_W

4 Estimate co-attention enhanced attention feature map as: $\Psi_F^{co} = \Psi_F \cdot \widetilde{A_{FM}} \in \mathbb{R}^{C \times hwd}$ (for fixed image) and $\Psi_M^{co} = \Psi_M \cdot \widetilde{A_{FM}} \in \mathbb{R}^{C \times hwd}$ (for moving image), and $\Psi_W^{co} = \Psi_F \cdot \widetilde{A_{FW}} \in \mathbb{R}^{C \times hwd}$ (for feedback attention).

5 Apply post-processing to those co-attention enhanced feature maps (unflattened $\in \mathbb{R}^{C \times h \times w \times d}$) using a convolutional kernel (γ), bias (Ω), and sigmoid activation (σ_2) with an output channel of 1: $\widetilde{\Psi^{co}} = \sigma_2(\gamma \Psi^{co} + \Omega) \in [0, 1]$, where $\widetilde{\Psi^{co}} \in \mathbb{R}^{h \times w \times d}$

6 Upsampling the refined attention maps using a trilinear interpolation (Υ) provides: $\xi_F = \Upsilon(\widetilde{\Psi_F^{co}}) \in \mathbb{R}^{H \times W \times D}$, $\xi_M = \Upsilon(\widetilde{\Psi_M^{co}}) \in \mathbb{R}^{H \times W \times D}$, and $\vartheta_W = \Upsilon(\widetilde{\Psi_W^{co}}) \in \mathbb{R}^{H \times W \times D}$

FBA for Self-supervision (block C) and Spatial Attention (block B). For the inputted fixed ($I_F \in \mathbb{R}^{H \times W \times D}$) and moving ($I_M \in \mathbb{R}^{H \times W \times D}$) images, we aim to find a φ to warp the I_M onto the I_F as $I_W = I_M \circ \varphi \in \mathbb{R}^{H \times W \times D}$, where H, W, and D indicate height, width, and depth of the 3D volume, respectively. In blocks **B** and **C**, we compute the 3D co-attention map [14] via feature embedding to direct our DLIR to emphasize coherent anatomical locations. This enables the DLIR to understand the association between two images in a global semantic feature embedded space. The embedding network ($\mathcal{E}$) consists of a series of 3D convolutions having filters of 64, 128, and 256 and an atrous spatial pyramid pooling [17] module to use parallel dilated convolutions with different rates (6, 12, and 18) to capture multi-scale contextual information. $\mathcal{E}$ returns an embedded features map of $\Psi = \mathcal{E}(I) \in \mathbb{R}^{C(=256) \times h \times w \times d}$, where $h, w, d = H/8, W/8, D/8$.

Algorithm 1 explains the procedures of spatial FBA (ϑ) and spatial co-attention (ξ) estimation. The obtained spatial co-attention maps (ξ_F and ξ_M) and the FBA co-attention map (ϑ_W) are then applied to the input images to highlight locations of importance. Since those co-attention maps are trainable to provide additional information to the backbones **A1** (generates initial field,

φ_i) and **A2** (generates final field, φ_f), their learning process is controlled by the NCC loss ($\mathcal{C}$), as defined in Eq. 2.

$$\mathcal{C}_{FBA} = \frac{\sum_{u=-w/2}^{w/2} \left(\xi_M \circ \varphi_f(p+u) - \mu_{\xi_M \circ \varphi_f}\right)\left(\vartheta_W(p+u) - \mu_{\vartheta_W}\right)}{\sqrt{\sum_{u=-w/2}^{w/2} \left(\xi_M \circ \varphi_f(p+u) - \mu_{\xi_M \circ \varphi_f}\right)^2} \sqrt{\sum_{u=-w/2}^{w/2} \left(\vartheta_W(p+u) - \mu_{\vartheta_W}\right)^2}}, \tag{2}$$

where p and μ are voxels and the mean of the attention maps in the cube window (w). To estimate the similar loss for the spatial co-attention block ($\mathcal{C}_{CoAT}$), Eq. 2 has been used with φ_i and ξ_F instead of φ_f and ϑ_W.

For the first registration in block **A1**, co-attention maps from block **B** are applied to the original images, I_F and I_M, as $I'_F = I_F \times \xi_F$, and $I'_M = I_M \times \xi_M \in \mathbb{R}^{H \times W \times D}$, while in the second registration in block **A2**, the co-attention map from block **C** is applied to the fixed image, as $I''_F = I_F \times \vartheta_W \in \mathbb{R}^{H \times W \times D}$, while the moving image retains its modification by the co-attention map from block **B**. Those refined volume pairs (I'_F & I'_M and I''_F & I'_M) are subjected to another constraint to control the learning process using the MSE loss ($\mathcal{M}$), as defined in Eq. 3.

$$\mathcal{M}_{FBA/CoAT} = \frac{1}{H \times W \times D} \sum_{i=1}^{H} \sum_{j=1}^{W} \sum_{k=1}^{D} \left(I''_F/I'_F(i,j,k) - I'_M(i,j,k) \circ \varphi_{f/i}\right)^2. \tag{3}$$

The final loss function to train our novel FBA-DLIR is defined in Eq. 4, where $\mathcal{C}_{FBA}$ and $\mathcal{M}_{FBA}$ provide self-supervision as feedback.

$$\begin{aligned} \widehat{\Theta} = \arg\max_{\Theta} \{ & \lambda_1 \mathcal{L}_{img}(I_F, I_M \circ \varphi_f) + \lambda_2 \times \mathcal{C}_{FBA} + \lambda_3 \times \mathcal{M}_{FBA} \\ & + \lambda_4 \times \mathcal{C}_{CoAT} + \lambda_5 \times \mathcal{M}_{CoAT} \}, \end{aligned} \tag{4}$$

where λ_{1-5} are the weighting factors, and all the loss terms are normalized to [0, 1]. Since all loss terms are normalized between [0, 1], the weighting factors are set to $\lambda_{1-5} = 1.0$ for this study, and it is sufficient to advocate for our novel FBA approach. Ahn et al. [13] reported that deformation fields were naturally smooth with spatial co-attention, and regularization ($\mathcal{L}_r(\varphi)$) of the DLIR block would not be needed. As such, we did not use regularization in our co-attention and FBA-based DLIR networks. Experimental results in Table 1 show the smoothness of the deformations fields.

2.2 Datasets and Training

Private Fetal 3D Echocardiography: Our fetal echocardiography dataset is a 4D (3D + time) multi-demographic fetal echocardiography collection of 26 fetal patients with 25 to 40 time frames, resulting in a total of 983 3D volumes. Images were acquired with the GE Volusion 730 ultrasound machine under the

STIC mode with the RAB $4-8L$ transducer and have an in-plane image resolution of $0.95\mu m \times 0.90\mu m$ (see details in [18]). Ethics approval was obtained under Kepler University Hospital IRB protocol 1009/2017 and the NUS DSRB protocol 2014/00056, and written informed consent was obtained from all patients. The endocardial boundaries and the MYO tissue space are manually labeled at the ES and ED time points and carefully smoothed with as little compromise to segmentation accuracy as possible, and a validated cardiac motion estimation algorithm (multi-scale Elastix with cyclic regularization) is used to propagate the labels to other time points [19]. A total of 78 fixed (ED frame) and moving (ES frame) pairs were selected from 983 volumes by various combinations of the time frames near the ES and those near the ED, ensuring a considerable deformation between fixed and moving volumes. The sample distribution for training/validation and testing of 78 fixed and moving pairs is 69 and 9, respectively.

Public Adult 3D Echocardiography: This MITEA dataset is a 3D adult echocardiography (18–84 years) comprising 134 patients with 268 sample acquisitions at the ED and ES times [20], resulting in 536 3D volumes. Images were acquired using a Siemens ACUSON SC2000 Ultrasound System and a 4Z1c matrix array transducer (Siemens Medical Solutions, CA, USA) with 36×48 elements. All the images are provided with the corresponding epicardial, endocardial, and MYO boundaries. The database comprises seven different classes with different numbers of samples: 164 healthy, 28 with LV hypertrophy, 24 with cardiac amyloidosis, 20 with aortic regurgitation, 16 with hypertrophic cardiomyopathy, 12 with dilated cardiomyopathy, and 4 with heart transplant recipients. The sample distribution for training/validation and testing of 268 fixed (ED frame) and moving (ES frame) pairs is 236 and 32, respectively.

Training and Evaluation: A batch of 4 and epochs of 200 were used with an Adam optimizer with a learning rate of 1e-5, and training was completed in 25 h with one NVIDIA RTX 3090 GPU. The volumetric images and masks were resampled to $64 \times 64 \times 64$ using bicubic interpolation for intensity images and nearest-neighbor interpolation for the masks. The results are evaluated using a dice similarity coefficient (DSC) for evaluating the overlapping, Hausdorff distance (HD) for measuring the anatomical border irregularity and coarseness, and percentage of the non-positive rates of the Jacobian determinant ($|\mathbf{J}_{\varphi_f}| \leq 0$) [21] of the deformation field (φ_f) for measuring its degree of smoothness.

3 Results and Discussions

Different DLIR configurations are used for an ablation study: **1. Vanilla DLIR** (Vanilla DLIR with only fixed and warped image similarity constraint using Eq. 1), **2. VoxelMorph** (Vanilla DLIR with an additional DSC loss constraint [16] with Eq. 1), **3. SCA-DLIR** (Vanilla DLIR with SCA [15] using Eq. 1), **4. CoVan-DLIR** (Vanilla DLIR with spatial co-attention [13]), **5. FBA+Vanilla DLIR** (Vanilla DLIR with FBA using Eq. 4), and **6. FBA+SCA-DLIR** (SCA-DLIR with FBA using Eq. 4).

Table 1. The quantitative registration results to show the non-overlapping benefits of each integrated module. Bold fonts indicate the best metrics.

Methods	Avg. DSC (↑)		Avg. HD (mm) (↓)		$\lvert\mathbf{J}_{\varphi_\mathbf{f}}\rvert \leq 0\,(\%)\ (\times\mathbf{10^{-2}})$ (↓)
	Myocardial	Endocardial	Myocardial	Endocardial	
3D Fetal Echocardiography					
Vanilla DLIR	0.832 ± 0.031	0.907 ± 0.009	5.66 ± 0.99	2.89 ± 0.64	7.5 ± 3.4
VoxelMorph [16]	0.839 ± 0.029	0.899 ± 0.013	5.44 ± 1.29	3.09 ± 0.84	$\mathbf{0.0 \pm 0.0}$
SCA-DLIR	0.840 ± 0.016	0.913 ± 0.008	5.61 ± 1.03	2.81 ± 0.56	3.2 ± 2.0
CoVan-DLIR [13]	0.845 ± 0.009	0.903 ± 0.005	5.51 ± 0.84	3.29 ± 0.74	$\mathbf{0.0 \pm 0.0}$
FBA with Vanilla	0.850 ± 0.025	0.911 ± 0.013	$\mathbf{3.52 \pm 2.18}$	2.96 ± 0.70	$\mathbf{0.0 \pm 0.0}$
FBA with SCA	$\mathbf{0.864 \pm 0.025}$	$\mathbf{0.921 \pm 0.015}$	4.87 ± 0.93	$\mathbf{2.78 \pm 0.60}$	$\mathbf{0.0 \pm 0.0}$
3D Adult Echocardiography [20]					
Vanilla DLIR	0.652 ± 0.076	0.749 ± 0.069	7.46 ± 1.72	7.59 ± 2.11	11.2 ± 0.945
VoxelMorph [16]	0.665 ± 0.079	0.749 ± 0.070	7.03 ± 1.72	6.80 ± 1.91	5.46 ± 0.496
SCA-DLIR	0.666 ± 0.076	0.751 ± 0.073	7.50 ± 1.52	6.82 ± 1.93	3.36 ± 0.520
CoVan-DLIR [13]	0.693 ± 0.072	0.760 ± 0.085	7.37 ± 1.65	5.79 ± 1.80	$\mathbf{0.0 \pm 0.0}$
FBA with Vanilla	0.680 ± 0.067	0.777 ± 0.075	6.39 ± 1.61	5.80 ± 1.45	4.89 ± 0.128
FBA with SCA	$\mathbf{0.702 \pm 0.048}$	$\mathbf{0.789 \pm 0.021}$	$\mathbf{5.99 \pm 1.91}$	$\mathbf{4.92 \pm 1.23}$	0.193 ± 0.319

The results of both adult and fetal echocardiography datasets in Table 1 demonstrate that adding SCA to the Vanilla DLIR improves anatomical benchmarks like DSC and HD and significantly ($p \ll 0.05$) improves $|\mathbf{J}_{\varphi_\mathbf{f}}|$ for both datasets. This reveals that SCA directs the DLIR to focus on where and what to learn during its training for cardiac motion estimation. However, adding DSC-based auxiliary loss to Vanilla DLIR (VoxelMorph) does not improve average DSC and HD for the fetal echocardiography dataset but mildly improves those metrics for the adult dataset. For both datasets, a significant ($p \ll 0.05$) improvement has been observed regarding the $|\mathbf{J}_{\varphi_\mathbf{f}}|$. This indicates that when the anatomical structures are tiny, as in 3D fetal echocardiography compared to 3D adult echocardiography, the successful VoxelMorph fails to result in good motion estimation in 3D fetal echocardiography. Our previous experience is that DSC-loss can improve anatomical measures of Vanilla DLIR in 2D echocardiography [12]; as such, this lack of improvement is likely due to a 3D image being the subject here and can demonstrate how successful strategies in 2D DLIR are often not effective in 3D volumes. When we add spatial co-attention to Vanilla DLIR (CoVan-DLIR), the average DSC and HD for the fetal echocardiography dataset mildly improve. In contrast, those metrics are significantly ($p \ll 0.05$) improved for the adult dataset. There are significant improvements in $|\mathbf{J}_{\varphi_\mathbf{f}}|$ for both datasets. These results reveal that CoVan-DLIR is mildly efficient in fetal echocardiography 3D volumes, which have poorer resolution and image quality than adult 3D volumes.

When our novel FBA is added to unsupervised SCA-DLIR, it significantly ($p \ll 0.05$) improves beyond vanilla DLIR, SCA-DLIR, co-attention, and VoxelMorph for both datasets, whereas FBA with Vanilla DLIR improves most of the metrics compared to those DLIRs. This suggests the genericity of our FBA module in supervising the DLIR networks. CoVan-DLIR has recently been proposed

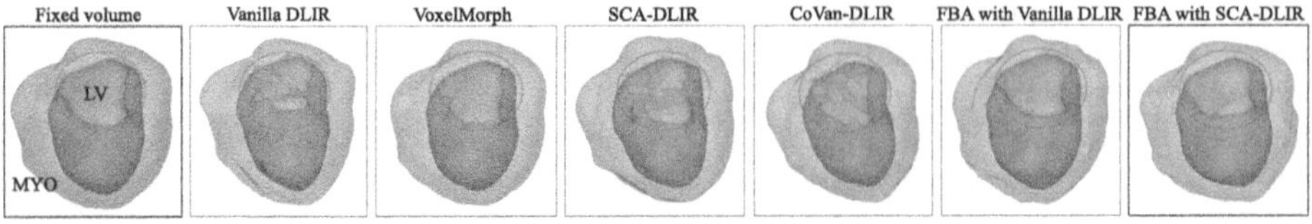

Fig. 2. Warped moving volumes from different DLIRs compared to the fixed volume for the same testing patient of the fetal echocardiography dataset, demonstrating endocardial and myocardial border irregularity and abnormal spikes (inside the red circle). (Color figure online)

in [13] as the first unsupervised 3D DLIR demonstrated using adult echocardiography and is thus arguably state-of-the-art. It enables the emphasis of specific spatial regions during DLIR to enable an enhancement from Vanilla DLIR. CoVan-DLIR also shows success in our experiments with the adult dataset, which has larger anatomical structures and better resolutions than fetal echocardiography volumes. However, when applied to 3D fetal echocardiography, it provides results similar or poor anatomical measures like DSC and HD compared to the SCA-DLIR. This suggests that SCA can similarly tap into the spatial co-attention effect regarding the anatomical criteria when applied to the 3D volumes having poorer resolution and image quality. In contrast, our FBA module with SCA-DLIR provides consistent results for both datasets. Again, the qualitative comparisons of those DLIRs have been exhibited in Fig. 2, showing the reconstructed warped moving volumes using our fetal echocardiography dataset. It can be observed that FBA results in warped volumes with fewer endocardial and myocardial border irregularity and abnormal spikes, with a visually good fit with the fixed volume, confirming our quantitative results of the fetal echocardiography dataset in Table 1.

However, our results show that enhancing CoVan-DLIR with our novel FBA can improve CoVan-DLIR even further. FBA adds a feedback co-attention block to CoVan-DLIR and repeats the registration process after the feedback. In FBA, a new co-attention map is computed between the warped and fixed images to be applied to the fixed image for the second DLIR (block **A2** in Fig. 1). The warped image is the transformation of the moving image towards the fixed image, and the DLIR is trained to match the warped image to the fixed image as much as possible. Our FBA co-attention map compares these two images that are supposed to match well; thus, it is a map to highlight locations with remaining registration matching errors. This thus gives FBA a new capability to more intelligently figure out where to focus further to minimize registration errors, which can be the reason for its better performance. In this sense, FBA can also be seen as a form of self-supervision for DLIR, where it provides feedback to the DLIR to highlight where there are still errors and to help guide the network to successful registration. In each iteration of the training, FBA uses the outputs of a first registration (block **A1** in Fig. 1) to compute a feedback co-attention map, which is used for a second registration (block **A2** in Fig. 1)

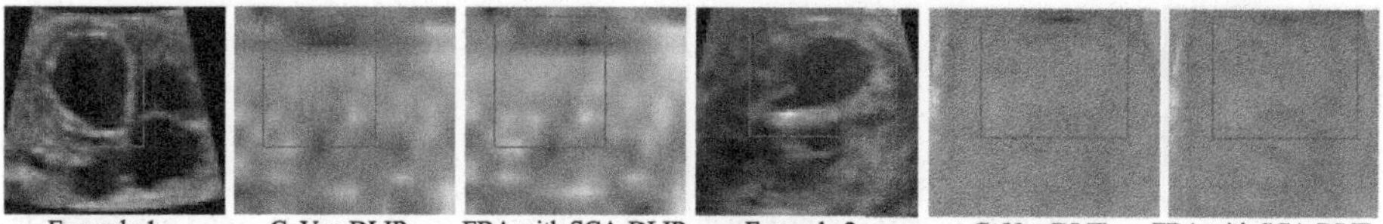

(a) Attention maps for the first epoch

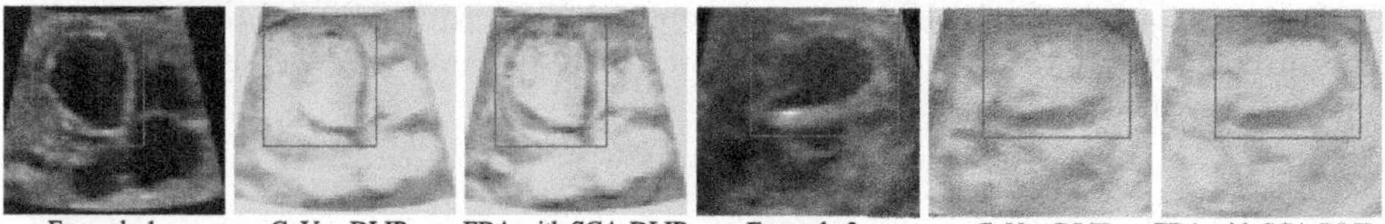

(b) Attention maps for the best epoch (final)

Fig. 3. Visualization of attention maps as heatmaps overlaid on the raw images, showing the DLIR's attention focus (Yellow: less focus and Blue: high focus). Figures (a) and (b) show the error convergence over the epoch of our novel FBA-based DLIR and CoVan-DLIR, comparing the first and best epochs' attention maps. The MYO structures are well visible (see red box) in our FBA with SCA-DLIR compared to CoVan-DLIR.

before losses are computed. In the subsequent iteration, the network starts with the second registration from the previous iteration and uses its errors to provide feedback for further registration. In this sense, the multiple iterations during FBA training represent an iterative registration error convergence process (see Fig. 3). Our results further show that FBA can improve SCA-DLIR similarly to the improvement it can provide to CoVan-DLIR. This suggests a non-overlapping benefit with SCA. However, FBA with SCA-DLIR has better DSC results than FBA with Vanilla DLIR, suggesting that FBA is more compatible with SCA-DLIR than CoVan-DLIR.

Figure 3 shows the co-attention map multiplied by the fixed image to visualize the attention adopted by the DLIRs, where the blue color presents strong emphasis, and the yellow color denotes a no emphasis.

Figure 3 (a) and Fig. 3 (b) show the error convergence over the epoch of our novel FBA-based DLIR and CoVan-DLIR, comparing the first and best epochs' attention maps. For the first epoch, both models focus on the entire image instead of a particular anatomy, i.e., the myocardial wall in cardiac motion. For the best final epoch, with CoVan-DLIR alone, the part of the MYO wall is blurred (and yellow, denoting a lack of emphasis), such as at the top of the images within the red box, which bounds the LV). This shows that CoVan-DLIR's attention map did not fully retrieve the MYO topology. In contrast, FBA's attention map recovers a well-visible MYO structure (in blue, which presents strong emphasis). This can enable better learning of the registration feature and, consequently, better unsupervised DLIR due to FBA's self-supervision. Figure 4 shows learned feature maps from Vanilla DLIR and our FBA with SCA-DLIR for our fetal

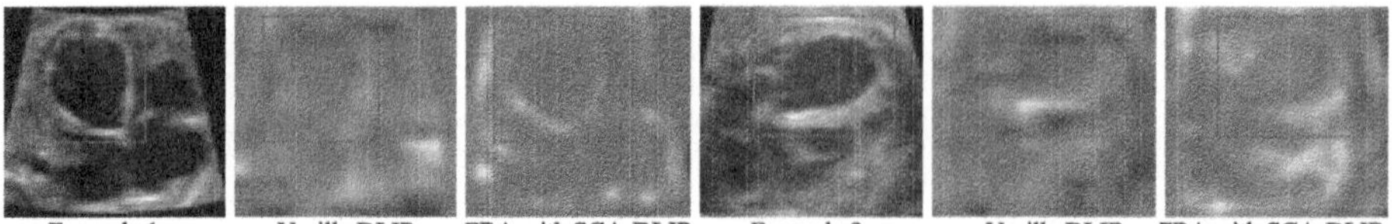

Fig. 4. Visualization of feature maps with corresponding raw images from identical DLIR layers for Vanilla DLIR and our proposed FBA with SCA-DLIR for the fetal echocardiography dataset, showing that our DLIR retrieves MYO's structural information in the feature maps (see red box).

echocardiography dataset. Here, we can observe that Vanilla DLIR feature maps do not preserve the MYO topology, but FBA with SCA-DLIR feature maps clearly displays the structure of the MYO, showing that our proposed feedback attention approach for the self-supervision of the unsupervised DLIR is successful at bringing emphasis to the relevant structures and that this is likely the basis for the observed improved results.

Our study thus confirms that this ability to place specific spatial and feature emphasis in conjunction with our novel FBA to guide unsupervised DLIR is essential in 3D echocardiography. The additional dimension in 3D images compared to 2D images means that the percentage of the image that contains the anatomical structures of interest becomes lower, and DLIR training will consequently be more difficult and less effective. The attention mechanism that guides DLIR training with feedback on the remaining registration error towards important locations and features can overcome this sparser region of interest problem and offer better and generic performance in 3D echocardiography.

Limitations and Future Works: The proposed FBA-based DLIR has the limitation of being complex and taking longer for inference, requiring 600–750 ms for a single inference, which is long compared to the 180 ms for Vanilla DLIR and VoxeMorph, 450ms for SCA-DLIR, and 250 ms for CoVan-DLIR. Future work can include: (1) optimizing the size and complexity of the network and the loss term weights in Eq. 4 to optimize performance while reducing computational time; (2) investigating if pre-trained self-supervised feature embedding network can give better performance; (3) testing the network on a larger dataset to confirm the robustness and scalability of the model; and (3) testing if FBA can be successfully applied to other modalities, such as cardiac MRI.

4 Conclusion

We proposed a novel FBA module to provide self-supervision to the 3D DLIR and showed that this improves DLIR. FBA enhances the guidance provided to DLIR by emphasizing locations where registration errors remain high to help the network improve accuracy, which is equivalent to an iterative error convergence. When FBA is combined with an SCA-DLIR backbone to enable both spatial and feature attention, results are further improved, and robust DLIR results are observed despite challenges in resolution, image quality, and small cardiac structures. Our results further show that co-attention feedback is an ideal approach to scaling up from 2D DLIR to 3D, which can be challenging for several successful 2D DLIR methods. Our proposed network can thus contribute to advancing state-of-the-art fetal echocardiographic motion estimation, potentially leading to a more accurate assessment of fetal and adult cardiac function.

Acknowledgments. M. K. Hasan was supported by the EPSRC-DTP studentship funds from the ICL, UK. G. Yang was supported in part by the ERC IMI (101005122), the H2020 (952172), the MRC (MC/PC/21013), the Royal Society (IEC/NSFC/211235), the NVIDIA Academic Hardware Grant Program, the SABER project supported by Boehringer Ingelheim Ltd, and the UKRI Future Leaders Fellowship (MR/V023799/1).

Disclosure of Interests. The authors have no competing interests to declare that are relevant to the content of this article.

References

1. Heyde, B., et al.: Elastic image registration versus speckle tracking for 2-D myocardial motion estimation: a direct comparison in vivo. IEEE Trans. Med. Imaging **32**(2), 449–459 (2012)
2. Azzam, M., Wasef, M., Khalaf, H., Al-Habbaa, A.: 3D-based strain analysis and cardiotoxicity detection in cancer patients received chemotherapy. BMC cancer **23**(1), 760 (2023)
3. Marwick, T.H., Shah, S.J., Thomas, J.D.: Myocardial strain in the assessment of patients with heart failure: a review. JAMA cardiol. **4**(3), 287–294 (2019)
4. Cincotta, R., Kumar, S.: Future directions in the management of twin-to-twin transfusion syndrome. Twin Res. Hum. Genet. **19**, 285–291 (2016)
5. Tulzer, A., et al.: Valvuloplasty in 103 fetuses with critical aortic stenosis: outcome and new predictors for postnatal circulation. Ultrasound. Obstet. Gynecol. **59**, 633–641 (2022)
6. Gayat, E., Ahmad, H., Weinert, L., Lang, R.M., Mor-Avi, V.: Reproducibility and inter-vendor variability of left ventricular deformation measurements by three-dimensional speckle-tracking echocardiography. J. Am. Soc. Echocardiogr. **24**, 878–885 (2011)
7. van Oostrum, N.H., de Vet, C.M., van der Woude, D.A., Kemps, H.M., Oei, S.G., van Laar, J.O.: Fetal strain and strain rate during pregnancy measured with speckle tracking echocardiography: a systematic review. Eur. J. Obstet. Gynecol. Reprod. Biol. **250**, 178–187 (2020)
8. Adriaanse, B., et al.: Interobserver agreement in detailed prenatal diagnosis of congenital heart disease by telemedicine using four-dimensional ultrasound with spatiotemporal image correlation. Ultrasound Obst. Gynecol. **39**, 203–209 (2012)
9. Yeo, L., Luewan, S., Romero, R.: Fetal intelligent navigation echocardiography (fine) detects 98% of congenital heart disease. J. Ultrasound Med. **37**, 2577–2593 (2018)
10. Østvik, A., et al.: Myocardial function imaging in echocardiography using deep learning. IEEE Trans. Med. Imaging **40**, 1340–1351 (2021)

11. Wang, Z., Yang, Y., Sermesant, M., Delingette, H.: Unsupervised echocardiography registration through patch-based MLPs and transformers. In: International Workshop on Statistical Atlases and Computational Models of the Heart, Springer, pp. 168–178 (2023). https://doi.org/10.1007/978-3-031-23443-9_16
12. Hasan, M.K., Zhu, H., Yang, G., Yap, C.H.: Multi-scale, data-driven and anatomically constrained deep learning image registration for adult and fetal echocardiography. arXiv:2309.00831 (2023).
13. Ahn, S.S., et al.: Co-attention spatial transformer network for unsupervised motion tracking and cardiac strain analysis in 3d echocardiography. Med. Image Anal. **84**, 102711 (2023)
14. Lu, X., Wang, W., Ma, C., Shen, J., Shao, L. and Porikli, F.: See more, know more: unsupervised video object segmentation with co-attention siamese networks. In: Proceedings of the IEEE/CVF Conference on Computer Vision and Pattern Recognition, pp. 3623–3632 (2019)
15. Woo, S., Park, J., Lee, J.-Y., Kweon, I. S.: CBAM: convolutional block attention module, In: Proceedings of the European conference on Computer Vision (ECCV), pp. 3–19 (2018)
16. Balakrishnan, G., Zhao, A., Sabuncu, M.R., Guttag, J., Dalca, A.V.: Voxelmorph: a learning framework for deformable medical image registration. IEEE Trans. Med. Imaging **38**, 1788–1800 (2019)
17. Chen, L.C., Papandreou, G., Kokkinos, I., Murphy, K. and Yuille, A.L.: DeepLab: semantic image segmentation with deep convolutional nets, atrous convolution, and fully connected CRFs. Ieee Trans. Pattern Anal. Mach. intell. **40**, 834–848 (2017)
18. Wiputra, H., Chan, W.X., Foo, Y.Y., Ho, S., Yap, C.H.: Cardiac motion estimation from medical images: a regularisation framework applied on pairwise image registration displacement fields. Sci. rep. **10**, 18510 (2020)
19. Chan, W.X., Zheng, Y., Wiputra, H., Leo, H.L., Yap, C.H.: Full cardiac cycle asynchronous temporal compounding of 3D echocardiography images. Med. Image Anal. **74**, 102229 (2021)
20. Zhao, D., et al.: MITEA: a dataset for machine learning segmentation of the left ventricle in 3d echocardiography using subject-specific labels from cardiac magnetic resonance imaging. Frontiers Cardiovasc. Med. **9**, 1016703 (2023)
21. Lian, C.,et al.: CoCycleReg: collaborative cycle-consistency method for multi-modal medical image registration, Neurocomputing **500**, 799–808 (2022)

Learning Deformable Intra-Patient Liver Registration with Graph Cross-Attention

Stefano Fogarollo[1(✉)], Yannick Scharll[2], Reto Bale[2], and Matthias Harders[1]

[1] Department of Computer Science Interactive Graphics and Simulation Group (IGS), University of Innsbruck, Technikerstraße 21, Innsbruck, Austria
stefano.fogarollo@uibk.ac.at

[2] Interventional Oncology-Microinvasive Therapy (SIP), Department of Radiology, Medical University Innsbruck, Innsbruck, Austria

Abstract. Liver registration is a critical component of computer-assisted hepatic interventions. Accurate alignment of the liver surface and internal vascular network is essential for planning and evaluating interventions, and follow-up monitoring. While conventional techniques achieve high precision in spatial alignment, they often incur high computation times. Therefore, deep learning methods have been proposed for fast registration, albeit with limited generalization to real-world scenarios. In this work, we introduce a novel learnable graph cross-attention mechanism specifically designed for the alignment of the liver vascular system. We build on related work, and extract feature pyramids from the inputs in a dual-stream fashion. Then, we represent the features at the coarsest-scale as graphs, learn critical structural information, and model the relationship between the two graphs through cross-attention. Our proposed module is a novel solution that can enhance the performance of various encoder-decoder architectures. To the best of our knowledge, this is the first application of a graph cross-attention mechanism for liver registration. We carried out the experimental validation on 20 procedures from the Innsbruck University Hospital, demonstrating that our graph module yields significant improvements, with achieved enhancements ranging from 5 % to 7 %.

Keywords: liver registration · graph cross-attention

1 Introduction

Deformable registration of liver images is a necessary step in computer-assisted hepatic interventions. Achieving precise spatial alignment of abdominal CT scans of the same patient acquired at different time points, is crucial for a number of clinical workflows, including ablation procedures and image-guided interventions.

Supplementary Information The online version contains supplementary material available at https://doi.org/10.1007/978-3-031-73480-9_6.

M. Modat et al. (Eds.): WBIR 2024, LNCS 15249, pp. 73–85, 2024.
https://doi.org/10.1007/978-3-031-73480-9_6

For example, in minimally invasive tumor ablations, the post-procedural scan is aligned with the pre-procedural scan to immediately verify sufficient tumor coverage. If the coverage is insufficient, interventional radiologists may perform an additional ablation session. During these interventions, the liver undergoes highly non-rigid deformations, with peripheral regions exhibiting significantly greater motion compared to central regions [25,34]. Moreover, the alignment of the liver surface alone is not enough, and accurate registration of the hepatic vasculature is essential for effective tumor tracking and precise hepatic image guidance [5]. Therefore, we focus on deformable registration, which involves aligning images that exhibit non-linear distortions due to variations in anatomy, pathology, or organ motion. Traditional techniques like SyN [1], Elastix [16], and deedsBCV [12] are renowned for their accurate alignment capabilities. However, they are usually computationally expensive and impractical for fast registration tasks in clinical scenarios, often taking minutes to register abdominal scans [35]. Recently, adapting conventional techniques to modern hardware (GPUs), as done in corrField [11] and iter_lbp [9] has significantly reduced inference times, while maintaining precise alignment. However, these methods rely on iteratively solving a different optimization problem for each new input, and are usually constrained by the discretization of the problem landscape.

Deep learning methods have been introduced for fast registration [2], but they often struggle to generalize, especially when limited labels are available for training. Existing approaches can be categorized into single-stage and multi-stage pipelines. Popular single-stage architectures include VoxelMorph [2], CycleMorph [14] and Transmorph [6]: they concatenate the inputs images, feed them into a backbone network (e.g. U-Net [26], SwinTransformer [17]), and regress the dense displacement field in a single step. Related to our work, XMorpher [28] introduces a two-stream architecture, and a cross-attention-based module for feature fusion, while we perform cross-attention on graph features. As an alternative, in multi-stage pipelines displacements are incrementally estimated, allowing for a more precise alignment. Cascaded pipelines [38,39] have been proposed to handle large deformations, but often require a large number of parameters ($>$ 100 millions), or require multi-stage training schemes [23,38]. Recently, lightweight and efficient multi-stage architectures have emerged, showing promise in bridging the gap with conventional techniques. NICE-Net [20] performs coarse-to-fine registration by successively warping feature maps, while NICE-Trans [21] enhances this approach by employing a Transformer-based architecture, and estimating affine parameters. PIViT [19] exploits a pyramid structure for better alignment, without affine pre-alignment. ModeT [32] introduces a motion decomposition Transformer to estimate the deformation field from feature correlations, via cross-attention. Differently from this work, we perform cross-attention on graph features, in between the encoding and decoding steps. Despite the accuracy of these methods on inter-patient registration, the limited availability of labeled intra-patient abdominal registration datasets restricts their applicability [13], and they do not yet match the performance of conventional techniques for complex and irregular structures such as the hepatic vasculature.

Since the liver vascular network is organized as a complex tree [18], we propose leveraging a graph prior to improve the search for correspondences during registration. While we are not the first to suggest this, our method addresses limitations observed in prior works. GraformerDIR [37], DIEGraph [33], and GraphRegNet [8] are notable works that employ graph neural networks (GNN) for dense medical image registration. The Graformer architecture is limited by the initial volume misalignment, as the inputs are merely concatenated, resulting in minimal improvements (less than 0.1 mm in the average symmetric surface distance). DIEGraph learns a purely convolutional framework for estimating the dense displacements, with a limited improvement of 0.007% DICE. Our graph module, in contrast, consistently delivers at least 5% improvement in DICE, across many backbones, including UNet, SwinTransformer, and PIViT. In GraphRegNet [8] the authors propose to detect Förstner keypoints, estimate displacements based on the cost volume, and regularize them with a graph convolutional neural network, however they also note that a more descriptive keypoint graph is necessary in abdominal images. Recently, the U-Net based learning method described in [36], has explored arterial to portal venous phase registration, but it assumes the liver mask as input.

In this work, we focus on the intra-patient registration of critical vascular structures of the liver. The input is a pair of CT volumes of the same patient, taken at different times, possibly before and after an intervention. Following related work, we extract features in a dual-stream fashion. Then, we perform feature fusion at the coarsest scale under a graph representation: we treat the feature tensors as graphs, we learn structural information with a graph transformer network (GTN), and explicitly model the search for intra-patient correspondences with cross-attention on the two graphs. Our key contribution is a learnable graph cross-attention module that can enhance single-stage backbones (e.g. UNet, SwinTransformer), but also multi-stage networks (PIViT, ModeT). In the following, we introduce our method, conduct experiments on 20 procedures from the Innsbruck University Hospital, and ablate the core components to justify the design choices. Further experiments were also conducted on other anatomies, see the Supplementary Information.

2 Method

3D deformable image registration entails predicting a dense displacement field establishing voxel-wise correspondences between the two input volumes. Given a moving I_m and a fixed I_f volume, the goal is to predict a dense displacement field ϕ that aligns I_m with I_f, optimizing the following loss function:

$$\mathcal{L}(I_m, I_f, \phi) = \mathcal{L}_{\text{similarity}}(I_m \circ \phi, I_f) + \lambda \mathcal{L}_{\text{regularization}}(\phi), \tag{1}$$

where $\circ$ is the warping operation, and λ is the regularization hyper-parameter. $\mathcal{L}_{\text{similarity}}$ commonly includes mean squared error [6] or local normalized cross-correlation (LNCC); $\mathcal{L}_{\text{regularization}}$ usually incorporates isotropic diffusion [2] or bending energy [6]. The alignment of segmentation masks is also included in

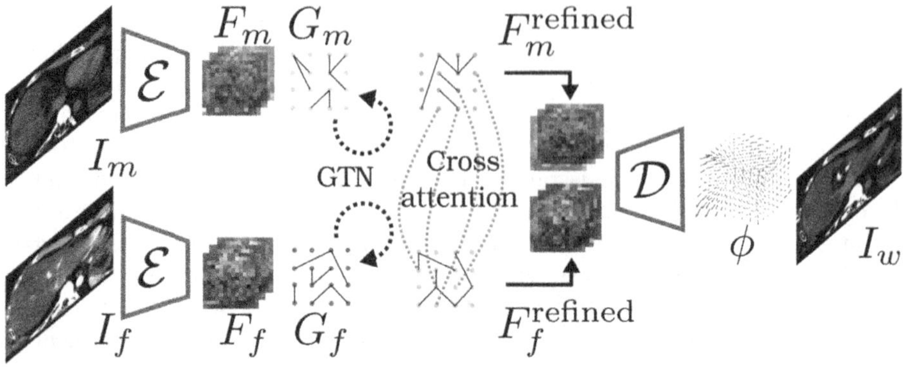

Fig. 1. Overview of our proposed method. We represent the feature tensors F_m and F_f as graphs G_m and G_f, respectively from moving and fixed image I_m, I_f, and perform feature fusion at the coarsest scale with a learnable graph transformer network (GTN). Then, we cross-attend the graphs and reshape them as tensors $F_m^{\text{refined}}, F_f^{\text{refined}}$. We proceed with the decoding steps, outputting the displacement field ϕ, that is used to warp the moving image into I_w.

the loss formulation, when available. Our architecture is depicted in Fig. 1, and described below.

Overview. Our module is in general agnostic to the choice of encoder-decoder pairs. Therefore, we instantiate it within a number of encoder-decoder backbones, including single-stage architectures such as VoxelMorph, Transmorph, and multi-stage ones like PIViT and ModeT. We assume a general encoder $\mathcal{E}$, that produces coarse feature tensors, followed by a decoder $\mathcal{D}$, that learns the fine-resolution dense displacement field $\phi \in \mathbb{R}^{3 \times D \times H \times W}$ from the coarse features. We encode the inputs I_m, I_f of size D, H, W in a dual-stream fashion, using the same encoder $\mathcal{E}$, producing two feature sets at coarsest scale $F_m, F_f \in \mathbb{R}^{C \times D_e \times H_e \times W_e}$, where C denotes the number of feature channels and D_e, H_e, W_e the spatial feature dimensions, respectively. Then, before applying the decoder $\mathcal{D}$, we enhance the feature representations with our proposed graph cross-attention module.

Graph Cross-attention. We reshape the feature sets F_m, F_f as $C \times (D_e \cdot H_e \cdot W_e)$, and interpret them as two graphs with $(D_e \cdot H_e \cdot W_e)$ nodes, each associated with a feature vector of size C. To retain topological information, we append positional encodings in the feature vector. Then, we form edges between each node and its k-nearest-neighbors. This process creates the graphs G_m and G_f from the moving and fixed feature sets, respectively. Empirically, we found that $k = 30$ is sufficient.

We process each graph separately with the same shared GTN. The GTN consists of a fully-connected embedding layer and L graph transformer layers (GTL) [29]: each layer performs a multi-head dot-product attention on the features of each node i based on the features of the neighboring nodes j. The updated features $\tilde{\mathbf{z}}$ of node $\mathbf{x}_i$, are given by

$$\tilde{\mathbf{z}}_i = \mathbf{W}_S\mathbf{x}_i + \sum_{j\in\mathcal{N}(i)} \alpha_{i,j}\mathbf{W}_V\mathbf{x}_j, \tag{2}$$

$$\alpha_{i,j} = \text{softmax}\left(\frac{(\mathbf{W}_Q\mathbf{x}_i)^\top(\mathbf{W}_K\mathbf{x}_j)}{\sqrt{d}}\right), \tag{3}$$

where $\mathbf{x}_j$ are the nodes in the neighborhood $\mathcal{N}(i)$ of node $\mathbf{x}_i$; $\alpha_{i,j}$ are the attention coefficients between nodes i, j; $\mathbf{W}_S, \mathbf{W}_Q, \mathbf{W}_K, \mathbf{W}_V$ are trainable linear weights representing self-propagation, query, key, and value, respectively; d is the feature dimensionality. In our experiments we set the number of layers to $L = 5$, and aggregate the features from neighboring nodes using a summation operation. Within each layer, we concatenate the outputs from $H = 8$ attention heads, and pass them to the next layer. In the final layer, however, we average the outputs.

Then, we perform the feature fusion step by searching for correspondences in the two graphs G_m and G_f using cross-attention. We compute the cross-attention coefficients by replacing the key and value embeddings with embeddings computed in the other graph:

$$\tilde{g}_m(i) = \sum_{k\in G_f} \text{softmax}\left(\frac{(\mathbf{C}_Q G_m(i))^\top(\mathbf{C}_K G_f(k))}{\sqrt{d}}\right)\mathbf{C}_V G_f(j), \tag{4}$$

$$\tilde{g}_f(i) = \sum_{k\in G_m} \text{softmax}\left(\frac{(\mathbf{C}_Q G_f(i))^\top(\mathbf{C}_K G_m(k))}{\sqrt{d}}\right)\mathbf{C}_V G_m(j), \tag{5}$$

where $\tilde{g}_m(i), \tilde{g}_f(i)$ denote the feature updates of node i in G_m, G_f, respectively; $\mathbf{C}_Q, \mathbf{C}_K, \mathbf{C}_V$ are trainable linear weights representing query, key, and value, respectively; and k are all nodes in the other graph. This step is critical, since it permits each node in one graph to exchange information with all nodes in the other one. Finally, we add the feature updates $\tilde{g}_m(i), \tilde{g}_f(i)$ to the original graph, and learn a fully-connected layer to output node-level embeddings. This graph processing does not modify the shape or size of the input tensors, but significantly enhances the feature representations, as demonstrated by our ablation studies. Then, we reshape the graphs as tensors $\in \mathbb{R}^{C\times D_e\times H_e\times W_e}$, producing $F_m^{\text{refined}}, F_f^{\text{refined}}$ and proceed with the steps required by the decoder $\mathcal{D}$. The pipeline is trained end-to-end, optimizing the loss function described in Eq. 1.

3 Experiments and Results

In the following, we report the evaluation of our graph module, when combined with both single-stage and multi-stage architectures. Given an encoder-decoder baseline, we instantiate our graph module within the same encoder-decoder pair.

We acknowledge the recent advances towards more efficient and effective frameworks for medical imaging tasks, and include MedNeXt [27] and SwinUNETR [10] in our experiments. To the best of our knowledge, this is the first application of such architectures in the context of medical image registration. For a fair comparison, we report the results of fast non-learning methods that take only a few seconds for inference in Table 3. Our main motivation is interventions for liver tumor ablation, therefore our focus (and limitation) to liver data. Nevertheless, we also tested on intra-patient lung registration data, compiled in the Supplementary Information.

Data Collection. We evaluate the methods on 20 hepatic stereotactic radiofrequency (SRFA) ablation procedures from the Innsbruck University Hospital, using the method described in [3]. Image registration is required to evaluate the effect of the procedure in removing the tumor. After the intervention, the post-procedural CT scan (portal venous phase) is registered to the pre-procedural one (arterial phase) to check immediately for sufficient tumor coverage. This is why the evaluation step should optimally take only a few seconds. The vasculature network is used by the interventional radiologists to check the correctness of the registration, therefore we collected manual segmentations of liver and hepatic vessels. The data has been annotated by an experienced radiologist, and further checked and corrected by a second experienced radiologist. We do not take landmark-based annotations into considerations, as these are difficult to obtain, and prone to inter-annotator variance. In the pre-processing steps, we rigidly pre-align the volumes, and normalize the intensities in $[0.0, 1.0]$.

Evaluation Metrics. We measure the alignment of liver and vascular masks (hepatic vessels and portal vein), in terms of DICE and vascularity clDICE [30], respectively, as well as the percentage of voxels with non-positive Jacobian determinant $|J_s|_{\leq 0}$. We employ clDICE for assessing vascular alignment since it takes into account also the topology of the mask. Both metrics take on values in $[0.0, 1.0]$, with 1.0 denoting two masks overlapping completely. Moreover, we measure the average surface distance (ASSD) on the vasculature mask, in millimeters; a lower distance is better. We perform five cross-validation experiments (14 training, 2 validation, and 4 test cases in each experiment), visualize selected qualitative results in Fig. 2, and report quantitative results below in Table 1 and in Table 2, including possible statistical significance. In the following tables, arrows indicate whether higher or lower values are better.

Training. We trained all the models with the Adam optimizer [15] with learning rate $1e-4$, optimizing the loss in Eq. 1; we follow related work and use LNCC for $\mathcal{L}_{\text{similarity}}$, and isotropic diffusion for $\mathcal{L}_{\text{regularization}}$ [2]. We also impose the DICE loss [22] on the available segmentation masks. We carried out the experiments on a `NVIDIA A40`, using PyTorch [24] and PyG [7] for the implementation of the methods. During training we employ dropout, and common data augmentation

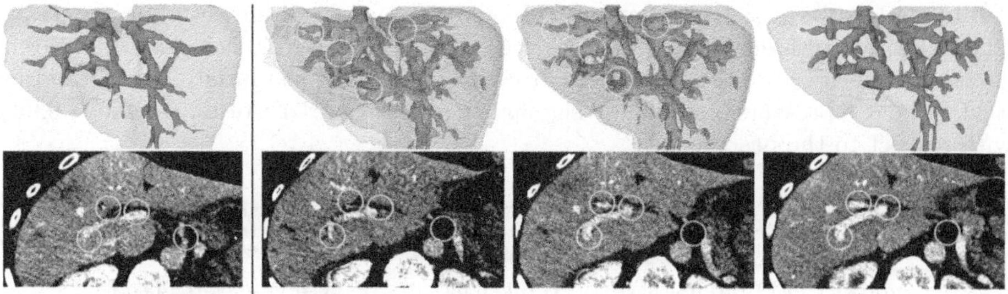

(a) Moving, vasc. ASSD: 3.22 | (b) TransMorph, vasc. ASSD: 2.48 | (c) *Ours*, vasc. ASSD: 2.33 | (d) Fixed

Fig. 2. Top: 3D renderings of Case # 1 from the Innsbruck University Hospital using 3D Slicer: the post-procedural volume (Moving) should be aligned with the pre-procedural volume (Fixed). Only liver and vascular segmentations are shown for clarity, in cyan and in red, respectively. The deformed liver is depicted in yellow, the vasculature in purple. Bottom: CT axial slices. The yellow circles mark critical points for assessing the registration quality, in the context of the clinical application. (Color figure online)

techniques: affine and deformable augmentations, Gaussian noise and blurring. The code for [33, 37] is not public, wherefore we developed our own implementation: we list the hyper-parameters in the Supplementary Information.

Table 1. Results using single-stage methods, grouped by backbone, respectively U-Net [26], SwinTransformer [17], MedNeXt [27], and SwinUNETR [10]; * denotes the baseline method in each group. In parenthesis, we specify the encoder-decoder backbone where we instantiate our method in. Text in bold represents statistically significant results (paired t-tests, p-value < 0.05, adjusted with Bonferroni correction), with respect to the baseline. XMorpher [28] is listed among U-Net methods because the architecture follows the U-Net structure, but it also includes cross-attention transformer blocks.

Method	Liver DICE ↑	Vasc. clDICE ↑	Vasc. ASSD ↓	$\|J_s\|_{\leq 0}$ ↓
VoxelMorph	$0.94_{\pm 0.08}$	$0.48_{\pm 0.04}$	$2.64_{\pm 0.21}$	$1.01_{\pm 0.1}$
CycleMorph	$0.94_{\pm 0.09}$	$0.49_{\pm 0.03}$	$2.66_{\pm 0.18}$	$0.96_{\pm 0.1}$
DIEGraph	$0.96_{\pm 0.06}$	$0.50_{\pm 0.04}$	$2.58_{\pm 0.17}$	$0.92_{\pm 0.1}$
GraformerDIR	$0.97_{\pm 0.08}$	$0.51_{\pm 0.05}$	$2.55_{\pm 0.16}$	$0.93_{\pm 0.2}$
XMorpher*	$0.97_{\pm 0.05}$	$0.53_{\pm 0.03}$	$2.54_{\pm 0.16}$	$0.93_{\pm 0.1}$
ours (U-Net)	$0.97_{\pm 0.07}$	$0.55_{\pm 0.01}$	$2.48_{\pm 0.19}$	$0.94_{\pm 0.1}$
TransMorph*	$0.95_{\pm 0.05}$	$0.54_{\pm 0.06}$	$2.51_{\pm 0.14}$	$1.52_{\pm 0.1}$
ours (SwinTransformer-CNN)	$\mathbf{0.98}_{\pm \mathbf{0.03}}$	$\mathbf{0.60}_{\pm \mathbf{0.05}}$	$2.44_{\pm 0.18}$	$\mathbf{1.03}_{\pm \mathbf{0.2}}$
MedNeXt*	$0.95_{\pm 0.04}$	$0.53_{\pm 0.04}$	$2.55_{\pm 0.16}$	$1.38_{\pm 0.1}$
ours (MedNeXt)	$0.97_{\pm 0.02}$	$\mathbf{0.60}_{\pm \mathbf{0.04}}$	$2.46_{\pm 0.15}$	$\mathbf{1.02}_{\pm \mathbf{0.1}}$
SwinUNETR*	$0.96_{\pm 0.03}$	$0.55_{\pm 0.05}$	$2.45_{\pm 0.16}$	$1.11_{\pm 0.1}$
ours (SwinUNETR)	$0.98_{\pm 0.02}$	$\mathbf{0.61}_{\pm \mathbf{0.06}}$	$\mathbf{2.32}_{\pm \mathbf{0.12}}$	$\mathbf{0.92}_{\pm \mathbf{0.1}}$

Table 2. Results using multi-stage methods, grouped by architecture, respectively PIViT [19], and ModeT [32]; * denotes the baseline method in each group. In parenthesis, we specify the encoder-decoder backbone where we instantiate our method in. Text in bold represents statistically significant results (paired t-tests, p-value < 0.05), with respect to the baseline.

Method	Liver DICE ↑	Vasc. clDICE ↑	Vasc. ASSD ↓	$\lvert J_s\rvert_{\leq 0}$ ↓
PIViT*	$0.97_{\pm 0.04}$	$0.58_{\pm 0.08}$	$2.34_{\pm 0.12}$	$0.41_{\pm 0.1}$
ours (PIViT)	$0.98_{\pm 0.02}$	$\mathbf{0.64}_{\pm 0.03}$	$2.25_{\pm 0.17}$	$0.44_{\pm 0.1}$
ModeT*	$0.98_{\pm 0.03}$	$0.58_{\pm 0.07}$	$2.32_{\pm 0.13}$	$0.52_{\pm 0.1}$
ours (ModeT)	$0.98_{\pm 0.01}$	$\mathbf{0.63}_{\pm 0.05}$	$\mathbf{2.14}_{\pm 0.17}$	$0.45_{\pm 0.2}$

Table 3. Results using fast non-learning methods (corrField [11], iter_lbp [9]), and instance-optimization based (convexAdam [31]). In these experiments we use a SwinTransformer-CNN backbone. Time is measured in seconds, using inputs of size $128 \times 224 \times 224$, and a `NVIDIA GTX 1660 Ti`.

Method	Liver DICE ↑	Vasc. clDICE ↑	Vasc. ASSD ↓	$\lvert J_s\rvert_{\leq 0}$ ↓	Time ↓
corrField	$0.98_{\pm 0.02}$	$0.59_{\pm 0.03}$	$2.32_{\pm 0.2}$	0.0	$1.41_{\pm 0.3}$
iter_lbp	$0.98_{\pm 0.01}$	$0.66_{\pm 0.04}$	$1.83_{\pm 0.2}$	0.0	$0.23_{\pm 0.2}$
convexAdam	$0.98_{\pm 0.01}$	$0.67_{\pm 0.06}$	$1.69_{\pm 0.1}$	0.0	$0.48_{\pm 0.1}$
ours	$0.98_{\pm 0.03}$	$0.60_{\pm 0.05}$	$2.44_{\pm 0.2}$	$1.03_{\pm 0.2}$	$0.18_{\pm 0.1}$

4 Ablation Studies and Discussion

Ablation Studies. We carry out ablation experiments for the graph construction (k-nearest-neighbors), the effect of the positional encoding, the cross-attention module, and the GNN architecture (for this, we replaced the proposed GTN with a graph convolutional network, GCN). We perform these experiments using the SwinTransformer-CNN backbone, and report the results in Table 4. Our proposed GCA module significantly improves the performance, while only a limited but not statistically significant improvement resulted from using a GTN, and positional information. Finally, no statistically significance was detected while varying the number of neighbours k during the graph construction (see the Supplementary Information).

Discussion. The empirical evidence supports the benefits of using the proposed cross-attention module: all the examined backbones, when paired with our module, exhibit improved vascular alignment. The experiments show that, even when using a U-Net backbone, we can improve on other cross-attention based methods regarding vascular alignment. No statistically significant differences have been detected regarding the displacement field metrics, nor the liver structure. We note that, despite using our module, the performance of deep learning methods still falls short of conventional techniques (see Table 3), showing remarkable

Table 4. Ablation studies. Text in bold denotes statistically significant results (paired t-tests, p-value < 0.05, adjusted with Bonferroni correction), when compared with the baseline method marked with *: removing our proposed graph module significantly lowers the performance.

Ablation Test	Liver DICE ↑	Vascula. clDICE ↑	$\|J_s\|_{\leq 0}$ ↓
ours*	0.98±0.03	0.60±0.05	1.03±0.2
no positional encoding	0.98±0.05	0.59±0.05	1.04±0.1
no CA	0.97±0.02	**0.56±0.06**	1.06±0.1
GTN → GCN	0.96±0.04	0.57±0.05	1.04±0.1
no GTN, no GCA	0.96±0.03	**0.55±0.02**	1.08±0.1

robustness against adverse noise conditions. This may be due to the training data available. However, learning methods are faster than conventional ones.

Limitations. In the experiments with our module, we encountered a 7% increase of GPU memory required during training, as well as a 15% increase in training time (one training iteration takes 2.31±0.2 seconds on a NVIDIA A40 GPU, when paired with a SwinTransformer-CNN backbone). Furthermore, due to the quadratic complexity of the cross-attention (see Eqs. 4 and 5), the time of each forward and backward pass is affected by the square of the spatial dimensions. To overcome this problem, an additional coarse layer in the encoder-decoder architecture may be designed, with the caveat of potentially losing fine-grained details. Moreover, in the graph-construction process the node voxels are connected based on the cosine distance in the feature tensors: this might lead to graph neighbourhoods containing a significant portion of local voxels. The neighbourhoods are re-computed at each graph transformer layer, but this might not be sufficient to extend the receptive field of the network. Higher number of neighbours (k), or different similarity measures should be investigated. Finally, while the ablation studies support our design choices, all the methods should be evaluated on larger datasets: to the best of our knowledge there is no public intra-patient liver dataset with vascular annotations. A comparison with commercial software would indicate the level of robustness in clinical scenarios. However, no hepatic evaluation software currently offers an open interface for obtaining the predicted displacement field.

5 Conclusions

Our work focuses on enhancing the performance of single-stage and multi-stage methods with a learnable graph cross-attention module, targeting intra-patient liver registration. It can be used to improve the alignment of the hepatic vascular network, which is necessary in a wide range of medical applications, from procedure planning to follow-up monitoring. We represent the latent features as graphs, and learn a graph cross-attention mechanism to promote alignment already in the latent code. The empirical evaluation and ablation studies show the efficacy of our module for achieving better results across diverse and challenging scenarios. In the future, we aim at solving the computational complexity aspect of the pipeline, as well as targeting multi-modal image fusion. Landmarks, like vascular bifurcations, should also be taken into consideration, especially when the image contains significant noise. Finally, we plan to evaluate our method on larger datasets, as soon as they become available.

Acknowledgements. The authors want to thank Nenad Radojicic for helping during the data annotation process. This project has been partially funded by Stiftung Propter Homines, Vaduz, Liechtenstein, and it has been supported by the Austrian Science Foundation (FWF, project DOC 110). The computational results presented have been achieved (in part) using the Vienna Scientific Cluster (VSC).

Conflict of Interest. The authors declare no conflict of interest. **Data, code and/or material availability.** The publicly available dataset DirLab [4] has been used. Patients at Innsbruck University Hospital gave informed consent for anonymous use of their data. Code is available at https://git.uibk.ac.at/informatik/igs/open/gca_pub.

References

1. Avants, B., Epstein, C., Grossman, M., Gee, J.: Symmetric diffeomorphic image registration with cross-correlation: Evaluating automated labeling of elderly and neurodegenerative brain. Medical Image Analysis **12**, 26–41 (2008). https://doi.org/10.1016/j.media.2007.06.004
2. Balakrishnan, G., Zhao, A., Sabuncu, M.R., Guttag, J., Dalca, A.V.: Voxelmorph: A learning framework for deformable medical image registration. IEEE Transactions on Medical Imaging **38**, 1788–1800 (2019). https://doi.org/10.1109/TMI.2019.2897538
3. Bale, R., Widmann, G., Haidu, M.: Stereotactic radiofrequency ablation. CardioVascular and Interventional Radiology **34**, 852–856 (2011). https://doi.org/10.1007/s00270-010-9966-z
4. Castillo, R., Castillo, E., Guerra, R., Johnson, V.E., McPhail, T., Garg, A.K., Guerrero, T.: A framework for evaluation of deformable image registration spatial accuracy using large landmark point sets. Physics in Medicine and Biology **54**, 1849–1870 (2009) https://doi.org/10.1088/0031-9155/54/7/001
5. Charnoz, A., Agnus, V., Malandain, G., Forest, C., Tajine, M., Soler, L.: Liver registration for the follow-up of hepatic tumors. In: Duncan, J.S., Gerig, G. (eds.) Medical Image Computing and Computer-Assisted Intervention – MICCAI 2005. pp. 155–162. Springer Berlin Heidelberg (2005)
6. Chen, J., Frey, E.C., He, Y., Segars, W.P., Li, Y., Du, Y.: Transmorph: Transformer for unsupervised medical image registration. Medical Image Analysis **82**, 102615 (2022) https://doi.org/10.1016/j.media.2022.102615
7. Fey, M., Lenssen, J.E.: Fast graph representation learning with pytorch geometric (2019), http://arxiv.org/abs/1903.02428
8. Hansen, L., Heinrich, M.P.: Graphregnet: Deep graph regularisation networks on sparse keypoints for dense registration of 3D lung cts. IEEE Transactions on Medical Imaging **40**, 2246–2257 (2021).https://doi.org/10.1109/TMI.2021.3073986

9. Hansen, L., Heinrich, M.P.: Revisiting iterative highly efficient optimisation schemes in medical image registration. In: MICCAI. pp. 203–212. Springer (2021)
10. He, Y., Nath, V., Yang, D., Tang, Y., Myronenko, A., Xu, D.: Swinunetr-v2: Stronger swin transformers with stagewise convolutions for 3D medical image segmentation. In: MICCAI. pp. 416–426. Springer (2023)
11. Heinrich, M.P., Handels, H., Simpson, I.J.A.: Estimating large lung motion in copd patients by symmetric regularised correspondence fields. In: Medical Image Computing and Computer-Assisted Intervention. pp. 338–345 (2015)
12. Heinrich, M.P., Jenkinson, M., Brady, M., Schnabel, J.A.: Mrf-based deformable registration and ventilation estimation of lung ct. IEEE Transactions on Medical Imaging **32**(7), 1239–1248 (2013).https://doi.org/10.1109/TMI.2013.2246577
13. Hering, A., Hansen, L., Mok, T.C.W., Chung, A.C.S., Siebert, H., Hager, S., Lange, A., Kuckertz, S., Heldmann, S., Shao, W., Vesal, S., Rusu, M., Sonn, G., Estienne, T., Vakalopoulou, M., Han, L., Huang, Y., Yap, P.T., Brudfors, M., Balbastre, Y., Joutard, S., Modat, M., Lifshitz, G., Raviv, D., Lv, J., Li, Q., Jaouen, V., Visvikis, D., Fourcade, C., Rubeaux, M., Pan, W., Xu, Z., Jian, B., Benetti, F.D., Wodzinski, M., Gunnarsson, N., Sjolund, J., Grzech, D., Qiu, H., Li, Z., Thorley, A., Duan, J., Grosbrohmer, C., Hoopes, A., Reinertsen, I., Xiao, Y., Landman, B., Huo, Y., Murphy, K., Lessmann, N., van Ginneken, B., Dalca, A.V., Heinrich, M.P.: Learn2reg: Comprehensive multi-task medical image registration challenge, dataset and evaluation in the era of deep learning. IEEE Transactions on Medical Imaging **42**, 697–712 (2023) https://doi.org/10.1109/TMI.2022.3213983
14. Kim, B., Kim, D.H., Park, S.H., Kim, J., Lee, J.G., Ye, J.C.: Cyclemorph: Cycle consistent unsupervised deformable image registration. Medical Image Analysis **71**, 102036 (2021) https://doi.org/10.1016/j.media.2021.102036
15. Kingma, D.P., Ba, J.: Adam: a method for stochastic optimization. In: ICLR (2015)
16. Klein, S., Staring, M., Murphy, K., Viergever, M., Pluim, J.: elastix: A toolbox for intensity-based medical image registration. IEEE Transactions on Medical Imaging **29**, 196–205 (2010). https://doi.org/10.1109/TMI.2009.2035616
17. Liu, Z., Lin, Y., Cao, Y., Hu, H., Wei, Y., Zhang, Z., Lin, S., Guo, B.: Swin transformer: Hierarchical vision transformer using shifted windows. In: 2021 IEEE/CVF International Conference on Computer Vision (ICCV). pp. 9992–10002 (2021) https://doi.org/10.1109/ICCV48922.2021.00986
18. Lorente, S., Hautefeuille, M., Sanchez-Cedillo, A.: The liver, a functionalized vascular structure. Scientific Reports **10**, 16194 (2020).https://doi.org/10.1038/s41598-020-73208-8
19. Ma, T., Dai, X., Zhang, S., Wen, Y.: Pivit: Large deformation image registration with pyramid-iterative vision transformer. In: MICCAI. pp. 602–612. Springer (2023)
20. Meng, M., Bi, L., Feng, D., Kim, J.: Non-iterative coarse-to-fine registration based on single-pass deep cumulative learning. In: MICCAI. pp. 88–97. Springer (2022)
21. Meng, M., Bi, L., Fulham, M., Feng, D., Kim, J.: Non-iterative coarse-to-fine transformer networks for joint affine and deformable image registration. In: MICCAI. pp. 750–760. Springer (2023)
22. Milletari, F., Navab, N., Ahmadi, S.A.: V-net: Fully convolutional neural networks for volumetric medical image segmentation. In: 2016 Fourth International Conference on 3D Vision (3DV). pp. 565–571 (2016).https://doi.org/10.1109/3DV.2016.79
23. Mok, T.C.W., Chung, A.C.S.: Large Deformation Diffeomorphic Image Registration with Laplacian Pyramid Networks. In: MICCAI. pp. 211–221. Springer (2020)

24. Paszke, A., Gross, S., Massa, F., Lerer, A., Bradbury, J., Chanan, G., Killeen, T., Lin, Z., Gimelshein, N., Antiga, L., Desmaison, A., Köpf, A., Yang, E., DeVito, Z., Raison, M., Tejani, A., Chilamkurthy, S., Steiner, B., Fang, L., Bai, J., Chintala, S.: Pytorch: an imperative style, high-performance deep learning library. In: 33rd International Conference on Neural Information Processing Systems (2019)
25. Paulsson, A., Yom, S., Anwar, M., Pinnaduwage, D., Sudhyadhom, A., Gottschalk, A., Chang, A., Descovich, M.: Is the liver a static organ: Intraorgan deformation quantified through fiducial displacement on 4DCT. International Journal of Radiation Oncology-Biology-Physics **93**(3, Supplement), E616 (2015) https://doi.org/10.1016/j.ijrobp.2015.07.2120
26. Ronneberger, O., Fischer, P., Brox, T.: U-net: Convolutional networks for biomedical image segmentation. In: Medical Image Computing and Computer-Assisted Intervention. pp. 234–241 (2015)
27. Roy, S., Koehler, G., Ulrich, C., Baumgartner, M., Petersen, J., Isensee, F., Jäger, P.F., Maier-Hein, K.H.: Mednext: Transformer-driven scaling of convnets for medical image segmentation. In: MICCAI. pp. 405–415. Springer (2023)
28. Shi, J., He, Y., Kong, Y., Coatrieux, J.L., Shu, H., Yang, G., Li, S.: Xmorpher: Full transformer for deformable medical image registration via cross attention. In: MICCAI. pp. 217–226. Springer (2022)
29. Shi, Y., Huang, Z., Feng, S., Zhong, H., Wang, W., Sun, Y.: Masked label prediction: Unified message passing model for semi-supervised classification. In: ICLR (2021)
30. Shit, S., Paetzold, J.C., Sekuboyina, A., Ezhov, I., Unger, A., Zhylka, A., Pluim, J.P.W., Bauer, U., Menze, B.H.: clDice - a novel topology-preserving loss function for tubular structure segmentation. In: 2021 IEEE/CVF Conference on Computer Vision and Pattern Recognition (CVPR). pp. 16555–16564 (2021) https://doi.org/10.1109/CVPR46437.2021.01629
31. Siebert, H., Hansen, L., Heinrich, M.P.: Fast 3D registration with accurate optimisation and little learning for learn2reg 2021. In: Biomedical Image Registration, Domain Generalisation and Out-of-Distribution Analysis. pp. 174–179 (2022)
32. Wang, H., Ni, D., Wang, Y.: Modet: Learning deformable image registration via motion decomposition transformer. In: MICCAI. pp. 740–749. Springer (2023)
33. Wang, L., Yan, Z., Cao, W., Ji, J.: Diegraph: dual-branch information exchange graph convolutional network for deformable medical image registration. Neural Computing and Applications **35**, 23631–23647 (2023). https://doi.org/10.1007/s00521-023-08979-y
34. Wise, P.A., Preukschas, A.A., Özmen, E., Bellemann, N., Norajitra, T., Sommer, C.M., Stock, C., Mehrabi, A., Müller-Stich, B.P., Kenngott, H.G., Nickel, F.: Intraoperative liver deformation and organ motion caused by ventilation, laparotomy, and pneumoperitoneum in a porcine model for image-guided liver surgery. Surgical Endoscopy **38**(3), 1379–1389 (2024) https://doi.org/10.1007/s00464-023-10612-x
35. Xu, Z., Lee, C.P., Heinrich, M.P., Modat, M., Rueckert, D., Ourselin, S., Abramson, R.G., Landman, B.A.: Evaluation of six registration methods for the human abdomen on clinically acquired ct. IEEE Transactions on Biomedical Engineering **63**(8), 1563–1572 (2016).https://doi.org/10.1109/TBME.2016.2574816
36. Xue, P., Zhang, J., Ma, L., Liu, M., Gu, Y., Huang, J., Liu, F., Pan, Y., Cao, X., Shen, D.: Structure-aware registration network for liver DCE-CT images. IEEE Journal of Biomedical and Health Informatics **28**(4), 2163–2174 (2024) https://doi.org/10.1109/JBHI.2024.3350166

37. Yang, T., Bai, X., Cui, X., Gong, Y., Li, L.: Graformerdir: Graph convolution transformer for deformable image registration. Computers in Biology and Medicine **147**, 105799 (2022).https://doi.org/10.1016/j.compbiomed.2022.105799
38. Zhao, S., Dong, Y., Chang, E., Xu, Y.: Recursive cascaded networks for unsupervised medical image registration. In: 2019 IEEE/CVF International Conference on Computer Vision (ICCV). pp. 10599–10609 (2019).https://doi.org/10.1109/ICCV.2019.01070
39. Zhao, S., Lau, T., Luo, J., Chang, E.I.C., Xu, Y.: Unsupervised 3D end-to-end medical image registration with volume tweening network. IEEE Journal of Biomedical and Health Informatics **24**, 1394–1404 (2020).https://doi.org/10.1109/JBHI.2019.2951024

Mamba? Catch The Hype Or Rethink What Really Helps for Image Registration

Bailiang Jian[1,2,3(✉)], Jiazhen Pan[1,2], Morteza Ghahremani[1,2,3], Daniel Rueckert[1,2,3,4], Christian Wachinger[1,2,3], and Benedikt Wiestler[1,2]

[1] Technical University of Munich, Munich, Germany
{bailiang.jian,jiazhen.pan}@tum.de
[2] Klinikum Rechts der Isar, Munich, Germany
[3] Munich Center for Machine Learning, Munich, Germany
[4] Imperial College London, London, UK

Abstract. VoxelMorph, proposed in 2018, utilizes Convolutional Neural Networks (CNNs) to address medical image registration problems. In 2021 TransMorph advanced this approach by replacing CNNs with Attention mechanisms, claiming enhanced performance. More recently, the rise of Mamba with selective state space models has led to MambaMorph, which substituted Attention with Mamba blocks, asserting superior registration. These developments prompt a critical question: does chasing the latest computational trends with "more advanced" computational blocks genuinely enhance registration accuracy, or is it merely hype? Furthermore, the role of classic high-level registration-specific designs, such as coarse-to-fine pyramid mechanism, correlation calculation, and iterative optimization, warrants scrutiny, particularly in differentiating their influence from the aforementioned low-level computational blocks. In this study, we critically examine these questions through a rigorous evaluation in brain MRI registration. We employed modularized components for each block and ensured unbiased comparisons across all methods and designs to disentangle their effects on performance. Our findings indicate that adopting "advanced" computational elements fails to significantly improve registration accuracy. Instead, well-established registration-specific designs offer fair improvements, enhancing results by a marginal 1.5% over the baseline. Our findings emphasize the importance of rigorous, unbiased evaluation and contribution disentanglement of all low- and high-level registration components, rather than simply following the computer vision trends with "more advanced" computational blocks. We advocate for simpler yet effective solutions and novel evaluation metrics that go beyond conventional registration accuracy, warranting further research across various organs and modalities.

1 Introduction

Efficient and accurate deformable registration is fundamental in neuroimaging analysis, facilitating the understanding of human brain dynamics and the disease

B. Jian and J. Pan—Equal contribution.
C. Wachinger and B. Wiestler—Equal Advising.

M. Modat et al. (Eds.): WBIR 2024, LNCS 15249, pp. 86–97, 2024.
https://doi.org/10.1007/978-3-031-73480-9_7

course assessment of various neurological disorders [36]. Cross-sectional brain image registration is pivotal when integrating data from a specific cohort to identify unique patterns or when aligning a new patient's scans with population atlases to apply pre-defined anatomical labels [4,8,14].

Improvements From Low-level Computational Blocks. Accurate, efficient, and rapid image registration remains a primary goal, both for research and clinical application. Over the past decade, numerous deep learning-based image registration methods have been proposed. Convolutional Neural Networks (CNNs), such as VoxelMorph-based methods [1,18], pioneered the field of learning-based registration methods by utilizing learned local convolutional priors to enhance registration accuracy and accelerate the inference process. Lately, Vision Transformer-based methods [3,9,26], such as Transmorph, have been introduced. These methods leverage the inherent attention mechanism to ensure global dependencies and registration coherence. Although these models claim higher registration accuracy than VoxelMorph, their inference speed is hindered by the quadratic computational complexity of the attention mechanism. Recently, Mamba [11], utilizing Selective State Space Models (SSMs), has been introduced in registration [12], claiming to balance both accuracy and efficiency by capturing long-range dependencies with nearly linear computational complexity. However, in our view, more experimental evidence with fair comparisons (same settings, training strategy, etc.) and more datasets are required to substantiate this argument.

Improvement From High-Level Registration-Specific Designs. In contrast to low-level computational blocks which enhance the registration performance at the lowest architecture level in a "brute force" alike way without clear justification or motivation, registration-specific designs address the problem at a higher architecture level considering the characteristics of registration tasks. **Coarse-to-fine optimization pyramid** [2,9,33,37] is a widely-used and effective technique that first estimates large deformations at a coarse scale and refines smaller motions at a finer scale. This approach is often combined with **iterative warping optimizations** [29,33,38,39], where large deformations are approximated through multiple estimation steps, allowing for fine-tuning during the registration process. Additionally, unlike U-Net-like architectures that concatenate fixed and moving images before feeding them into networks, the **dual-branch** approach with two separate encoders has gained popularity in registration and optical flow estimation [19,25,37,38]. This method can explicitly distinguish the dissimilarity between two images. Finally, calculating the regional **correlations** between fixed and moving images to establish voxel-wise correspondences [15,19,27,30] is vital and beneficial for accurate image registration.

As academic researchers specializing in medical image registration for several years, we have observed a recurring trend: the adaptation of low-level computation blocks, such as Transformers, Multilayer Perceptrons (MLPs), and Mamba, often without clear justification and motivation in the core of the registration problem. These methods have garnered significant attention. On the contrary, registration-specific designs have gained much less attention. We argue that techniques like coarse-to-fine approaches, correlation and iterative optimization hold substantial potential for achieving better results. Moreover, registration studies are typically proposed with multiple novel components, wherein advanced low-level computational blocks and registration-specific designs are **intertwined**. The performance contributions of individual components within these complex models, under controlled settings (same data domain and training presets), remain unclear.

The major contributions of our work can be summarized as follows:

1. **Modularized Component Analysis**: We conducted a comprehensive study using modularized components for every aforementioned block and performed fair comparisons across all blocks. This approach allowed us to disentangle, isolate, and identify the contribution of each component to the final registration results, providing clarity on which elements are most impactful.
2. **Extensive Dataset Evaluation**: Our experiments were carried out on five open-source brain MRI datasets with consistent experimental settings for all runs. This ensured a fair and controlled evaluation of Transformers, Mamba, and CNN-based methods, enabling a reliable performance comparison. We demonstrate the qualitative results with random selections to avoid any bias.
3. **Recommendation**: Our findings reveal that "advanced" low-level computational blocks (Transformers and Mamba) offer worse results compared to CNN-based methods in Brain MRI registration tasks. The primary contributions to registration performance stem from high-level registration-specific designs such as correlation layers and iterative warping pyramids. Based on these insights, we encourage the community to focus more on registration-specific designs instead of simply transplanting the "advanced" computational blocks without considering the registration characteristics. We advocate for focusing on simpler, more effective solutions and novel evaluation metrics that move beyond the conventional registration accuracy score, as our results reveal nuanced differences among the tested models.

2 Methodology

Preliminaries. Given a target image and a source image with their anatomical labels, (I_t, L_t) and (I_s, L_s), respectively, we want to build a network parametrized by θ to predict a dense deformation field ϕ which establishes the voxel-wise spatial correspondence between target and source images $F_\theta(I_s, I_t) = \phi$. To optimize the parameters θ, we define the training loss as:

$$\mathcal{L} = \mathcal{L}_{\text{sim}}(I_t, I_s \circ \phi) + \gamma \mathcal{L}_{\text{seg}}(L_t, L_s \circ \phi) + \lambda \mathcal{L}_{\text{reg}}(\phi), \quad (1)$$

where $I_s \circ \phi$ represents the warped source image, $\mathcal{L}_{\text{sim}}$ measures the similarity between images, $\mathcal{L}_{\text{seg}}$ computes the dice loss between segmentation labels, and $\mathcal{L}_{\text{reg}}$ is the smoothness regularization function which penalizes the irregularities of ϕ. γ and λ are pre-determined hyperparameters for Dice loss and regularization, respectively.

We apply VoxelMorph [1] as our registration vanilla method to model θ and estimate the deformation field ϕ. It utilizes U-Net [32] as the architecture backbone and takes concatenated target and source images as the network inputs. U-Net forward-passes them to the encoders with CNN-based low-level computational blocks and reduces the image size when reaching the U-Net bottleneck, and upsamples to the original image size in its decoders. It should be noted that instead of using max-pooling and trilinear upsampling to down-/upsample the feature maps, we apply convolutional kernels with stride size 2 and transposed convolutional kernels for down-/upsampling to boost the networks' performance.

2.1 Improvement Using Advanced Low-Level Computational Blocks

Vision Transformer [6], introduced in computer vision in 2021, has demonstrated superior performance over CNN-based methods [13] in various tasks, such as image classification, detection, and segmentation. It leverages Attention mechanisms that can model global dependencies, unlike local kernels used in CNNs. Based on that, Swin-Transformer was proposed [22], using the shift-window mechanism and achieved linear computational complexity with respect to image size. Swin-Transformer was further introduced into the image registration field [3]. In TransMorph, the authors reported improved registration performance and more coherent deformation fields compared to VoxelMorph due to leveraging these global dependencies. In our work, we replace the computational building blocks (CNNs in the baseline encoder) with Transformers to attempt to improve the registration performance, and this network is dubbed as "TM".

Mamba [11] improves the efficiency of State Space Models (SSMs), which are linear time-invariant systems that map stimulation to response with linear ordinary differential equations. The continuous-time model is first discretized to enable integration into deep learning algorithms. Further, Mamba introduces a selective scan mechanism and hardware-accelerated algorithm, incorporating time-dependent information and significantly boosting SSM efficiency. Concurrently introduced to the registration field, Mamba has shown strong capabilities in handling long sequences with nearly linear complexity [12]. In this work, we attempt to improve the registration accuracy by replacing the computational elements from CNNs with Mamba blocks. This replaced network version is denoted as "Mam".

Large-Kernel Convolution [5] shows the capability of long-range dependencies as well by scaling up the kernel size. LKU-Net [17] claimed that the effective receptive field of a parallel convolutional block with the largest kernel size of 5 is sufficient to address brain MRI registration without global dependencies. They achieve similar dice score as TransMorph while with more implausible deformations. In this work, we evaluate the influence of replacing computational elements from CNNs with large-kernel CNNs and denote it as "LKU".

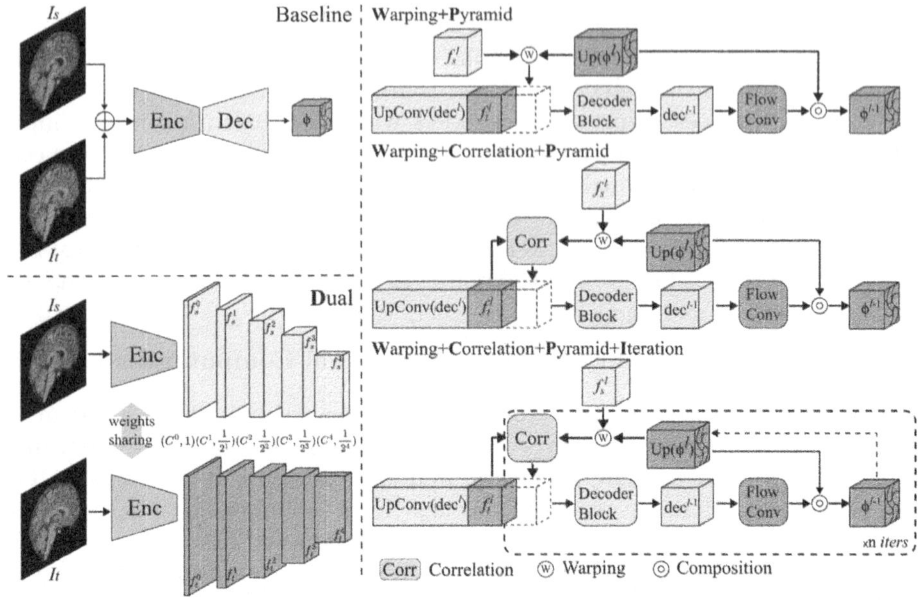

Fig. 1. Overview of the baseline and the modularized components. (a) The upper left shows the U-Net-based *Baseline* methods, which concatenates the source and target images and predicts the final deformation field directly. (b) The bottom left shows the ***D****ual*-stream encoder variant. (c) The right figure presents the detailed workflow of each registration-specific element (***P****yramid*, ***W****arping*, ***C****orrelation* and ***I****teration*) at the specific level ℓ. Level ℓ corresponds to $2^{-\ell}$ resolution. The width of the cubes in the right diagram does not correspond to the exact number of channels of the features, but only for display convenience.

2.2 Improvement Using High-Level Registration-Specific Designs

Dual Stream Encoders. Dual stream encoders utilize two separate branches with shared weights to encode the target and source images independently [25, 35]. Compared to the widely-applied U-Net-based architectures [1,3,12], which concatenate the two images before feeding them into the encoder, dual stream encoders are favored in modern optical flow works [37,38]. This approach constrains the network to produce meaningful representations of each image separately before combining them at a higher level. This process resembles the standard matching approach, akin to standard matching process where features are extracted from image patches and then compared. Additionally, the dual stream encoder architecture facilitates the straightforward introduction of motion pyramid, warping, and correlation layers (see the following paragraphs for more details). This technique is denoted as "**D**" in this work.

Motion Pyramid and Warping. The Motion Pyramid architecture, also known as coarse-to-fine deformation field estimation, is a classic and widely used technique for motion tracking and registration [2,10,33,37]. In registration networks, a pyramid of feature maps of the target and source images are initially generated. The feature maps at a coarse scale (typically 1/8 or 1/16) are used to predict global and large deformations. The finer moving feature is then warped using this estimated deformation to reduce dissimilarity. By estimating the motions bottom up, the deformation field are progressively refined at finer scales. We denote the motion pyramid and warping as "**PW**" in this work.

Correlation Layers. Correlation layers can be used in registration to explicitly measure region dissimilarity and guide deformation field estimation [7,19,28, 38]. In this work, we first flatten the spatial dimensions of the feature maps to obtain feature vectors $\mathbf{f}_t, \mathbf{f}_s \in \mathbb{R}^{HWD \times N}$, where N denotes the embedding feature length. We then compute the correlation as the inner product between these two feature vectors:

$$\mathbf{C} = \mathrm{Corr}(\mathbf{f}_t, \mathbf{f}_s) = \frac{1}{N}\mathbf{f}_t^\top \mathbf{f}_s, \tag{2}$$

resulting in $\mathbf{C} \in \mathbb{R}^{HWD \times HWD}$. To reduce the computational complexity, we compute a partial correlation within a neighborhood volume of size d^3 and indicate the correlation layers as "**C**" in this study.

Iterative Optimization. is applied in many recent works to improve the registration accuracy [23,29,31,38,39]. The direct deformation prediction is decomposed by the iterative refinement in which the registered image is optimized step by step. We apply this registration optimization strategy in this work, and the exact architecture is depicted in Fig. 1. This technique is abbreviated as "**I**".

3 Experiments and Results

Datasets. We use OASIS [24], ADNI [16], IXI[1], LPBA [34], and Mindboggle [20] to conduct experiments on mono-modal cross-sectional brain MRI deformable registration. The train/test split are as follows: OASIS (330/84), ADNI (234/43), IXI (301/115), LPBA (0/40, zero-shot), Mindboggle (0/100, zero-shot). During training, random pairs are taken within each datasets. Importantly, using identical random seed and dataloader settings, we ensure that all methods register the same pair of images in each training iteration. For testing, 200 separate random pairs are sampled from each dataset.

Evaluation Metrics. We compute the Dice score (DSC) and the 90th percentile of the Hausdorff distance (HD90) using the anatomical segmentation labels as metrics of registration accuracy. To assess the deformation field plausibility and smoothness, we report the standard deviation of the logarithm of the Jacobian determinant (SDlogJ) and the percentage of non-diffeomorphic voxels (NDV) [21] within the brain area.

[1] https://brain-development.org/ixi-dataset/.

Table 1. The architectural details of each method. We report the number of trainable parameters (Params) in the unit of million (M), the number of floating point operations (FLOPs) in the unit of GB, the maximum allocated GPU memory during training (Max. Mem Train) in the unit of GB, and the GPU runtime of inference on a pair of images (Runtime Infer) in the unit of second.

	Params (M)	FLOPs (GB)	Max.Mem Train (GB)	Runtime Infer (s)
VXM	2.6	235.2	8.54	5.27
Mam-VXM	2.4	231.4	10.86	7.66
TM	46.6	379.1	9.22	5.09
Mam-TM	39.0	289.1	9.63	5.44
LKU	8.3	517.4	10.74	3.68
Dual	1.9	216.5	8.55	6.76
DWP	1.9	169.5	8.59	6.31
DWCP	8.4	214.6	9.12	9.94
DWCPI	8.1	380.1	9.56	12.42

General Setup**.** The layer to predict the flow field(s) at any level in any methods, i.e., the *FlowConv*, is a $k = 3, s = 1, out\_channel = 3$ convolution with weights initialized from normal distribution $(0, 10^{-5})$ and bias initialized as zero. The final predicted displacement field of each method is at half-resolution. It is then upsampled and upscaled to warp the full-size source image and label. The weights γ, λ of $\mathcal{L}_{\text{seg}}$ and $\mathcal{L}_{\text{reg}}$ are set to 0.5. For the motion pyramid, at each sub-level, we only compute the image similarity and smoothness regularization. The losses are scaled in proportion to the resolution of the output level. For example, at level 4 where the resolution is 2^{-4}, the computed loss is $2^{-4} * (\mathcal{L}_{\text{sim}} + \lambda\mathcal{L}_{\text{reg}})$. The motion pyramid is computed at resolutions [1/16, 1/8, 1/4]. The random seed of training and testing is set to 2023. All methods are trained with 100 epochs with learning rate 10^{-4} with exponential decay rate 0.996.

Method-Wise Setup**.** For baseline methods, the channel number of the vanilla *VoxelMorph (VXM)* model is set to [16,32,32] respectively. There are two remaining convs with 32 channels before *FlowConv*. *Mam-VXM* replaces the CNNs in VXM's encoder with Mamba blocks while keeping the rest identical. *TransMorph (TM)* uses $embed\_dim = 96$ as referred in the paper [3]. Similar to *Mam-VXM*, *Mam-TM* replace the Swin-Transformers in TM's encoder with Mamba blocks. *LKU-Net* use $large\_kernel\_size = 5, feat\_chan = 16$ as referred in the paper [17]. The ***D****ual*-stream variant of VXM halves the encoder feature channel numbers, while the motion ***P****yramid* and ***W****arping* variant's decoder is using the same number of channels as VXM without the remaining convolutions. The ***C****orrelation* variant computes correlation with different number of radius at different level. At the bottom 1/16 resolution, the global correlation is computed, while radii of [3,2,1] are used at the rest resolutions [1/8,1/4,1/2]. From the preliminary experiment, we found that only iterations at the last two levels will

Table 2. Cross-sectional registration results of brain T1-MRI. Separate 200 pairs are randomly sampled from 100 subjects in Mindboggle and 40 subjects in LPBA. SDlogJ is in the scale $\times 10^2$. The top two results are marked in bold.

	LPBA				MindBoggle			
	DSC ↑	HD90 ↓	SDlogJ ↓	NDV(%) ↓	DSC ↑	HD90 ↓	SDlogJ ↓	NDV(%) ↓
affine	54.0±4.8	–	–	–	51.5±4.3	–	–	–
VXM	67.0±3.6	7.54±0.81	7.71±0.79	0.35±0.14	68.8±2.3	5.41±0.73	8.51±0.48	0.34±0.07
Mam-VXM	67.5±3.3	7.45±0.76	7.93±0.88	0.40±0.10	69.2±2.3	5.37±0.69	9.07±0.54	0.45±0.09
TM	67.3±3.3	7.43±0.75	7.77±0.81	0.35±0.10	68.8±2.3	5.35±0.70	8.94±0.51	0.40±0.07
Mam-TM	66.8±3.5	7.53±0.77	7.92±0.80	0.37±0.10	68.6±2.3	5.37±0.70	8.88±0.54	0.39±0.08
LKU	67.2±3.2	7.42±0.73	8.43±1.31	0.61±0.38	68.8±2.2	5.35±0.69	11.61±1.50	0.85±0.16
Dual	66.4±3.7	7.57±0.79	8.43±1.11	0.44±0.25	68.6±2.2	5.38±0.71	8.69±0.51	0.42±0.08
DWP	70.4±2.6	6.91±0.60	7.19±0.64	0.24±0.06	69.4±2.2	5.33±0.69	8.47±0.52	0.35±0.06
DWCP	**70.4±2.7**	**6.98±0.64**	**6.59±0.71**	**0.07±0.05**	**69.6±2.2**	**5.28±0.69**	**7.47±0.40**	**0.08±0.02**
DWCPI	**71.3±2.3**	**6.88±0.59**	**6.82±0.43**	**0.01±0.01**	**69.8±2.2**	**5.28±0.69**	**7.68±0.42**	**0.01±0.00**

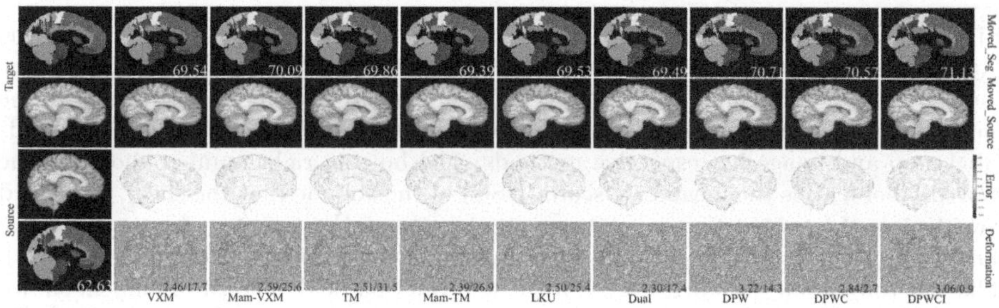

Fig. 2. The sagittal-viewed visualization results of a randomly sampled pair from the LPBA dataset. The first column shows the target label map and image, and the source image and label map from top to bottom. The other columns correspond to the methods to be compared. The first and second rows display the warped source (moved) label map and image by respective methods. The bottom right number shown in the moved label map is the mean dice score (DSC) of the volume (not the slice). The third row depicts the subtraction map (error map) between the target and moved images. The value is within the range [-1,1] since the image intensities are normalized to [0,1]. The last row shows the warped grid of the deformation field. The two numbers shown are the mean foreground displacement magnitude and the percentage of foreground non-diffeomorphic voxels, both statistics are computed for the entire image volume.

contribute to performance improvement. Therefore, the ***Iteration*** variant sets number of iterations $n = 2$ at resolution [1/4, 1/2]. The numerical architecture details can be found in Table 1.

Results. The zero-shot evaluation on LPBA and MindBoggle, with quantitative results, is presented in Table 2. The qualitative visualizations on LPBA and MindBoggle are shown in Fig. 2 and 3. Comprehensive qualitative (sagittal and axial views) and quantitative evaluations on OASIS, ANDI, and IXI are available in the **Supplementary materials**. In a controlled experimental environment,

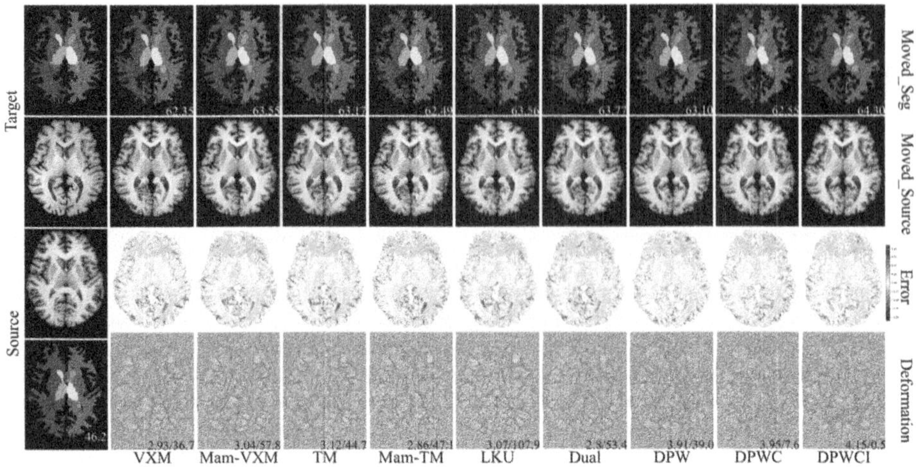

Fig. 3. The axial-viewed visualization result of a randomly sampled pair from the Mindboggle dataset. The first column shows the target label map and image, and the source image and label map from top to bottom. The other columns correspond to the methods to be compared. The first and second rows show the warped source (moved) label map and image by respective methods. The bottom right number shown in the moved label map is the mean dice score (DSC) of the volume (not the slice). The third row depicts the subtraction map (error map) between the target and moved images. The value is within the range [-1,1] since the image intensities are normalized to [0,1]. The last row shows the warped grid of the deformation field. The two numbers shown are the mean foreground displacement magnitude and the percentage of foreground non-diffeomorphic voxels, both statistics are computed for the entire image volume.

it is observed across all datasets that using more advanced computational blocks, such as Mamba or Transformers, does not improve and may even worsen performance. In contrast, utilizing registration-specific designs on top of VXM[2] can enhance registration performance, with improvements of approximately 1.5% and 5% in LPBA's zero-shot evaluation. Additionally, a qualitative evaluation was performed on randomly selected subjects from all five datasets. The differences in registered images, segmentations, residual errors, and deformation fields between different methods are subtle across all datasets, with the registration performance of VXM (baseline method) already being satisfactory.

4 Discussion and Conclusion

In this work, we set out to answer two critical questions: 1. Do "more advanced" low-level computational blocks, such as Transformer or Mamba, genuinely enhance medical imaging registration? 2. If "chasing trends" does not lead to substantial improvements, what truly can enhance medical image registration?

[2] All the registration-specific designs is only applied on top of VXM in this work.

Through comprehensive and rigorous experiments on five brain MR datasets, our findings revealed that employing "more advanced" computational blocks can barely improve, and in most cases, even worsen, registration accuracy. In contrast, incorporating high-level registration-specific designs, such as coarse-to-fine and correlation layers, can improve registration performance, albeit marginally, with a 1.5% increase in dice score compared to the baseline. Notably, the vanilla *VoxelMorph* already produced competitive qualitative registration results, nearly indistinguishable from the best-performing models. Therefore, we encourage the community to apply the simplest *VoxelMorph* to handle the most brain image registration problems, and to use *DPWCI* networks in the cases if the marginal higher registration accuracy is demanded. We release all these models and relevant codes at `github.com/rethink-reg`.

Based on our findings, we advise the community to approach the development of complex, trend-driven registration architectures **with caution**. Simplifying designs can often yield comparable or even superior results. While the pursuit of innovative techniques is essential, it is equally crucial to understand why a particular method is suitable for registration and what factors contribute to its success. We also advocate for the development of new evaluation metrics, moving beyond the conventional registration accuracy score, as our results show nuanced differences among the tested models. Moreover, we suggest focusing on more substantive and impactful areas, including real-time and data-efficient registration, robustness, generalizability, patient-specific adaptation, and the interpretability of deep learning models.

Acknowledgments. This work was supported by BMWi (project "NeuroTEMP") research funding and the Munich Center of Machine Learning (MCML). We gratefully thank the review and feedback provided by Yawei Li.

Disclosure of Interests. The authors declare no competing interests.

References

1. Balakrishnan, G., Zhao, A., Sabuncu, M.R., Guttag, J., Dalca, A.V.: Voxelmorph: a learning framework for deformable medical image registration. IEEE Trans. Med. Imaging **38**(8), 1788–1800 (2019)
2. Brox, T., Bruhn, A., Papenberg, N., Weickert, J.: High accuracy optical flow estimation based on a theory for warping. In: Computer Vision-ECCV 2004, pp. 25–36. Springer (2004). https://doi.org/10.1007/978-3-540-24673-2_3
3. Chen, J., Frey, E.C., He, Y., Segars, W.P., Li, Y., Du, Y.: TransMorph: transformer for unsupervised medical image registration. Med. Image Anal. **82**, 102615 (2022)
4. Coffey, C., et al.: Quantitative cerebral anatomy of the aging human brain: a cross-sectional study using magnetic resonance imaging. Neurology **42**(3), 527–527 (1992)
5. Ding, X., Zhang, X., Han, J., Ding, G.: Scaling up your kernels to 31x31: revisiting large kernel design in CNNs. In: Proceedings of the IEEE/CVF Conference on Computer Vision and Pattern Recognition, pp. 11963–11975 (2022)

6. Dosovitskiy, A., et al.: An image is worth 16x16 words: transformers for image recognition at scale. ICLR (2021)
7. Dosovitskiy, A., Fischer, P., Ilg, E., et al.: FlowNet: learning optical flow with convolutional networks. In: Proceedings of the IEEE International Conference on Computer Vision, pp. 2758–2766 (2015)
8. Fotenos, A.F., Snyder, A., Girton, L., Morris, J., Buckner, R.: Normative estimates of cross-sectional and longitudinal brain volume decline in aging and ad. Neurology **64**(6), 1032–1039 (2005)
9. Ghahremani, M., Khateri, M., Jian, B., Wiestler, B., Adeli, E., Wachinger, C.: H-ViT: a hierarchical vision transformer for deformable image registration. In: Proceedings of the IEEE/CVF Conference on Computer Vision and Pattern Recognition, pp. 11513–11523 (2024)
10. Ghoul, A., et al.: Attention-aware non-rigid image registration for accelerated MR imaging. IEEE Trans. Med. Imaging (2024)
11. Gu, A., Dao, T.: Mamba: linear-time sequence modeling with selective state spaces. International Conference on Machine Learning (ICML) (2024)
12. Guo, T., Wang, Y., Meng, C.: MambaMorph: a mamba-based backbone with contrastive feature learning for deformable MR-CT registration. arXiv preprint arXiv:2401.13934 (2024)
13. He, K., Zhang, X., Ren, S., Sun, J.: Deep residual learning for image recognition. In: Proceedings of the IEEE Conference on Computer Vision and Pattern Recognition, pp. 770–778 (2016)
14. Hering, A., Hansen, L., Mok, T.C., Chung, A.C., et al.: Learn2Reg: comprehensive multi-task medical image registration challenge, dataset and evaluation in the era of deep learning. IEEE Trans. Med. Imaging **42**(3), 697–712 (2022)
15. Hosni, A., Rhemann, C., othters: Fast cost-volume filtering for visual correspondence and beyond. IEEE Trans. Pattern Anal. Mach. Intell. **35**(2), 504–511 (2012)
16. Jack Jr, C.R., et al.: The alzheimer's disease neuroimaging initiative (ADNI): MRI methods. J. Magn. Reson. Imaging An Official J. Int. Soc. Mag. Reson. Med. **27**(4), 685–691 (2008)
17. Jia, X., Bartlett, J., Zhang, T., Lu, W., Qiu, Z., Duan, J.: U-Net vs transformer: is u-net outdated in medical image registration? In: International Workshop on Machine Learning in Medical Imaging, pp. 151–160. Springer (2022). https://doi.org/10.1007/978-3-031-21014-3_16
18. Jian, B., Navab, N., Wendler, T., et al.: Weakly-supervised biomechanically-constrained CT/MRI registration of the spine. In: International Conference on Medical Image Computing and Computer-Assisted Intervention, pp. 227–236. Springer (2022). https://doi.org/10.1007/978-3-031-16446-0_22
19. Kang, M., Hu, X., Huang, W., et al.: Dual-stream pyramid registration network. Med. Image Anal. **78**, 102379 (2022)
20. Klein, A., Mensh, B., Ghosh, S., Tourville, J., Hirsch, J.: Mindboggle: automated brain labeling with multiple atlases. Bmc Med. Imaging **5**, 1–14 (2005)
21. Liu, Y., Chen, J., Wei, S., Carass, A., Prince, J.: On finite difference Jacobian computation in deformable image registration. Int. J. Comput. Vis. 1–11 (2024)
22. Liu, Z., Lin, Y., Cao, Y., et al.: Swin transformer: hierarchical vision transformer using shifted windows. In: Proceedings of the IEEE/CVF International Conference on Computer Vision, pp. 10012–10022 (2021)
23. Ma, T., Zhang, S., Li, J., Wen, Y.: IIRP-Net: iterative inference residual pyramid network for enhanced image registration. In: Proceedings of the IEEE/CVF Conference on Computer Vision and Pattern Recognition, pp. 11546–11555 (2024)

24. Marcus, D.S., Wang, T.H., Parker, J., et al.: Open access series of imaging studies (oasis): cross-sectional MRI data in young, middle aged, nondemented, and demented older adults. J. Cogn. Neurosci. **19**(9), 1498–1507 (2007)
25. Meng, M., Bi, L., Feng, D., Kim, J.: Non-iterative coarse-to-fine registration based on single-pass deep cumulative learning. In: International Conference on Medical Image Computing and Computer-Assisted Intervention, pp. 88–97. Springer (2022). https://doi.org/10.1007/978-3-031-16446-0_9
26. Meng, M., Bi, L., Fulham, M., Feng, D., Kim, J.: Non-iterative coarse-to-fine transformer networks for joint affine and deformable image registration. In: International Conference on Medical Image Computing and Computer-Assisted Intervention, pp. 750–760. Springer (2023). https://doi.org/10.1007/978-3-031-43999-5_71
27. Meng, M., Feng, D., Bi, L., Kim, J.: Correlation-aware coarse-to-fine MLPs for deformable medical image registration. In: Proceedings of the IEEE/CVF Conference on Computer Vision and Pattern Recognition, pp. 9645–9654 (2024)
28. Pan, J., Huang, W., Rueckert, D., Küstner, T., Hammernik, K.: Reconstruction-driven motion estimation for motion-compensated MR CINE imaging. IEEE Trans. Med. Imaging (2024)
29. Pan, J., Rueckert, D., Küstner, T., Hammernik, K.: Efficient image registration network for non-rigid cardiac motion estimation. In: Machine Learning for Medical Image Reconstruction, MICCAI 2021, pp. 14–24 (2021)
30. Pan, J., Rueckert, D., Küstner, T., Hammernik, K.: Learning-based and unrolled motion-compensated reconstruction for cardiac MR CINE imaging. In: International Conference on Medical Image Computing and Computer-Assisted Intervention, pp. 686–696 (2022)
31. Qiu, H., Hammernik, K., Qin, C., Chen, C., Rueckert, D.: Embedding gradient-based optimization in image registration networks. In: International Conference on Medical Image Computing and Computer-Assisted Intervention, pp. 56–65. Springer (2022). https://doi.org/10.1007/978-3-031-16446-0_6
32. Ronneberger, O., Fischer, P., Brox, T.: U-net: Convolutional networks for biomedical image segmentation. In: International Conference on Medical Image Computing and Computer-Assisted Intervention. pp. 234–241. Springer (2015). https://doi.org/10.1007/978-3-319-24574-4_28
33. Rueckert, D., Aljabar, P.: Nonrigid registration using free-form deformations: application to breast MR images. IEEE Trans. Med. Imaging **18**(8), 712–721 (1999)
34. Shattuck, D.W., Mirza, M., Adisetiyo, V., et al.: Construction of a 3D probabilistic atlas of human cortical structures. Neuroimage **39**(3), 1064–1080 (2008)
35. Shi, J., He, Y., Kong, Y., et al.: XMorpher: full transformer for deformable medical image registration via cross attention. In: International Conference on Medical Image Computing and Computer-Assisted Intervention, pp. 217–226. Springer (2022). https://doi.org/10.1007/978-3-031-16446-0_21
36. Sotiras, A., Davatzikos, C., Paragios, N.: Deformable medical image registration: A survey. IEEE Trans. Med. Imaging **32**(7), 1153–1190 (2013)
37. Sun, D., Yang, X., Liu, M.Y., Kautz, J.: PWC-Net: CNNs for optical flow using pyramid, warping, and cost volume. In: Proceedings of the IEEE Conference on Computer Vision and Pattern Recognition, pp. 8934–8943 (2018)
38. Teed, Z., Deng, J.: RAFT: recurrent all-pairs field transforms for optical flow. In: Computer Vision–ECCV 2020, pp. 402–419. Springer (2020). https://doi.org/10.1007/978-3-030-58536-5_24
39. Zhao, S., Dong, Y., Chang, E.I., Xu, Y., et al.: Recursive cascaded networks for unsupervised medical image registration. In: Proceedings of the IEEE/CVF International Conference on Computer Vision, pp. 10600–10610 (2019)

Robustness

Assessing the Robustness of Image Registration Models Under Domain Shifts with Learnable Input Images

Iris D. Kolenbrander[1,2](✉), Vidya Prasad[3], Leanne Zikken[1], Maureen A. J. M. van Eijnatten[1,2], Matteo Maspero[4,5], and Josien P. W. Pluim[1,2]

[1] Medical Image Analysis Group, Department of Biomedical Engineering, Eindhoven University of Technology, Eindhoven, The Netherlands

[2] Eindhoven Artificial Intelligence Systems Institute, Eindhoven University of Technology, Eindhoven, The Netherlands

i.d.kolenbrander@tue.nl

[3] Visual Analytics Group, Department of Mathematics and Computer Science, Eindhoven University of Technology, Eindhoven, The Netherlands

[4] Computational Imaging Group for MR Diagnostics and Therapy, University Medical Center Utrecht, Utrecht, The Netherlands

[5] Department of Radiotherapy, University Medical Center Utrecht, Utrecht, The Netherlands

Abstract. Deep learning models have revolutionized image registration but their accuracy can degrade under unforeseen data variations (domain shifts). It is crucial to assess model robustness under such shifts, often accomplished using simulated domain shifts and expert annotations, e.g., landmarks. This work presents ProactiV-Reg, an annotation-free approach that utilizes a learnable image mapping: it iteratively adjusts a moving image to align with a fixed image under simulated domain shifts. The distances between the perturbed and the optimized images reveal model robustness. We evaluate ProactiV-Reg on three models, showcasing its ability to detect robustness differences, identify dominant perturbations, and provide insights into the model's input requirements.

Keywords: deformable image registration · deep learning · robustness

1 Introduction

Many image registration methods now use deep learning models to enable time-critical medical applications [1,4,20]. However, a major challenge remains: These

---

Supplementary Information The online version contains supplementary material available at https://doi.org/10.1007/978-3-031-73480-9_8.

M. Modat et al. (Eds.): WBIR 2024, LNCS 15249, pp. 101–111, 2024.
https://doi.org/10.1007/978-3-031-73480-9_8

models are susceptible to domain shifts, i.e., where the data used for training differs from real-world applications in terms of contrast, quality, resolution, or patient population. Such discrepancies can lead to inaccurate registrations [11,14], potentially jeopardizing downstream clinical tasks. This limitation underscores the need to assess the robustness of deep learning models to domain shifts to ensure their reliable application.

In parallel, others have addressed model reliability by quantifying model uncertainty using techniques such as Bayesian neural networks, Monte Carlo dropout, and ensemble models. These techniques may help clinicians evaluate the trustworthiness of predictions [8,12]. In addition, methods such as test-time optimization, where the model is fine-tuned on a single image pair during testing, can reduce the inaccuracies caused by domain shifts [2,18]. While these approaches are valuable during deployment, our work focuses on assessing model robustness under domain shifts during the validation phase.

Model robustness is often assessed by simulating domain shifts and evaluating accuracy using landmarks or segmentations, which require expert annotators [5, 6]. *ProactiV*, a recently proposed technique, uses visual analytics to proactively investigate model behavior under domain shifts [19]. Image perturbations are applied to an image and the method then adapts the image until it produces desired outputs. This input image is referred to as the "optimized image". When significant adaptations are required to achieve the desired output with a specific model, this indicates potential susceptibility to the perturbation. By analyzing the distances between perturbed and optimized images, users can assess the model's robustness to domain shifts. However, ProactiV is designed for single-input tasks with available ground truth predictions, such as MRI reconstruction, and cannot be directly applied to image registration with two input images and no readily available ground truth.

This work presents an adaptation of ProactiV to make it suitable for image registration, offering a model-agnostic method for assessing model robustness under domain shifts. We eliminate the ground truth dependency and investigate the hypothesis that the distances between perturbed and optimized images are reliable indicators of model robustness. Finally, we demonstrate the method on three state-of-the-art registration models and apply it in two use cases.

2 Methods

2.1 Image Registration Robustness

Deformable image registration aligns a moving image (M) to a fixed image (F) using a dense displacement vector field (DVF: ϕ), predicted by a registration model (R_θ). Robustness refers to the model's ability to maintain accuracy when encountering data different from its training data. We formulate two concepts of model robustness. Both are concerned with registration failures due to input variations, but differ in how this might be assessed.

1. **Sensitivity to input variations:** A non-robust model shows output changes in response to input variations. This issue is particularly relevant in the context of predictable domain shifts where the model should remain invariant. For example, the output DVF (and resulting image alignment) of a model changes as a result of deteriorating image quality.
2. **Implicit input requirements:** A non-robust model might need input modifications to achieve accurate registration, a concept relevant for unknown domain shifts, where data discrepancies are unclear. Understanding these input modifications may help identify the specific image characteristics (the input requirements) needed for optimal model performance. For instance, if a new image pair contains an artifact not present in the training data, the model might perform better if this artifact is corrected beforehand.

2.2 ProactiV-Reg

We use both concepts (sensitivity to input variations and implicit input requirements) to gain insights into the model's robustness under domain shifts. To achieve this, we propose ProactiV-Reg, an adaptation of ProactiV tailored to deformable image registration (Fig. 1). Our code is publicly available at https://github.com/iriskolenbrander/ProactiV-Reg-WBIR.git.

Step 1. Image Perturbation. This step simulates domain shifts by introducing controlled modifications to both the fixed and moving images. The perturbations can be applied separately or in combination, with varying severity levels (L) as detailed in the Supplementary Materials (A.1). The perturbed moving and fixed images are denoted by M^T and F^T.

Step 2. Learnable Image Mapping. Given a specific registration model and a chosen perturbation, ProactiV-Reg aims to find the image ($\hat{M^T}$) that, when fed to the model together with the perturbed fixed image (F^T), produces a DVF ($\hat{\phi^T}$) that results in good image alignment (Eqs. 1 and 2).

In the optimization process, the moving image is made learnable and the model's parameters are frozen. The pixels of the learnable image (q) are iteratively updated using gradient descent to minimize an objective function. The objective function is the image similarity between the perturbed fixed (F^T) and the image resulting from applying the predicted DVF ($\hat{\phi^T}$) to a non-learnable copy of the moving image (q'). Specifically, the similarity is the weighted average of the local normalized cross-correlation (NCC) (window size=7) at three resolutions [16]. The optimization problem is formulated as follows:

$$\hat{M^T} = argmin_q(-NCC(R_\theta(q, F^T) \circ q', F^T)) \quad (1)$$

Here, "$\circ$" indicates spatial transformation, q is the learnable image (initialized as $q_0 = M^T$), and q' is the non-learnable copy of q (updated in each iteration). The voxel values of q at step n are updated as follows:

$$q_{n+1} = q_n - \alpha * \nabla_{q_n} NCC(R_\theta(q_n, F^T) \circ q'_n, F^T) \quad (2)$$

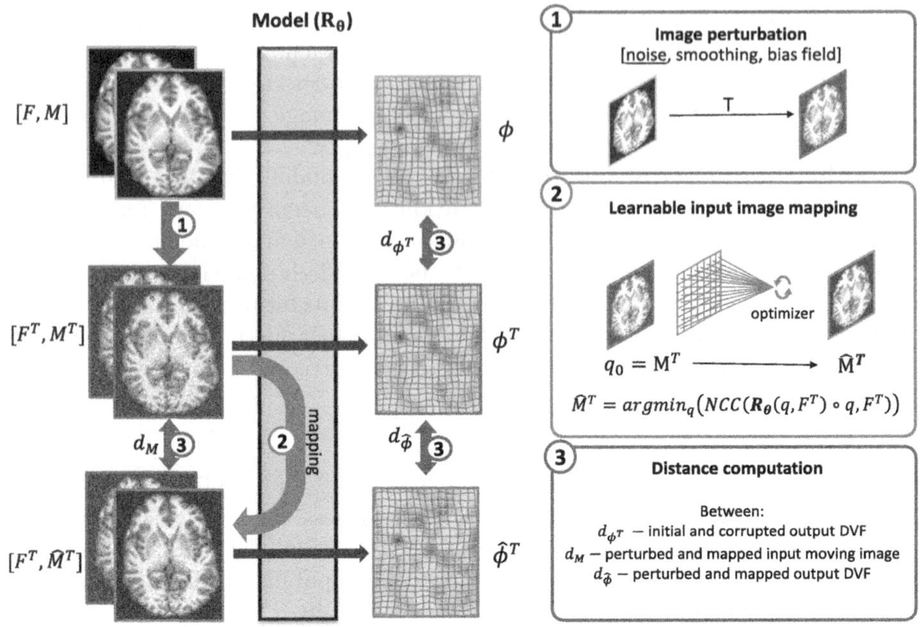

Fig. 1. The ProactiV-Reg method: (1) The fixed (F) and moving images (M) are perturbed by T, (2) The perturbed moving image (M^T) is used as learnable input (q) and mapped to the image ($\hat{M}^T$) that produces the desired DVF ($\hat{\phi}^T$) when fed to the registration model (R_θ), and (3) the distances $d_{\hat{M}}$, d_{ϕ^T}, and $d_{\hat{\phi}}$ are computed.

Using a non-learnable copy for the spatial transformation ensures that gradients are propagated through the model and prevents the optimization from simply converging to $\hat{M}^T = F^T$. The optimization is performed using the Adam optimizer with a fixed learning rate of 1e-3 for each model (details on models in Sect. 2.4) on a PC with an Nvidia Tesla V100-PCIe 32GB GPU. The process continues until the loss function converges (change in value below 1e-4 for 20 consecutive iterations) or a maximum of 100 iterations is reached.

Step 3. Distance Computation. Finally, ProactiV-Reg computes three distances using the mean squared error (MSE) (Eq. 3). These distances aim to quantify the changes induced by the perturbation and the learnable mapping.

- $d_{\hat{M}}$: Represents the input adaptations introduced by the learnable mapping.
- $d_{\hat{\phi}}$: Represents the change in the model's output caused by the mapping.
- d_{ϕ^T}: Represents the changes induced by the domain shift (if available when simulating domain shifts).

$$d_{\hat{M}} = MSE(M^T, \hat{M}^T); \qquad d_{\hat{\phi}} = MSE(\phi^T, \hat{\phi}^T); \qquad d_{\phi^T} = MSE(\phi, \phi^T) \tag{3}$$

The original ProactiV also included the distances relative to the ground truth, which we excluded because ground truth is not commonly available in image registration. We chose the MSE from the image metrics MSE, mean absolute error, and structural similarity image metric, based on the stronger correlation observed between MSE and registration accuracy (Supplementary materials: Table S1).

2.3 Brain MRI Dataset

We evaluate ProactiV-Reg on a dataset of 51 3D brain T1-weighted magnetic resonance images (MRI) from the open access OASIS dataset [15], which includes adult subjects diagnosed with mild to moderate Alzheimer's disease. We use Hoopes *et al.*'s processed dataset [10], consisting of images of size $180 \times 192 \times 180$ after skull-stripping, affine alignment to a common atlas (FreeSurfer [7]), resampling of voxel sizes to $1 \times 1 \times 1$ mm, and gray value scaling to the range [0–1]. Semi-automatic labels are available for 35 brain structures. A single, randomly selected subject serves as the fixed image to which all other subjects (moving images) are registered, resulting in 50 image pairs.

Synthetic Deformations. In the experiments described in Sects. 3.1 and 3.2 (Sect. 3.2: nonlinear deformations), we use synthetically deformed brain MRIs, applying generated DVFs to the original MRIs [3]. While this approach allows us to validate the distances using ground truth DVFs, they are not part of the ProactiV-Reg method (and thus are not required to assess robustness).

We apply two DVFs to each MRI in the original dataset: a low-resolution DVF (on a $5 \times 6 \times 5$ control point grid) obtaining the moving image, followed by a statistical model-based DVF ($20 \times 24 \times 20$ grid), obtaining the fixed image. The statistical model captures true variations between subjects and is built using principal component analysis (PCA) capturing 80% of the variations in deformations produced by iterative registration (in Elastix [13] using mutual information, a three-resolution scheme, 100 iterations, and a final B-spline grid spacing of 8 mm). The magnitude of the deformations is controlled by σ_ϕ, which was set to 4500 unless otherwise specified (Supplementary Materials: A.3). The resulting synthetic images exhibit realistic variations between the brains of different subjects.

2.4 Deep Learning Models

We select three pre-trained convolutional neural network models based on their proven registration performance on T1-weighted brain MRIs and the availability of their pre-trained weights. As the purpose of this study is to demonstrate ProactiV-Reg, not to evaluate the three models, we will refer to them as models A, B, and C.

- Model A [1]: This unsupervised U-Net architecture has four/three convolutional blocks in the encoder/decoder (32 filters), skip connections, and four

convolutional blocks in the registration head (16 filters). It is trained using MSE loss and regularization.
- Model B [9]: This weakly supervised U-Net architecture adopts a similar structure to model A but with more filters (256 filters in both encoder/decoder and registration head). It is trained using the soft Dice loss (of binary segmentations) and regularization.
- Model C [17]: This unsupervised model incorporates a conditional architecture with three levels of convolutional encoders, conditional modules, and decoders (28 filters). The model is conditioned on a regularization weight (RW=0.4 unless otherwise specified) and was trained using a pyramid local NCC and regularization.

2.5 Evaluation Metrics

Registration accuracy is measured by propagating the moving segmentation using the predicted deformation and computing the 95th percentile Hausdorff distance (95%HD) of the cerebral cortex, cerebellar cortex, brainstem, lateral ventricle, and hippocampus (calculated as the average of 95%HD values of bilateral structures). In addition, the synthetically deformed dataset allows the (mean absolute) error computation of the predicted and ground-truth deformations.

3 Experiments and Results

3.1 Distances as Indicators of Registration Robustness

This section investigates whether the distances ($d_{\hat{M}}$, d_{ϕ^T}, and $d_{\hat{\phi}^T}$) can serve as measures of registration robustness. Using a synthetically deformed dataset, we study how the distances and the registration error are affected by different intensity-based perturbations, including noise, smoothing, or bias field (severity levels $L = 10$; see Supplementary Materials: A.1). We use model A to minimize the number of iterative optimizations.

The distances increase with noise and bias field but remain constant with smoothing (Fig. 2). These results correlate with the registration error, supporting our hypothesis that the distances reflect registration robustness. Notably, d_{ϕ^T} and $d_{\hat{\phi}^T}$ increase substantially under bias fields, while $d_{\hat{M}}$ increases only slightly. In other words, the image mapping introduces small changes to the input image that significantly affect the predicted DVF. These results suggest that multiple distances are essential for a more complete understanding.

3.2 Models' Robustness Under Simulated Domain Shifts

We demonstrate ProactiV-Reg in two use cases, focusing on understanding model behavior under simulated domain shifts. We validate our findings by examining the impact of these shifts on the alignment of brain structures.

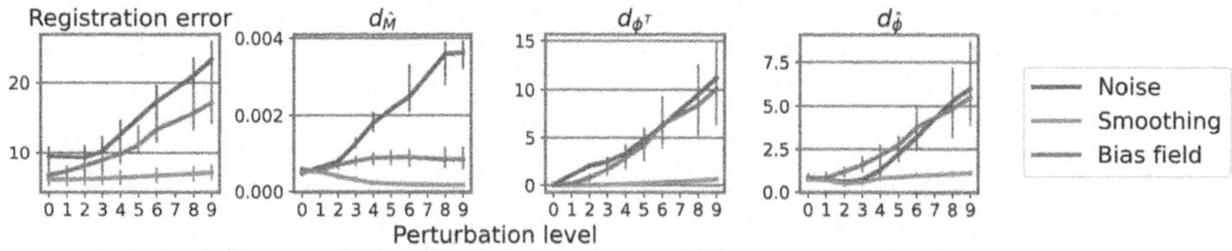

Fig. 2. The effect of intensity-based perturbations on the registration error and distances of model A evaluated on a synthetically deformed dataset. The distances $d_{\hat{M}}$ and $d_{\hat{\phi}}$ represent the input and output change needed for optimal registration; d_{ϕ^T} represents the output change caused by the domain shift.

Intensity-Based Perturbations. We investigate the robustness of three models (A, B, and C) to intensity-based domain shifts, including noise, smoothing, and bias field (with varying severity levels, $L = 5$) on the brain MRI dataset. Our goal is to identify the most robust model and the most impactful domain shift, assessed using three visualization panels in ProactiV-Reg. The *distances panel (i)* shows the distances under a single perturbation for low-level analysis. The *global panel (ii)* summarizes the distances under combined perturbations of smoothing and bias field (25 combinations for $L = 5$), reduced to two dimensions via PCA. The *instance panel (iii)* shows examples of perturbed and optimized images, useful for analyzing local changes induced by the learnable mapping.

Figure 4i reveals that model A is the most sensitive to noise and bias field, with substantial increases in $d_{\hat{M}}$ and d_{ϕ^T}. Figure 4ii supports this, showing more dispersed data points for model A (arrow q) compared with the tightly clustered points of models B and C (arrows r-s). All models display relatively constant distances under smoothing perturbations (Fig. 4i), indicating good robustness to this domain shift. These findings align with the observed brain structure alignment after registration (Supplementary Materials: Fig. S1). Interestingly, the optimized images of model A contained additional fine-grained details and a low-resolution, inhomogeneous signal resembling a bias field (Fig. 4iii), suggesting that model A relies on bias field correction for optimal registration.

Nonlinear Deformations. We explore the use of ProactiV-Reg for optimizing the regularization weight (RW) in model C. We assess model C's robustness under nonlinear deformations while setting the RW to 0.1, 0.25, 0.5, 0.75, or 1.0. We create six datasets with different deformation magnitudes ($\sigma_\phi \in$[3000, 4000, 5000, 6000, 7000, 8000]; Supplementary Materials: A.3).

Figure 3b shows that higher DVF constraints (higher RW) result in poorer structure alignment for larger deformations. This finding is also reflected in the distances (Fig. 3a), indicating that these distances can be used instead of segmentations to optimize the RW for robust registration across varying deformations.

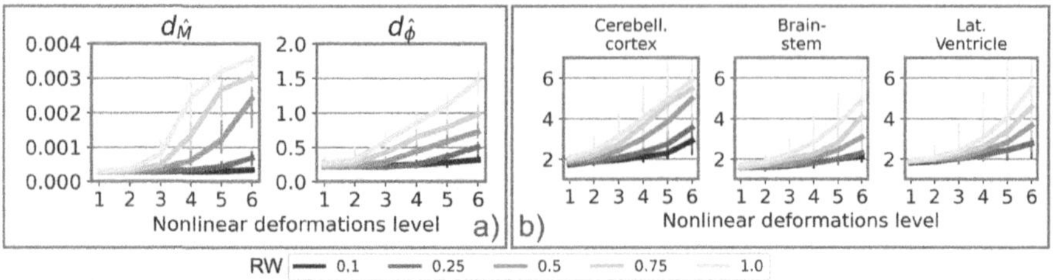

Fig. 3. The impact of nonlinear deformations on the distances (a) and the brain structure alignment (b) of model C conditioned on different regularization weights (RW). $d_{\hat{M}}$ and $d_{\hat{\phi}}$ represent the input and output change needed for optimal registration.

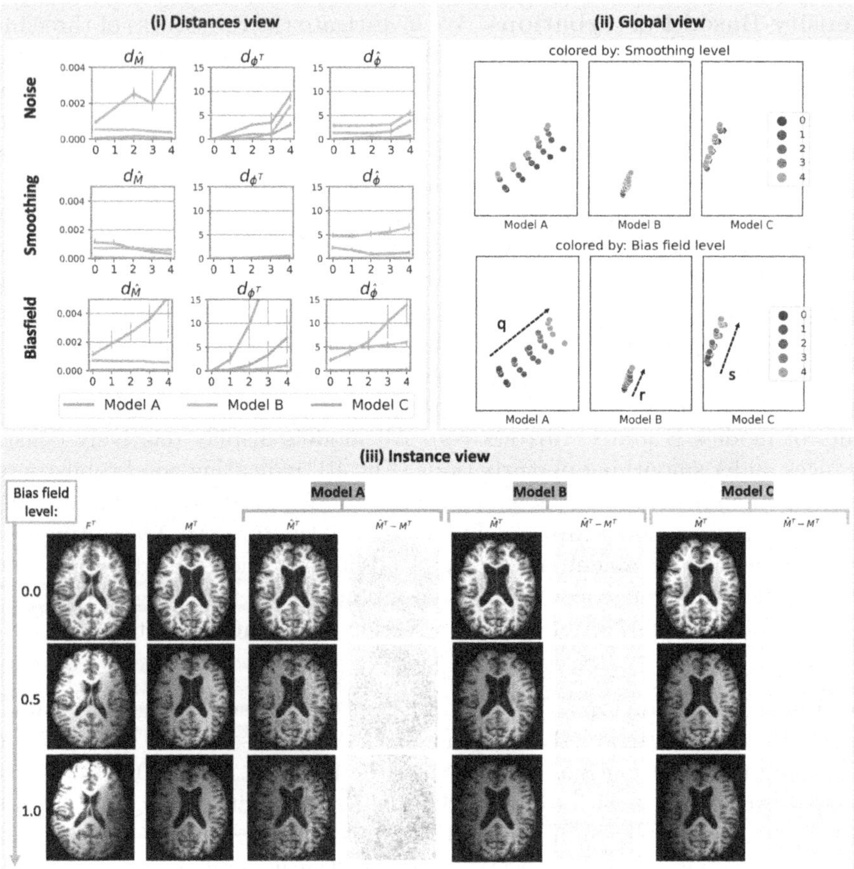

Fig. 4. ProactiV-Reg for assessing the robustness of registration models. *Distances view (i)*: shows the distances influenced by a single perturbation, where distances $d_{\hat{M}}$ and $d_{\hat{\phi}}$ represent the input and output change needed for optimal registration, and d_{ϕ^T} the output change caused by the domain shift. *Global view (ii)*: summarizes the distances, where each data point represents a combination of perturbations. *Instance view (iii)*: shows the local changes added by the mapping. $[F^T, M^T]$ = perturbed images; $\hat{M}^T$ = optimized moving image; $\hat{M}^T - M^T$ = difference image.

3.3 Discussion and Conclusion

This paper presents ProactiV-Reg, a model-agnostic method that provides valuable insights into the robustness of deep learning-based deformable image registration methods under domain shifts. Our results demonstrate ProactiV-Reg for detecting differences in the robustness of three registration models, identifying impactful domain shifts, and determining the optimal regularization weight.

The observed robustness levels are consistent with previously reported findings in the literature. In particular, our results confirm the relatively weaker generalizability of model A and the higher robustness of models B and C across different datasets, as reported by Hoffmann *et al.* [9] and Mok *et al.* [18]. The instance-level analysis provide valuable insights into the implicit input requirements of model A: it relied on the correction of bias fields for optimal registration.

The current implementation of ProactiV-Reg uses fixed design choices to serve as a proof of concept. Our results highlight that some insights can only be derived from computing multiple distances, e.g., a model may require only minor input adjustments to achieve significant improvements. Future work will involve revisiting some of the design choices. For example, mapping both moving and fixed images to desired inputs could be explored. This approach might encounter convergence problems due to the increased optimization complexity.

While this work focused on three models and four simulated domain shifts, ProactiV-Reg is universally applicable to other models and domain shifts. It can also be used to assess whether a particular registration model performs well on new data from different distributions. In this context, the simulation of perturbations is not necessary. Expert annotations and ground truth deformations, although used for validation in this study, are not required for the application of ProactiV-Reg. Ultimately, this method can guide the development of robust registration models by informing the redesign of network architectures or training procedures. In the future, we will demonstrate ProactiV-Reg to guide the design of data augmentation for improving model robustness.

In conclusion, ProactiV-Reg is a versatile tool for assessing the robustness of registration models. It is model agnostic, does not rely on expert annotations or other ground truth, and can incorporate any domain shift, making it a promising method for guiding the development of deep learning models towards robust image registration.

Acknowledgements. This work was supported by the Irene Curie Fellowship program and the Eindhoven Artificial Intelligence Systems Institute.

Conflict of Interest. The authors have no relevant conflicts to disclose.

References

1. Balakrishnan, G., Zhao, A., Sabuncu, M.R., Guttag, J., Dalca, A.V.: Voxelmorph: A learning framework for deformable medical image registration. IEEE Transactions on Medical Imaging **38**(8), 1788–1800 (2019). https://doi.org/10.1109/TMI.2019.2897538
2. Baum, Z.M.C., Hu, Y., Barratt, D.C.: Meta-registration: Learning test-time optimization for single-pair image registration. In: Aylward, S., Noble, J.A., Hu, Y., Lee, S.L., Baum, Z., Min, Z. (eds.) Simplifying Medical Ultrasound. pp. 162–171 (2022). https://doi.org/10.1007/978-3-031-16902-1_16
3. Corral Acero, J., Zacur, E., Xu, H., Ariga, R., Bueno-Orovio, A., Lamata, P., Grau, V.: SMOD - data augmentation based on statistical models of deformation to enhance segmentation in 2D Cine Cardiac MRI. In: Coudière, Y., Ozenne, V., Vigmond, E., Zemzemi, N. (eds.) Functional Imaging and Modeling of the Heart. pp. 361–369 (2019).https://doi.org/10.1007/978-3-030-21949-9_39
4. De Vos, B., Berendsen, F., Viergever, M., Sokooti, H., Staring, M., Išgum, I.: A deep learning framework for unsupervised affine and deformable image registration. Medical Image Analysis **52**, 128–143 (2019). https://doi.org/10.1016/j.media.2018.11.010
5. Elmahdy, M.S., Jagt, T., Zinkstok, R.T., Qiao, Y., Shahzad, R., Sokooti, H., Yousefi, S., Incrocci, L., Marijnen, C., Hoogeman, M., Staring, M.: Robust contour propagation using deep learning and image registration for online adaptive proton therapy of prostate cancer. Medical Physics **46**(8), 3329–3343 (2019). https://doi.org/10.1002/mp.13620
6. Eppenhof, K., Maspero, M., Savenije, M., de Boer, J., van der Voort van Zyp, J., Raaymakers, B., Raaijmakers, A., Veta, M., van den Berg, C., Pluim, J.: Fast contour propagation for MR-guided prostate radiotherapy using convolutional neural networks. Medical Physics **47**(3), 1238–1248 (2020). https://doi.org/10.1002/mp.13994
7. Fischl, B.: Freesurfer. NeuroImage **62**(2), 774–781 (2012). https://doi.org/10.1016/j.neuroimage.2012.01.021
8. Grzech, D., Kainz, B., Glocker, B., le Folgoc, L.: Image registration via stochastic gradient Markov Chain Monte Carlo. In: Sudre, C.H., et al. (eds.) Uncertainty for Safe Utilization of Machine Learning in Medical Imaging, and Graphs in Biomedical Image Analysis. pp. 3–12 (2020). https://doi.org/10.1007/978-3-030-60365-6_1
9. Hoffmann, M., Billot, B., Greve, D.N., Iglesias, J.E., Fischl, B., Dalca, A.V.: SynthMorph: Learning contrast-invariant registration without acquired images. IEEE Transactions on Medical Imaging **41**(3), 543–558 (2022). https://doi.org/10.1109/TMI.2021.3116879
10. Hoopes, A., Hoffmann, M., Greve, D., Fischl, B., Guttag, J., Dalca, A.: Learning the effect of registration hyperparameters with HyperMorph. Machine Learning for Biomedical Imaging (IPMI 2021 special issue), 1–30 (2022)
11. Ketcha, M.D., Silva, T.S.D., Han, R., Uneri, A., Vogt, S., Kleinszig, G., Siewerdsen, J.H.: Learning-based deformable image registration: effect of statistical mismatch between train and test images. Journal of Medical Imaging **6**(4), 044008 (2019). https://doi.org/10.1117/1.JMI.6.4.044008
12. Khawaled, S., Freiman, M.: NPBDREG: Uncertainty assessment in diffeomorphic brain MRI registration using a non-parametric Bayesian deep-learning based approach. Computerized Medical Imaging and Graphics **99**, 102087 (2022). https://doi.org/10.1016/j.compmedimag.2022.102087

13. Klein, S., Staring, M., Murphy, K., Viergever, M.A., Pluim, J.P.W.: Elastix: A toolbox for intensity-based medical image registration. IEEE Transactions on Medical Imaging **29**(1), 196–205 (2010). https://doi.org/10.1109/TMI.2009.2035616
14. Kolenbrander, I.D., Maspero, M., Hendriksen, A.A., Pollitt, R., van der Voort van Zyp, J.R.N., van den Berg, C.A.T., Pluim, J.P.W., van Eijnatten, M.A.J.M.: Deep-learning-based joint rigid and deformable contour propagation for magnetic resonance imaging-guided prostate radiotherapy. Medical Physics **51**(4), 2367–2377 (2024). https://doi.org/10.1002/mp.17000
15. Marcus, D.S., Wang, T.H., Parker, J., Csernansky, J.G., Morris, J.C., Buckner, R.L.: Open Access Series of Imaging Studies (OASIS): Cross-sectional MRI Data in Young, Middle Aged, Nondemented, and Demented Older Adults. Journal of Cognitive Neuroscience **19**(9), 1498–1507 (2007). https://doi.org/10.1162/jocn.2007.19.9.1498
16. Mok, T.C.W., Chung, A.C.S.: Large deformation diffeomorphic image registration with laplacian pyramid networks. In: Martel, A.L., et al. (eds.) International Conference on Medical Image Computing and Computer Assisted Intervention – MICCAI 2020. pp. 211–221 (2020).https://doi.org/10.1007/978-3-030-59716-0_21
17. Mok, T.C.W., Chung, A.C.S.: Conditional deformable image registration with convolutional neural network. In: de Bruijne, M., et al. (eds.) International Conference on Medical Image Computing and Computer Assisted Intervention – MICCAI 2021. pp. 35–45 (2021). https://doi.org/10.1007/978-3-030-87202-1_4
18. Mok, T.C.W., Li, Z., Xia, Y., Yao, J., Zhang, L., Zhou, J., Lu, L.: Deformable medical image registration under distribution shifts with neural instance optimization. In: Cao, X., et al. (eds.) International Conference on Machine Learning in Medical Imaging – MLMI 2023. pp. 126–136 (2024). https://doi.org/10.1007/978-3-031-45673-2_13
19. Prasad, V., van Sloun, R.J.G., Vilanova, A., Pezzotti, N.: ProactiV: Studying deep learning model behavior under input transformations. IEEE Transactions on Visualization and Computer Graphics **30**(8), 1–14 (2023). https://doi.org/10.1109/TVCG.2023.3301722
20. Terpstra, M.L., Maspero, M., Bruijnen, T., Verhoeff, J.J., Lagendijk, J.J., van den Berg, C.A.: Real-time 3D motion estimation from undersampled MRI using multi-resolution neural networks. Medical Physics **48**(11), 6597–6613 (2021). https://doi.org/10.1002/mp.15217

Challenging the Robustness of Image Registration Similarity Metrics with Adversarial Attacks

Robin Rexeisen and Xiaoyi Jiang(✉)

Faculty of Mathematics and Computer Science, University of Münster, Münster, Germany
{r.rexeisen,xjiang}@uni-muenster.de

Abstract. Multi-modal image registration is a crucial task in various medical applications. A typical technique here is iterative optimization, whose success depends on the reliability of the used similarity metric. In this work, we systematically challenge the robustness of two such popular metrics, Mutual Information and Cross-Cumulative Residual Entropy, by employing adversarial techniques from the deep learning field. Our experiments show resistance to small perturbations, while indicating a higher vulnerability as the perturbations increase. Furthermore, our results indicate that certain structural patterns emerge during this process. Additionally, we examine the functional landscape of both metrics. Consequently, this work emphasizes the robustness of these metrics while also offering a starting point for more insights into the underlying patterns that contribute to their failure.

Keywords: Image registration · similarity metrics · robustness · adversarial attacks

1 Introduction

Medical image registration is a fundamental pre-processing step in numerous clinical applications, such as image fusion, organ atlas creation and tumor growth monitoring [2,7]. A substantial proportion of these tasks processes multi-modal images, for which intensity-based approaches were found to be more effective than landmark-based solutions [3,5]. The typical procedure here involves an iterative optimization of a similarity metric, on which the efficacy of the alignment process highly depends on. Defining such a metric is particularly challenging in the context of multi-modality. Notwithstanding, several proposals emerged over the last decades [8,10,25,26]. The de facto standard of the field is Mutual Information (MI) with its variants such as Normalized Mutual Information (NMI) and conditional Mutual Information (cMI) [9,13,21,22]. Nevertheless, over time certain patterns were found, such as thin structures and soft tissues, which negatively impact the performance of MI [17,19]. Furthermore, in [20,26], the authors

M. Modat et al. (Eds.): WBIR 2024, LNCS 15249, pp. 112–126, 2024.
https://doi.org/10.1007/978-3-031-73480-9_9

note the improvable noise robustness and proposed Cross-Cumulative Residual Entropy (CCRE) as a viable alternative.

Noise sensitivity is also a general issue in deep learning, as demonstrated by Szegedy et al. [23]. They were capable of generating adversarial examples, perturbations invisible to the human eye, which changed the prediction of neural networks. Meanwhile, deep learning has also established for medical problems including image registration [7,9], which also has consequences for their robustness in this field [14].

Although the aforementioned work indicates a potential weakness for similarity metrics such as MI for certain perturbations, no prior work has systematically investigated this in a computable fashion analogous deep learning, to the best of our knowledge. Consequently, the objective of this study is to automatically generate potential counterexamples via adversarial techniques. In particular, we employ the Fast Gradient Sign Method (FGSM) [6] and Projected Gradient Descent (PGD) [15] approaches. Given that these adversarial examples were specifically designed to deceive the corresponding similarity metric, they are expected to challenge the metric in the most rigorous manner. An effectively generated counterexample fails then to be aligned correctly, since such a task requires a reliable similarity metric. Thus, in this work, we don't focus directly on the image registration task, but on the properties of the similarity metric.

1.1 Prior Work

Noise robustness is a typical quality measure for a registration similarity metric. E.g. [8] and [10] use it as a criterion for supporting their proposals. Some authors conduct a comprehensive examination by evaluating the robustness of their proposals against MI at varying noise levels [20,26]. However, to the best of our knowledge, no prior work has analyzed robustness to adversarial examples that are explicitly generated with the goal to deceive the similarity metrics.

Adversarial methods are also applied before in the context of image registration. Primarily in the field of generative adversarial networks (GAN), in which the generator network contents with a separate discriminator by generating realistic data samples. A common application of GAN is modality translation [24,27], which reduces the multi-modal problem into a mono-modal registration by synthesizing a distinctive modality from another. An alternative method is to train the generator to predict deformation fields for a realistic alignment [4,28]. Although GANs and adversarial attacks both rely on the generation of contrasting images, they are fundamentally distinct. The purpose of adversarial training is to yield a network that represents a specific distribution, rather than to challenge its resistance from perturbations.

In the field of deep learning, however, several existing techniques have been developed with precisely this objective. Besides FGSM [6] and PGD [15], some noteworthy works include Basic Iterative Method [12], Carlini and Wagner [1], and DeepFool [16]. However, to the best of our knowledge, these methods have not been applied to arbitrary image registration similarity metrics. Therefore, this work will demonstrate a methodology for applying the established methods to this task.

2 Methods

The following sections begin with a brief introduction to the registration problem, followed by a presentation of our adversarial approach for challenging the robustness of similarity metrics. Subsequently, we define the Intrinsic Registration Error on which our evaluation is based. We then describe further evaluation methods and end with details about our implementation.

2.1 Registration Problem

The classical registration task consists of a moving image I_{moving} that is transformed by $\mathcal{T}_\theta(I_{\text{moving}})$ with specific parameters θ. The goal is to find the parameters $\overline{\theta}$ that align the moving to a given fixed image I_{fixed}. In medical context, the fixed and moving images are usually of different modalities. For simplification, we assume in this work that I_{fixed} and I_{moving} are 2D images, while $\mathcal{T}$ represents a 2D rigid transformation. In this context $\theta = (\theta_x, \theta_y, \theta_r) \in \mathbb{R}^3$ with θ_x, θ_y being the pixel translation in x- and y-direction, respectively, and θ_r represents the rotation in degrees.

The quality of the alignment is supposed to be expressed through a registration metric $\mathcal{O}$, such that the overall problem is to solve:

$$\overline{\theta} = \arg\max_{\theta} \mathcal{O}(I_{\text{fixed}}, \mathcal{T}_\theta(I_{\text{moving}})). \tag{1}$$

This is classically handled in an iterative scheme. In this study, we focus mainly on the similarity metrics MI and CCRE.

2.2 Adversarial Registration Examples

The objective of this work is to examine counterexamples that challenge the robustness of the similarity metrics when used for image registration. To achieve this, we intend to apply established adversarial generation techniques from the deep learning field. For this purpose, let I be a certain image that belongs to a certain class $\hat{y}$. In context of deep learning, an adversarial example of I for a classification network f, that aims $f(I) = \hat{y}$, is described by a bounded perturbation δ, that is supposed to change the outcome, i.e. $f(I + \delta) \neq \hat{y}$.

To generate such examples, different methods were proposed. White-box approaches, for example, exploit the gradient of the targeted network for this purpose. They exist in an untargeted and targeted version.

In an untargeted white-box attack, the perturbation δ is often searched within a specific boundary described by $\varepsilon > 0$ in the following way:

$$\arg\max_{\|\delta\|_p \leq \varepsilon} \mathcal{L}(F(I + \delta),\ \hat{y}). \tag{2}$$

Here, $\mathcal{L}$ represents the loss function the network is trained with, e.g. cross-entropy. The utilized norm that bounds δ is defined by the choice of $p \in (0, \infty]$. For FGSM and PGD, this is typically $p = \infty$. Given that $\mathcal{L}$ describes the discrepancy of the network, maximizing this function minimizes the probability that $I + \delta$ is in the correct class $\hat{y}$, which simultaneously increases the probability of $I + \delta$ being included in any of the incorrect classes.

To increase the likelihood that $I+\delta$ gets classified as a specific incorrect class $y' \neq \hat{y}$, a targeted white-box attack often solves:

$$\underset{\|\delta\| \leq \varepsilon}{\arg\min}\, \mathcal{L}(F(I + \delta),\ y'). \tag{3}$$

As we do not seek to address the registration problem directly, but rather to develop a methodology for computing adversarial examples, we propose a discrete approach. The goal is to utilize (2) and (3) to enable the application of previously proposed adversarial methods for deep learning. To achieve this, we specify a discrete set of fixed transformation parameters:

$$\Theta := \{\theta_1, \ldots, \theta_K\}, \quad K \in \mathbb{N}. \tag{4}$$

Each of the θ_k can be interpreted as a specific class. Consequently, we need to ensure that for each registration that we conduct, the truth parameter $\hat{\theta}$ is part of this fixed parameter set, i.e. $\tilde{\theta} \in \Theta$.

Next, the probability needs to be computed for a registration outcome to be in a specific "transformation class". For this, we compute the similarity metric value o_k for each set of parameters:

$$o_k(I) := \mathcal{O}(I_{\text{fixed}}, I(\theta_k)),\ k = 1, \ldots, K. \tag{5}$$

According to Eq. (1), the overall task is then to find arg max $\{o_1, \ldots, o_K\}$. Thus, a subsequent softmax among all o_k results in a probability distribution $p(I) := (p_1(I), \ldots, p_K(I))$ for the registration classes of Θ:

$$p_k(I) := \frac{e^{o_k}(I)}{\sum_{j=1}^{K} e^{o_j(I)}},\ k = 1, \ldots, K. \tag{6}$$

Equation (2) can thus be applied in the context of image registration:

$$\overline{\delta} := \underset{\|\delta\|_p \leq \varepsilon}{\arg\max}\, \mathcal{L}(p(I_{\text{moving}} + \delta),\ \hat{p}), \tag{7}$$

where $\hat{p}$ is the one-hot encoding of $\hat{\theta}$ within Θ. Similarly, one can reformulate Eq. (3) for the targeted variant. In the following, we denote the generated adversarial moving image as $I_{\text{moving}}^{\text{adv}} := I_{\text{moving}} + \overline{\delta}$.

2.3 Intrinsic Registration Error

Since the similarity metric is supposed to be highest for the right transformation parameters (Eq. (1)), the overall goal of an adversarial attack is to change this global maximum by altering the moving image.

We define the distance between this global maximum $\theta \in \mathbb{R}^3$ and a given truth alignment $\hat{\theta} \in \mathbb{R}^3$ as Intrinsic Registration Error (IRE) $\mathcal{E}$, which is, in fact, the L^1 distance between them:

$$\mathcal{E}(\theta, \hat{\theta}) := \left|\theta_x - \hat{\theta}_x\right| + \left|\theta_y - \hat{\theta}_y\right| + \left|\theta_r - \hat{\theta}_r\right|. \tag{8}$$

It corresponds to the lowest possible error a registration algorithm can yield when solving Eq. (1) and is thus a measurement of the reliability of the similarity metric rather than of a specific optimization technique.

IRE adopts a conventional registration assessment methodology by considering the discrepancies in the transformation parameters. For more complex transformation models, an adaption of the corresponding evaluation method is conceivable. We note that Eq. (8) mixes the pixel and degree distance of the parameters, which may make it difficult to interpret. However, we assume that this is sufficient for our purposes, since we are primarily interested in the change of this error introduced by adversarial techniques.

More specifically, we define the change of the IRE between the global maximum θ_{orig} that comes from the original image and the maximum θ_{adv} from the adversarial image by:

$$\Delta_{\hat{\theta}}(\theta_{\text{adv}}, \theta_{\text{orig}}) := \mathcal{E}(\theta_{\text{adv}}, \hat{\theta}) - \mathcal{E}(\theta_{\text{orig}}, \hat{\theta}). \tag{9}$$

Accordingly, we call this the Intrinsic Registration Error Distance (IRED). We also omit the truth parameter $\hat{\theta}$ to shorten the expression $\Delta(\theta_1, \theta_2)$ for clarity if it is evident from the context. We define an adversarial attack as successful if and only if it increases the IRED. Consequently, an unsuccessful attack causes no change in this value, or even decreases it.

2.4 Evaluation Methodology

In our analysis, we seek to evaluate the effect of an adversarial attack on the similarity metric. To be free from side effects, e.g. the quality of an optimization algorithm, we operate directly on a discrete version of the function graph

$$G_{\mathcal{O}}(I) := \{(\theta, \mathcal{O}(I_{\text{fixed}}, I(\theta))) \mid \theta \in \Theta_G\} \subseteq \mathbb{R}^4. \tag{10}$$

Specifically, we compare this graph between the original image pair $G_{\mathcal{O}}(I_{\text{moving}})$ and its adversarial counterpart $G_{\mathcal{O}}(I^{\text{adv}}_{\text{moving}})$ on the fixed discrete grid $\Theta_G := ([-5, 5] \cap \mathbb{Z})^3$. The size of Θ_G was limited due to the high computational cost. Furthermore, for $\mathcal{O}$ we set either MI or CCRE in this work. This approach allows us to study the variations in the local structure of the similarity metric, induced by an adversarial example, by examining the changes in local maxima. Additionally, the global maxima needed to determine the IRED (9) are also located on this grid.

2.5 Implementation Details

In our examination of class-based adversarial attacks, we leveraged both Fast Gradient Sign Method (FGSM) and Projected Gradient Descent (PGD) approaches in the targeted and untargeted version. Specifically, for the targeted variant, we optimized for the transformation class with the maximum distance from the true transformation, as defined by Eq. (8). Consequently, this version can be conceptualized as a "quality" attack, as it seeks to achieve the most critical perturbations. In contrast, the untargeted version can be regarded as a "quantity" of errors, as it seeks for any possible error-inducing perturbation. Furthermore, we conducted experimental evaluations with perturbation bounded by $\varepsilon \in \{0.01, 0.05\}$. Notably, for PGD, we employ a step size of 20 and assigned $\alpha = 0.1$.

Our implementation is written in Python, where we utilize the torchattack [11] as well as torchio [18] package. For the sake of reproducibility, all source code will be made available online[1].

3 Results

In this section, we present the results of our study on the adversarial robustness of the aforementioned registration metrics, as introduced in Subsect. 2.2. The first part specifies the experimental configuration employed. We proceed by comparing the Signal-to-Noise Ratio of the generated adversarial moving image with that of the original to gain insight into the induced perturbation. Next, we concentrate on the variation of the number of local maxima in the landscape of the similarity metric to explore the effect of the adversarial generations on it. In the final experiment, we evaluate the impact of the attacks on the performance of image registration within a discrete area. Last but not least, we discuss these outcomes.

3.1 Experimental Settings

To assess the adversarial robustness of the similarity metrics, we leveraged the IXI dataset[2]. Specifically, from a subset of 30 T1- and T2-weighted MR volume pair, we extracted three random axial slices per volume, resulting in a total of 90 slices at an original resolution of 256 × 256 pixels without interpolation or resampling. Following quality control measures to eliminate images without substantial structure, our final test set comprised 81 slices. Subsequently, these slices underwent rigid transformations, defined by $\theta = (\theta_x, \theta_y, \theta_r) \in \mathbb{R}^3$, where $\theta_x, \theta_y, \theta_r$ are each uniformed sampled from $[-2, 2] \cap \mathbb{Z}$. As before, θ_r denotes rotations in degrees, while θ_x and θ_y represent translations in pixels. This ensures that this truth parameter can be localized on the function graph (see (10)) as well as in the fixed parameter set Θ (see (4)). Consequently, we also set $\Theta = [-2, 2] \cap \mathbb{Z}$.

[1] https://zivgitlab.uni-muenster.de/ag-pria/adversarial_registration_attack.

[2] https://brain-development.org/ixi-dataset/.

3.2 Signal-to-Noise Ratio

In the first instance, we conducted an investigation into the disparity in Signal-to-Noise Ratio (SNR) between the original and adversarial images. This analysis provides insight into the structural properties of the introduced perturbations. A decrease in SNR suggests that the added perturbations resemble unstructured noise, while an elevated SNR implies that these perturbations exhibit a specific pattern.

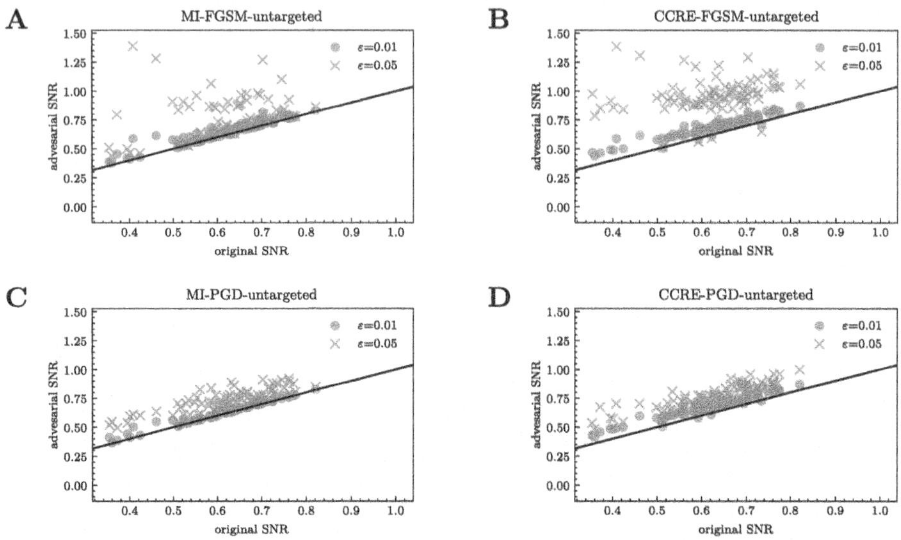

Fig. 1. Signal-to-noise ratio of the original moving image and its adversary counterpart for different combinations of registration similarity metrics and attack variants (A-D). The plots illustrate the attack outcomes for varying ϵ values. Points above the diagonal line exhibit an increased SNR after perturbing, whereas points below the diagonal line demonstrate a decreased SNR. The targeted variants are omitted for brevity, as they follow a comparable distribution to their untargeted counterparts.

The findings are illustrated in Fig. 1. Some results of the targeted version are omitted for the sake of brevity, as their distribution is found to be quite similar to that of the untargeted version. Some generated adversarial images with their SNR are also listed in Fig. 2. Notably, increasing ε corresponds with an elevated SNR for both methods. This increase in SNR is observed in 79% of instances for MI and 97% for CCRE. The results indicate that CCRE requires even more frequent application of structured perturbations to achieve successful deception compared to MI. Moreover, the FGSM approach typically results in a higher SNR increase, which can also be interpreted in terms of a more substantial structural perturbation for higher ε. A visual inspection of Fig. 2 supports this interpretation. Instead of mere random noise, an enhanced contrast along with greater detail are discernible in the picture with increased SNR. This is especially apparent for higher ε.

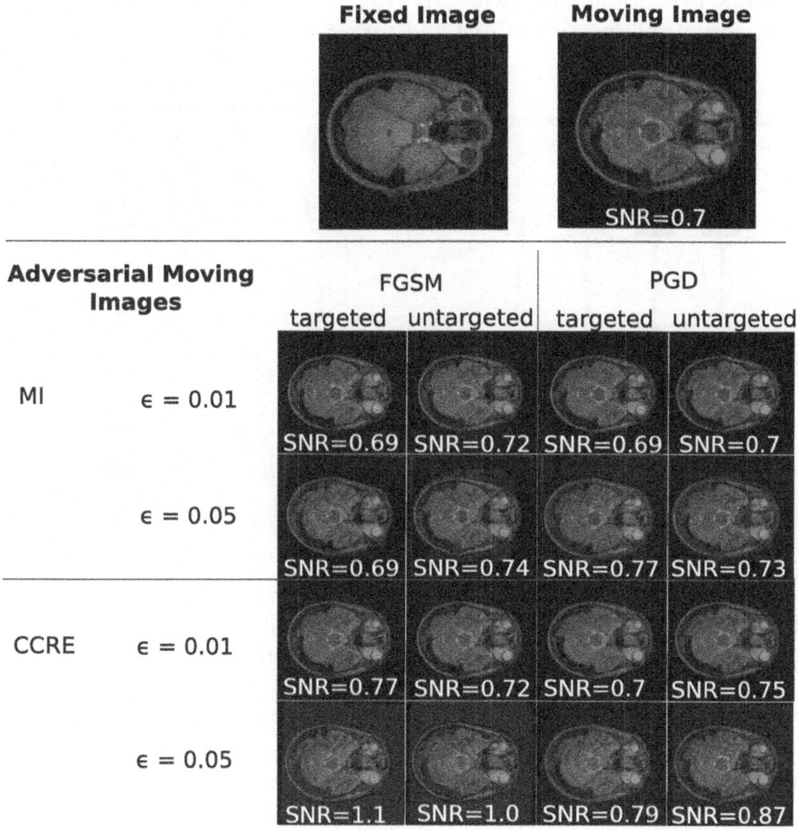

Fig. 2. Examples where each attack variant for each ε increases the IRE compared to the original moving image (top right) as intended. The fixed and moving image are displayed on the top. The different attack outcomes on the moving image are visible in the bottom part. The corresponding Signal-to-Noise Ratio is stated below each (adversarial) moving image.

3.3 Local Maxima

Next, we examined the frequency of local maxima occurring in the functional graphs. As we operate on a discrete grid (see (10)), these are easily to localize. This analysis serves as a means of characterizing the deformation induced by adversarial attacks. Specifically, the emergence of novel optima corresponds to regions where the deformation exhibits localized peaks, whereas the disappearance of existing optima occurs when the graph becomes smoother.

A comparison of the pre- and post-perturbation maxima for the different methodologies and parameters is provided in Fig. 3. Our findings indicate that the number of maxima remains largely invariant across most scenarios. However, notable exceptions arise with MI (Fig. 3), particularly under conditions of increased acceptable perturbations and targeted attacks.

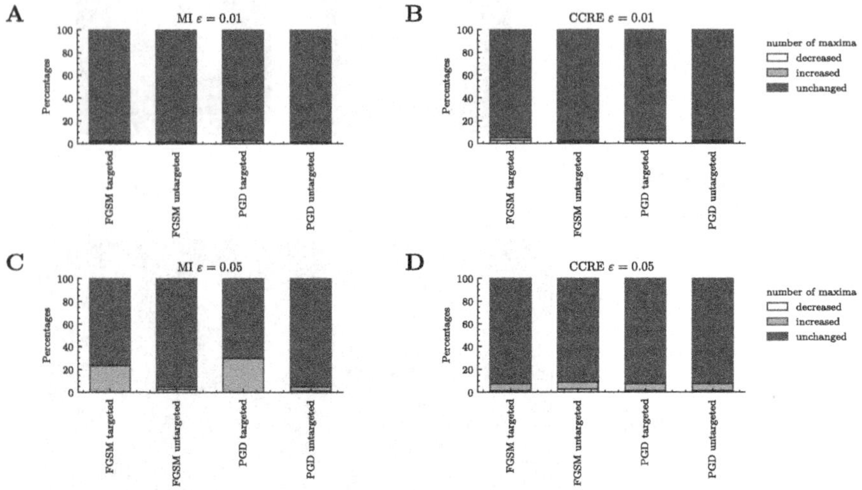

Fig. 3. The fraction of changes in the number of local maxima between the original image pair and its adversarial counterpart for registration with MI (A,C) and CCRE (B,D) within the region Θ and different admissible perturbations, expressed by ε. Each plot illustrates the differences for the various applied adversarial techniques.

3.4 Intrinsic Registration Error

In the last analysis step, we focus on $\Delta(\overline{\theta}_{\text{adv}}, \overline{\theta}_{\text{orig}})$, i.e. the IRED between the global maxima of the adversarial $\overline{\theta}_{\text{adv}}$ and the original $\overline{\theta}_{\text{orig}}$ similarity metric landscape. Recall that an adversarial attack is supposed to increase this error. Thus, by analyzing success frequency, we gather an idea of the robustness of their corresponding similarity metric.

Upon examination of Fig. 5, several observations concerning the IRED for various attack methods can be identified. For $\varepsilon = 0.01$, the attacks did not generate adversarial examples in the majority of cases. However, they were still able to do so to a certain extent. Notably, for the MI model, FGSM untargeted demonstrated the highest capability with a success rate of approximately 32%, while PGD untargeted had the highest rate for CCRE at around 19.8%. Figure 2 provides an illustration of some the successfully generated adversarial examples. For $\varepsilon = 0.01$, the additional perturbation is barely discernible to the human eye.

Increasing ϵ to 0.05, both targeted and untargeted attacks demonstrated a notable increase in success rates. Specifically, for the MI model, PGD targeted

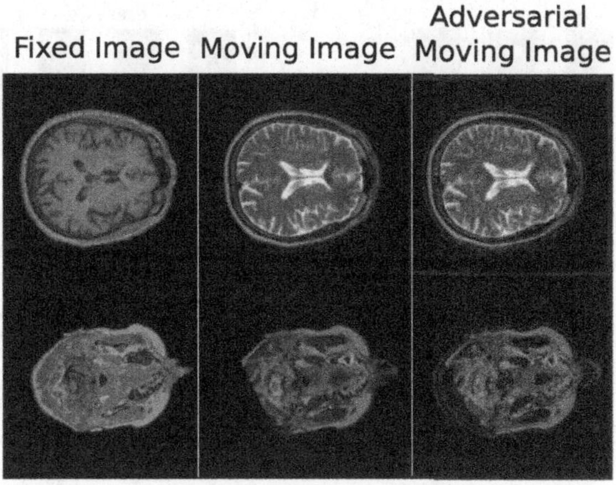

Fig. 4. Examples where the IRE went down, contrary to the anticipated outcome of an adversarial approach. The left column presents the fixed image, the middle column the moving image, and the right columns exhibits the outcome of the untargeted FGSM attack on the moving image for MI with $\varepsilon = 0.05$.

displayed the highest success rate of approximately 75.3%, while FGSM untargeted yielded the highest rate for CCRE at around 76.5%. Again, Fig. 2 illustrates some successful outcomes of these attacks. It is noteworthy that FGSM untargeted also led to a reduction of the IRE in 37% of cases for MI, in contrast to the intended purpose of the method. Two characteristic examples can be found in Fig. 4. In the upper image, it appears that the introduced perturbation may have enhanced the IRE by chance. In contrast, the lower image reveals that the structural elements from the fixed image were replicated in the adversarial generation, particularly in the lower-left corner. The introduction of this new, coherent information naturally has a positive impact on the registration task.

For MI, Fig. 5 shows that the targeted approaches produced more successful adversarial example for $\varepsilon = 0.05$, whereas untargeted variants were more effective for $\varepsilon = 0.01$. In contrast, CCRE demonstrated a tendency to achieve slightly higher success rates in untargeted attacks across the different ε values.

An analysis of the IRED distribution depicted in Fig. 6 reveals several notable findings. First, as anticipated, the targeted version exhibits the largest IRE change compared to their untargeted counterpart. Moreover, our findings suggest that for $\varepsilon = 0.05$ for the targeted attacks, a higher IRE increase is statistically more prevalent for MI when compared to CCRE.

In the case of an untargeted FGSM attack in MI and at $\varepsilon = 0.05$, the IRED distribution appears symmetrically distributed around zero. This implies that IRE increases and decreases are equally probable to occur in comparable magnitudes. In the context of targeted attacks in MI with $\varepsilon = 0.05$, FGSM produces a higher IRE increase more frequently than PGD. In contrast, CCRE appears to

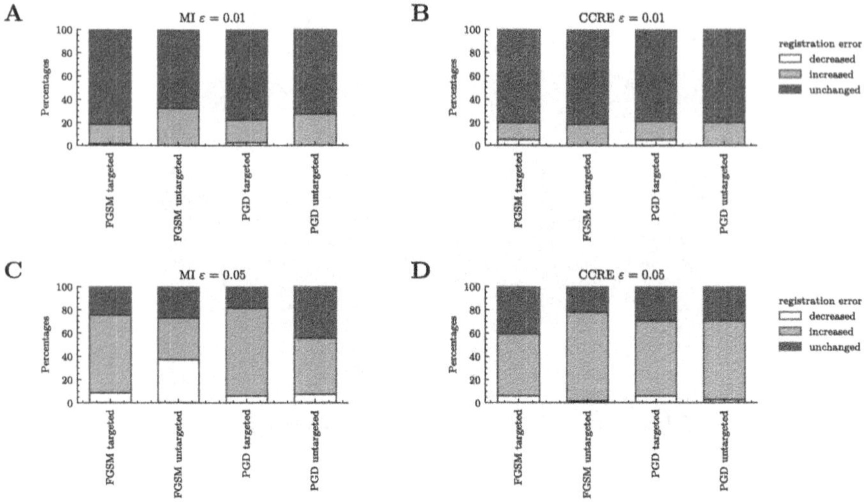

Fig. 5. Percentages of the different outcomes for the IRED as defined in Eq. (9) when comparing the similarity metric landscape of the perturbed moving image with the original. The figures illustrate this difference between the various attack approaches for MI (A,C) and CCRE (B,D) as well as for different perturbation ranges (A,B and C,D).

yield comparable distributions for FGSM and PGD, particularly for the targeted version.

The results can be summarized as follows: for smaller changes to the input data, both MI and CCRE appear to resist adversarial attacks in most cases. CCRE demonstrates even greater resilience during these instances. However, this resistance weakens for larger perturbations, making both MI and CCRE vulnerable to successful attacks utilizing the appropriate techniques. Another significant observation is that targeted attacks appear to result more frequently in higher IRE increases, whereas untargeted ones produce more reliable perturbations. It is noteworthy that MI for higher perturbations, particularly for FGSM, is an exception to this trend, as the effect of the adversarial example can be unpredictable for higher perturbations.

3.5 Discussion

The results of the registration error tests indicate that both MI and CCRE are susceptible to deception (Sect. 3.4), though this is unlikely to occur in the absence of visible perturbations. For perceivable modification, both metrics demonstrate a notable resilience.

With regard to the induced perturbations, the SNR experiments (Sect. 3.2) indicate that these exhibit a specific structural pattern, particularly in the case of CCRE. Moreover, our maxima study (Sect. 3.3) indicates that the global form of the similarity metrics remains relatively stable with respect to the neighboring

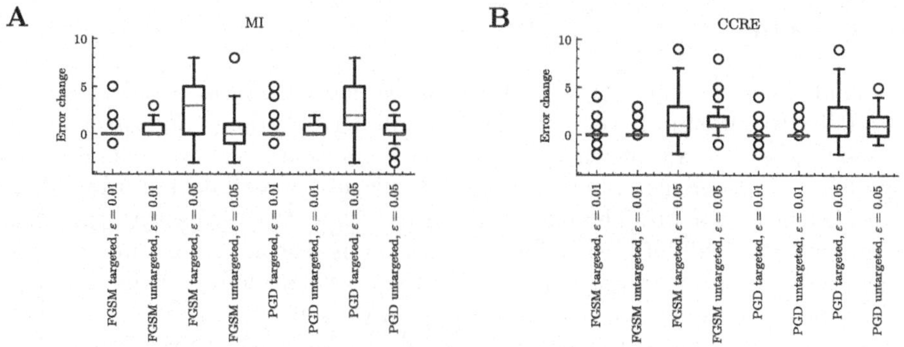

Fig. 6. Distribution of IRED as defined in Eq. (9) visualized with box plot diagrams. (A) and (B) visualize the distribution for MI and CCRE, respectively, across the different adversarial techniques and perturbations.

rigid registration parameters range we focused on. This suggests that the optima are primarily shifted rather than generated anew. Consequently, identifying the correct registration transformation for the adversarial example becomes challenging, even for advanced optimization methods. An exception to this observation is MI, where targeted attacks are capable of yielding new local maxima. This finding may be indicative of MI's susceptibility towards deformations in its function plot. Considering the success rate (Sect. 3.4), these deformations may be essential for a successful adversarial generation for MI.

Regarding this MI success rate, the targeted attack is demonstrably more effective than the untargeted version for higher perturbations. In particular, the results for FGSM are more volatile there than for any other variant. This phenomenon is not observed for untargeted attacks on CCRE or lower perturbations, where they appear to be more successful than their targeted counterparts. This efficacy of the untargeted method generally aligns with expectations, as it imposes fewer restrictions on the resulting perturbations, placing primary emphasis on success than on large IRED. The counterintuitive outcome for MI probably indicates that the particular attack variant does not align with this registration metric when ε increases. Especially FGSM entails a linearity assumption within a local area, which has stronger impact for higher ε. In this context, imposing more constraints in form of a particular target class could assist in guiding towards a viable solution despite this weakness, which could elucidate the result. For all other combinations of ε and similarity metrics, each untargeted attack exhibits approximately the same effectiveness and, as already noted, is more successful than its targeted counterpart. This indicates that they are equally suitable and suggests that CCRE is sufficiently linear even for large perturbations.

4 Conclusion

Our findings underscore the robustness of MI and CCRE, as demonstrated by the ability to withstand even adversarial small perturbations. Conversely, in the presence of larger adversarial changes, both similarity metrics exhibited vulnerability. However, here, our findings also suggest potential limitations in the attack's efficacy for MI. The results of our experiments indicate that the adversarial examples exhibit a specific pattern that is distinct from random noise. This pattern seems to work by preserving the overall structure of the functional landscape of the similarity metrics, causing only a shift of it. The findings of this study have the potential to contribute to a deeper comprehension of registration metrics.

This study represents a first attempt to gain insight into the impact of the adversarial approach on similarity metrics. In future works, we intend to reinforce our findings through a more comprehensive investigation. In particular, we aim to conduct a more thorough evaluation of different attacks with varying parameters, examine the changes in the functional landscape through the use of more sophisticated methods, and investigate a broader range of perturbation levels. Additionally, we aim to identify certain visual patterns that are particularly conducive to generating effective adversarial examples. Furthermore, we intend to extend the applicability of these methods to other similarity metrics, such as NMI. Finally, further investigation into more intricate transformations, such as deformation fields, in conjunction with three-dimensional volumes instead of slices, represents another viable direction for future research. Our ultimate goal is to deepen the understanding of these metrics and optimize their utilization for registration tasks.

Acknowledgements. This work was supported by the Deutsche Forschungsgemeinschaft (DFG) - CRC 1450 - 431460824.

References

1. Carlini, N., Wagner, D.: Towards Evaluating the Robustness of Neural Networks. In: IEEE Symposium on Security and Privacy (SP). pp. 39–57 (2017)
2. Chen, J., Liu, Y., Wei, S., Bian, Z., Subramanian, S., Carass, A., Prince, J.L., Du, Y.: A survey on deep learning in medical image registration: New technologies, uncertainty, evaluation metrics, and beyond. CoRR **abs/2307.15615** (2023)
3. El-Baz, A., Acharya, R., Mirmehdi, M.: Multi Modality State-of-the-Art Medical Image Segmentation and Registration Methodologies. Springer (2011)
4. Fan, J., Cao, X., Xue, Z., Yap, P.T., Shen, D.: Adversarial similarity network for evaluating image alignment in deep learning based registration. In: MICCAI. pp. 739–746. Springer (2018)
5. Ferrante, E., Paragios, N.: Slice-to-volume medical image registration: A survey. Medical Image Analysis **39**, 101–123 (2017)
6. Goodfellow, I.J., Shlens, J., Szegedy, C.: Explaining and harnessing adversarial examples. In: ICLR (2015)

7. Haskins, G., Kruger, U., Yan, P.: Deep learning in medical image registration: A survey. Machine Vision and Applications **31**(1-2), 8 (2020)
8. Heinrich, M.P., Jenkinson, M., Bhushan, M., Matin, T., Gleeson, F.V., Brady, S.M., Schnabel, J.A.: MIND: Modality independent neighbourhood descriptor for multi-modal deformable registration. Medical Image Analysis **16**(7), 1423–1435 (2012)
9. Jiang, X., Ma, J., Xiao, G., Shao, Z., Guo, X.: A review of multimodal image matching: Methods and applications. Information Fusion **73**, 22–71 (2021)
10. Kasiri, K., Fieguth, P., Clausi, D.A.: Self-similarity measure for multi-modal image registration. In: IEEE International Conference on Image Processing (ICIP). pp. 4498–4502 (2016)
11. Kim, H.: Torchattacks : A pytorch repository for adversarial attacks. CoRR **abs/2010.01950** (2020)
12. Kurakin, A., Goodfellow, I.J., Bengio, S.: Adversarial examples in the physical world. In: ICLR (2017)
13. Loeckx, D., Slagmolen, P., Maes, F., Vandermeulen, D., Suetens, P.: Nonrigid Image Registration Using Conditional Mutual Information. IEEE Transactions on Medical Imaging **29**(1), 19–29 (2010)
14. Ma, X., Niu, Y., Gu, L., Wang, Y., Zhao, Y., Bailey, J., Lu, F.: Understanding adversarial attacks on deep learning based medical image analysis systems. Pattern Recognition **110**, 107332 (2021)
15. Madry, A., Makelov, A., Schmidt, L., Tsipras, D., Vladu, A.: Towards deep learning models resistant to adversarial attacks. In: ICLR (2018)
16. Moosavi-Dezfooli, S.M., Fawzi, A., Frossard, P.: DeepFool: A simple and accurate method to fool deep neural networks. In: CVPR. pp. 2574–2582 (2016)
17. Penney, G.P., Weese, J., Little, J.A., Desmedt, P., Hill, D.L.G., Hawkes, D.J.: A comparison of similarity measures for use in 2-D-3-D medical image registration. IEEE Transactions on Medical Imaging **17**(4), 586–595 (1998)
18. Pérez-García, F., Sparks, R., Ourselin, S.: TorchIO: a Python library for efficient loading, preprocessing, augmentation and patch-based sampling of medical images in deep learning. Computer Methods and Programs in Biomedicine p. 106236 (2021)
19. Pluim, J., Maintz, J., Viergever, M.A.: Mutual-information-based registration of medical images: A survey. IEEE Transactions on Medical Imaging **22**(8), 986–1004 (2003)
20. Rao, M., Chen, Y., Vemuri, B.C., Wang, F.: Cumulative Residual Entropy: A New Measure of Information. IEEE Transactions on Information Theory **50**(6), 1220–1228 (2004)
21. Sengupta, D., Gupta, P., Biswas, A.: A survey on mutual information based medical image registration algorithms. Neurocomputing **486**, 174–188 (2022)
22. Studholme, C., Hill, D., Hawkes, D.J.: An overlap invariant entropy measure of 3D medical image alignment. Pattern Recognition **32**(1), 71–86 (1999)
23. Szegedy, C., Zaremba, W., Sutskever, I., Bruna, J., Erhan, D., Goodfellow, I.J., Fergus, R.: Intriguing properties of neural networks. In: ICLR (2014)
24. Tanner, C., Özdemir, F., Profanter, R., Vishnevsky, V., Konukoglu, E., Göksel, O.: Generative adversarial networks for MR-CT deformable image registration. CoRR **abs/1807.07349** (2018)
25. Viola, P., Wells, W.: Alignment by maximization of mutual information. In: ICCV. pp. 16–23 (1995)
26. Wang, Vemuri, Rao, Chen: Cumulative residual entropy, a new measure of information & its application to image alignment. In: ICCV. pp. 548–553 (2003)

27. Wei, D., Ahmad, S., Huo, J., Huang, P., Yap, P.T., Xue, Z., Sun, J., Li, W., Shen, D., Wang, Q.: SLIR: Synthesis, localization, inpainting, and registration for image-guided thermal ablation of liver tumors. Medical Image Analysis **65**, 101763 (2020)
28. Yan, P., Xu, S., Rastinehad, A.R., Wood, B.J.: Adversarial image registration with application for MR and TRUS image fusion. In: MLMI@MICCAI. pp. 197–204 (2018)

Comparative Study on Co-registration Techniques for Diffusion-Weighted Breast MRI and Improved ADC Mapping

Luise Brock[1,2](✉), Andrzej Liebert[1], Hannes Schreiter[1], Dominika Skwierawska[1], Chris Ehring[1], Jessica Eberle[1], Frederik B. Laun[1], Michael Uder[1], Lorenz Kapsner[1], Sabine Ohlmeyer[1], Dominique Hadler[1], Florian Knoll[1], and Sebastian Bickelhaupt[1]

[1] Radiologisches Institut, Universitätsklinikum Erlangen, Friedrich-Alexander-Universität Erlangen-Nürnberg (FAU), Erlangen, Germany
Luise.Brock@uk-erlangen.de

[2] Department of Artificial Intelligence in Biomedical Engineering (AIBE), Friedrich-Alexander-Universität Erlangen-Nürnberg (FAU), Erlangen, Germany

Abstract. Breast magnetic resonance imaging (MRI) increasingly incorporates diffusion-weighted imaging (DWI) sequences. DWI allows to calculate apparent diffusion coefficient (ADC) maps which can supplement differentiation between malignant and benign breast lesions. However, artifacts in DWI are not infrequent, e.g., due to patient motion, pulsation or other sources, which can cause shifts between the different b-value acquisitions and affect the accuracy of the ADC map. This IRB-approved, retrospective study includes $n = 141$ women with breast lesions undergoing breast MRI including an ultra-high b-value DWI acquisition ($b = 1500 s/mm^2$). Thirteen different rigid and non-rigid co-registration methods facilitated by the ANTs library were evaluated. The targets for co-registration were different b-value DWI. ADC maps were calculated from the co-registered DWI images and analyzed quantitatively, measuring mean, standard deviation, and within-lesion coefficient of variance (CoV), using repeated-measures ANOVA alongside Dunnett post-hoc tests in manual segmentations of the target lesions performed on ultra-high b-value images. In addition, we evaluated between-patient CoV and area under the curve (AUC) for differentiation of benign and malignant lesions, as well as sensitivity, specificity, and accuracy to measure clinical efficacy.

Amongst the $n = 13$ co-registration techniques, the iterative SyNCC co-registration method with ultra-high b-value DWI images as target demonstrated the most notable improvement in standard deviation and within-lesion CoV while maintaining consistent mean ADC values. However, none of the co-registration method significantly improved the differentiation of lesion types. Nonetheless, the study offers a detailed comparison of co-registration methods for DWI images in ADC map generation.

Keywords: Breast MRI · ADC · Co-registration

Supplementary Information The online version contains supplementary material available at https://doi.org/10.1007/978-3-031-73480-9_10.

M. Modat et al. (Eds.): WBIR 2024, LNCS 15249, pp. 127–136, 2024.
https://doi.org/10.1007/978-3-031-73480-9_10

1 Introduction

Breast MRI is a pivotal imaging method in breast cancer diagnostics, e.g., for screening in high-risk individuals and specific indications in preoperative staging of breast cancer and post-therapeutic monitoring [1]. Dynamic contrast-enhanced (DCE) MRI offers a high sensitivity in detecting suspicious lesions after contrast injection. However, its ability to characterize detected lesions can be limited in some cases, leading to false-positives and unnecessary biopsies in retrospect. In this context, diffusion-weighted imaging (DWI) has emerged as a potential complementary non-contrast-enhanced MRI technique that is increasingly integrated into multiparametric breast MRI protocols [2–5]. DWI allows depicting local alterations of tissue microstructure by mapping Brownian molecular motion of water molecules. The scope of these effects can be quantified via the apparent diffusion coefficient (ADC), calculated from the different DWI b-value acquisitions acquired during an examination. In the context of breast MRI, ADC has been suggested to aid in differentiating between benign and malignant lesions and to potentially aid in predicting treatment response in breast cancer patients undergoing neoadjuvant chemotherapy [3, 6]. A recent review of the literature in 2022 highlights the potential of DWI and its features to enable earlier and more accurate diagnosis, thereby potentially improving treatment outcomes [5, 7].

However, technical challenges such as distortion artifacts, fat saturation issues, and lengthy examination times potentially leading to motion artifacts, can impact image quality and pose challenges to DWI [8–10]. Inter-b-value misalignment and distortion can impair the creation of accurate ADC maps, thus compromising the ability to distinguish between malignant and benign lesions. To address this, this study evaluates 13 different co-registration methods from the ANTS library to enhance ADC maps, conducting both technical and clinical performance assessments.

2 Materials and Methods

2.1 Dataset and Imaging Sequences

This IRB-approved retrospective study conducted at University Hospital Erlangen, Germany, included breast MRI examinations from 2015 to 2020. A total of 141 cases were included, comprising 68 benign and 73 malignant lesions, with a mean age of 52 years ($\pm$12).

MRI examinations were conducted using 1.5 ($n = 7$) and 3 T ($n = 134$) MRI systems (Aera (1.5 T) and Skyra Fit (3 T); Siemens Healthineers, Erlangen, Germany) equipped with dedicated multi-channel breast coils. The multiparametric MRI protocol encompassed pre- and post-contrast T1-weighted sequences, T2-weighted sequences, and multi-b-value diffusion-weighted imaging (DWI) sequences with axial orientation and b-values of 50 (b50), 750 (b750), and 1500 (b1500) s/mm2 (acquisition parameters: Supplementary Table 1).

Lesion segmentation, both multi-slice and single-slice, was carried out on b1500 images by a medical student (J.E.) with one year of breast DWI experience under the supervision of a board-certified radiologist (S.B., $>$ 10 years of experience in breast DWI).

2.2 Experimental Setup

The aim of this study was to evaluate different co-registration methods and their effectiveness in aligning different b-value DWIs to facilitate a more accurate calculation of ADC maps. Co-registration was performed using ANTs, a leading medical image registration and segmentation toolkit [11].

Thirteen co-registration methods were compared using either b50 or b1500 as target images, employing various registration techniques from the ANTs library. Given the deformable nature of breast tissue and the prevalence of distortion artifacts in high-b-value DWIs, non-rigid registration was used trying to address all types of motion effectively. Additionally, rigid registration, known for its robustness and reliability, was also employed for comparison. For all targets, Rigid registration (rotation and translation only), Similarity registration (scaling, rotation, and translation), and Affine registration (Rigid registration with additional scaling) were utilized, with additional non-rigid transformations evaluated for b1500 targets, employing SyN and SyNCC methods. SyN involves symmetric normalization and affine and deformable transformations, using mutual information as the optimization metric [12]. Similar to SyN, SyNCC, uses cross-correlation as the metric, resulting in increased computation time.

Furthermore, in cases where b1500 was the target, co-registration was performed iteratively and non-iteratively. In other cases, all co-registrations were performed non-iteratively. Iterative co-registration involved aligning the b50 image with the b750 image after the latter had already been aligned with the b1500 image.

In cases where b1500 was not the target, segmentations performed on b1500 images also required co-registration. In these cases, the b1500 image was co-registered with the target, and the segmentation was then shifted using the same transformation to ensure its alignment with the lesion.

We prioritized comparing the methods sequencing, target and type of transformation from the ANTs documentation, rather than optimizing complex co-registration methods. Table 1 lists the 13 different types of co-registration and their abbreviations for future reference. The method without co-registration is abbreviated as *no-coreg*.

A recent study suggests that using only two b-values for ADC calculations maintains comparable predictive value and repeatability to using four b-values, thereby ensuring accuracy in treatment response assessment [6, 13]. Hence, our study exclusively focused on two-b-value ADC calculations, computed voxel-wise using Eq. (1) where S(b1) and S(b2) represent the signal intensities corresponding to two different b-values: 750 s/mm2 and 50 s/mm2, respectively [14]. Subsequently, mean, standard deviation (Std), and within-lesion coefficient of variation (CoV) of the ADC values within each segmentation were calculated from the generated ADC maps.

$$ADC = \frac{\ln\left(\frac{S(b_2)}{S(b_1)}\right)}{b_1 - b_2} \quad (1)$$

The data processing, registration, and statistical analysis were implemented in python (v3.9.13) using ANT's library (antspyx 0.3.8). The experiments were performed on a single workstation (Intel i9-12900K 3.20 GHz, 64GB RAM).

Table 1. Overview over 13 different co-registration methods with the co-registration type, target and whether it is an iterative or non-iterative method specified

	Co-registration type	Target for Co-registration	Iterative
Non-iterative			
b50-Affine	Affine	b50	no
b50-Rigid	Rigid	b50	no
b50-Similarity	Similarity	b50	no
b1500-Affine	Affine	b1500	no
b1500-Rigid	Rigid	b1500	no
b1500-Similarity	Similarity	b1500	no
b1500-SyN	SyN	b1500	no
b1500-SyNCC	SyNCC	b1500	no
Iterative			
b1500-iter-Affine	Affine	b1500	yes
b1500-iter-Rigid	Rigid	b1500	yes
b1500-iter-Similarity	Similarity	b1500	yes
b1500-iter-SyN	SyN	b1500	yes
b1500-iter-SyNCC	SyNCC	b1500	yes

2.3 Performance Evaluation

Technical Performance Evaluation The technical evaluation aimed to identify the best performing methods. Initially, mean, standard deviation (Std), and within-lesion Coefficient of Variation (CoV) of ADC values in benign and malignant lesions were assessed. For accurate diagnostic thresholds, minimal variance within the lesion is crucial. Statistical significance was determined using repeated measures ANOVA with Dunnett's post hoc analysis [15], considering results with p-values < 0.05 as significant. Subsequently, the top four performing methods were chosen for further clinical investigation. As an additional technical assessment, performance time was calculated by averaging the time for ten co-registration calculations.

Diagnostic Performance Evaluation Several metrics were calculated for the best performing methods. To quantify the heterogeneity of lesion ADC means within a patient cohort, we calculated the CoV between patients by dividing the mean of the patient mean ADC per method by its Std. The goal of the between-patient CoV was to evaluate how well the ADC threshold can be defined across the patient cohort. To assess whether the number of outliers reduces after co-registration, they were quantified using the interquartile range (IQR) method and compared between methods [16]. Furthermore, the area under the curve (AUC), sensitivity, specificity and accuracy for the optimal threshold of the individual methods were calculated using the ROC curve. Statistical significance was assessed by comparison with the *no-coreg* method using McNemar's test [17].

Additionally, we compared performance using both optimal ($1.53\text{x}10^{-3}$ mm2/sec) and conservative ($1.68\text{x}10^{-3}$ mm2/sec) ADC cutoffs from the A6702 trial, which indicated a potential 21% reduction in biopsies without compromising sensitivity [5, 18].

3 Results

This study aimed to enhance ADC map accuracy by co-registering b50 and b750 images. Figure 1 demonstrates the impact of co-registration, particularly the *b1500-iter-SyNCC* method, on individual b-value images and ADC maps in severe misalignment. It shows improved alignment of lesion segmentation from b1500 with lesions in b50 and b750 after co-registration. Additionally, the ADC map from co-registered images exhibits enhanced lesion coverage by the newly calculated ADC. Figure 1 shows an example of the value of co-registration transformations.

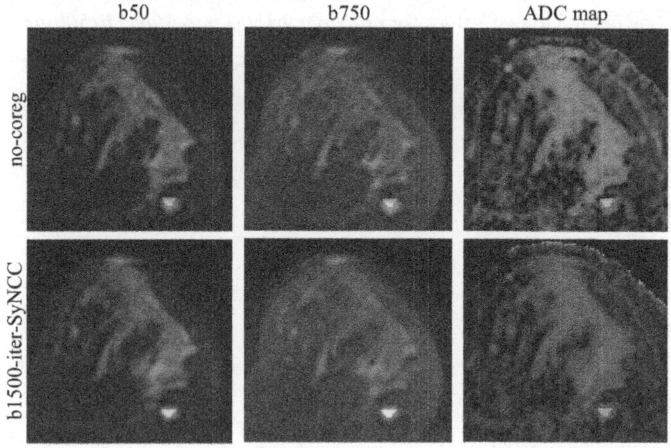

Fig. 1. DWI images with b values b = 50 s/mm2 (b50) and b = 750 s/mm2 (b750) before and after co-registration, using the b1500-iter-SyNCC method and the ADC map, created from the DWI images, showing a better alignment of segmentations and lesions after co-registration

3.1 Technical Performance Comparison

The best-performing methods according to within-lesion CoV are *no-coreg, b1500- iter-Affine, b1500-iter-Rigid, b1500-iter-Similarity,* and *b1500-iter-SyNCC* as they showed the lowest CoV for both benign and malignant lesions (see to Fig. 2).

As these four methods, together with the *no-coreg* method, had the lowest within-lesion CoV, they are further examined in detail with respect to this metric, as well as the mean and Std for benign and malignant lesions. The box plots show that the mean remains consistently stable overall, and for the SynCC method, the Std and within-lesion CoV decrease compared to *no-coreg*, especially in malignant cases (see Fig. 3).

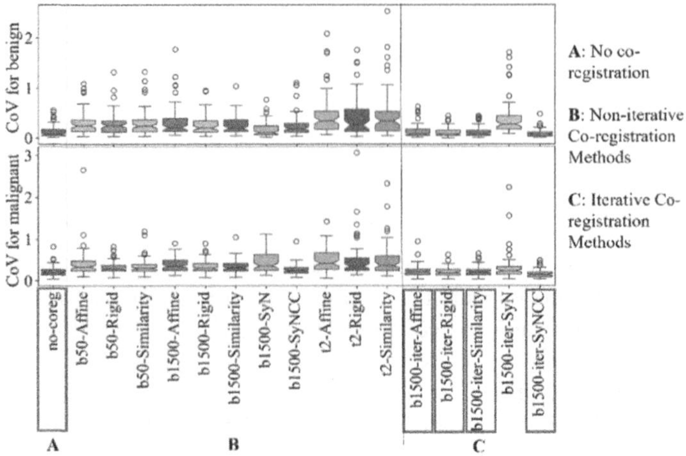

Fig. 2. Within-lesion Coefficient of Variance (CoV) for benign and malignant lesions for ADC maps from 15 different co-registration methods and not co-registered method (*no-coreg*) visualized in a boxplot, showing the lowest values for *no-coreg, b1500-iter-Affine, b1500-iter-Rigid, b1500-iter-Similartiy,* and *b1500-iter-SyNCC* (marked red)

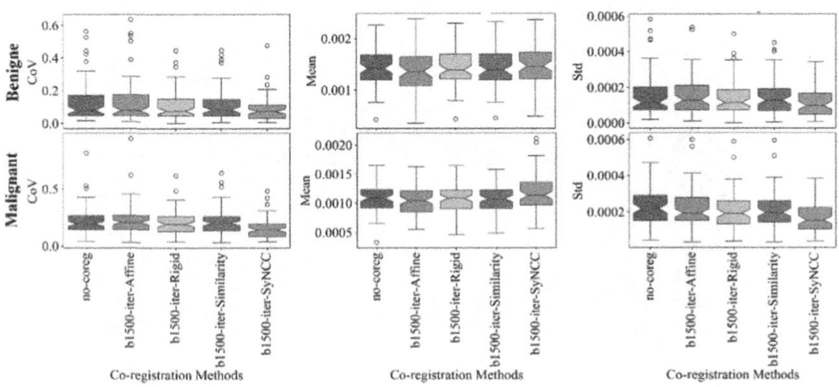

Fig. 3. Within-lesion Coefficient of Variance (CoV), Mean and standard deviation (Std) of ADC maps from the four best performing co-registration methods and not co-registered method (*no-coreg*) in a boxplot for benign (top) and malignant (bottom) data. Mean stays relatively consistent, CoV and Std decrease for *b1500-iter-SynCC*, especially in malignant data

To assess statistical significance, a repeated measures ANOVA with Dunnett post-hoc test was conducted to compare the values of the different co-registration methods with *no-coreg*. This revealed significant changes in CoV and Std for *b1500-iter-SyNCC* compared to *no-coreg*, particularly in malignant cases where the changes were highly significant for Std ($P < 0.001$) and CoV ($P < 0.001$). Even the mean showed a slightly significant change compared to *no-coreg* in malignant lesions ($P = 0.5$) (see Table 2).

Table 2. P-values obtained from Dunnett post-hoc tests assessing statistical significance in Mean, Standard Deviation (STD), and within-lesion Coefficient of Variation (CoV) alterations across various co-registration methods compared to no co-registration

	p-value for Mean	p-value for Std	p-value for CoV
Benign			
b1500-iter-Affine	0.320	0.816	0.677
b1500-iter-Rigid	0.936	0.424	0.471
b1500-Similarity_iter	0.804	0.426	0.555
b1500-iter-SyNCC	0.543	**0.030**	**0.026**
Malignant			
b1500-iter-Affine	0.367	0.656	0.860
b1500-iter-Rigid	0.836	0.310	0.394
b1500-iter-Similarity	0.665	0.426	0.609
b1500-iter-SyNCC	**0.050**	**0.5e-3**	**0.2e-3**

3.2 Diagnostic Performance Comparison

Lesion ADC heterogeneity, measured by between-patient CoV, remained consistent across all co-registration methods and *no-coreg* (0.25 ± 0.03) (Supplementary Table 2).

Few outliers were observed in the *no-coreg* method. Malignant cases showed minimal outliers, suggesting limited room for improvement through co-registration. In benign cases, there were fewer mean outliers but more Std outliers. Notably, the *b1500-iter-SyNCC* method significantly reduced outliers from 4 to 0 (Supplementary Table 3).

Clinical performance evaluation via ROC curves revealed poorer performance regarding the AUC for the technically proficient *b1500-iter-SyNCC* method, with marginal improvements for other methods (Fig. 4).

For most methods, the time to perform the co-registration method was short (t = 4.7 ± 0.2 s), but notably longer for *b1500-iter-SyNCC* (t = 24.38 ± 13 min), which makes this method less suitable for clinical practice.

Sensitivity, specificity, and accuracy were assessed for optimal thresholds but no statistically significant difference between performances was noticed. *b1500-iter-Rigid* performed best and moderately better than *no-coreg*. Similarly, *b1500-iter-Similarity* performed marginally better than *no-coreg*, while *b1500-iter-SyNCC* exhibited notably poorer in lesion differentiation (Table 3 and Supplementary Fig. 1).

Performance was assessed using thresholds from the A6702 study. While *b1500-iter-Rigid* showed slightly better accuracy than *no-coreg*, no significant differences were observed between most co-registration methods and *no-coreg*, except for *b1500-iter-SyNCC*, which performed worse (see Table 3 and Supplementary Table 4) .

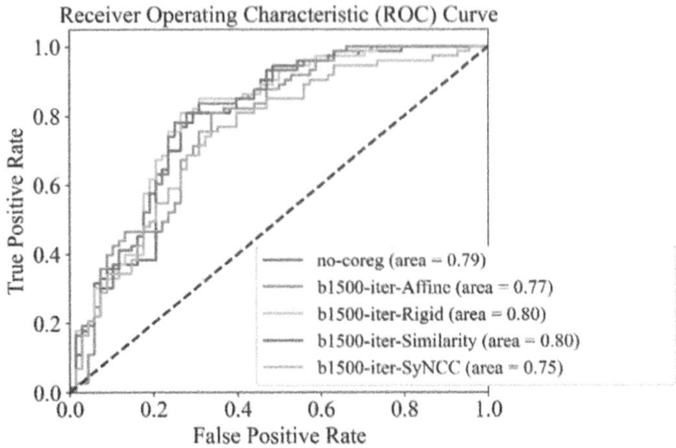

Fig. 4. ROC-curve for four best performing co-registration methods and the no-co-registration method (*no-coreg*), indicating additionally the AUC for each method in the legend

Table 3. Sensitivity, specificity and accuracy for individual threshold and A6702 trial threshold

	Sensitivity	Specificity	Accuracy	Threshold
Individual thresholds [mm²/sec]				
no-coreg	0.774	0.747	0.759	1.31×10^{-3}
b1500-iter-Affine	0.763	0.720	0.738	1.26×10^{-3}
b1500-iter-Rigid	0.781	0.766	0.773	1.27×10^{-3}
b1500-iter-Similarity	0.761	0.770	0.766	1.25×10^{-3}
b1500-iter-SyNCC	0.707	0.710	0.709	1.33×10^{-3}
Optimal thresholds according to A6702 trial				
no-coreg	0.933	0.634	0.695	1.53×10^{-3}
b1500-iter-Affine	0.962	0.626	0.688	"
b1500-iter-Rigid	0.933	0.640	0.702	"
b1500-iter-Similarity	0.897	0.625	0.681	"
b1500-iter-SyNCC	0.75	0.624	0.660	"

4 Discussion and Conclusion

In this study, various co-registration strategies for DWI acquisitions were investigated to improve ADC quantification in breast MRI. The focus was on different orders, as well as trying different non-rigid and rigid registration techniques. For future comparison it might be worth looking at different registration tools, such as Elastix [19], although that would be beyond the scope of this study. Amongst the assessed co-registration techniques four methods showed promising technical performance. In particular, iterative methods outperformed non-iterative methods, suggesting iterative co-registration to be superior due to its incremental approach, minimizing error-prone adjustments.

Amongst, *b1500-iter-SyNCC* excelled in technical evaluation, although with longer performance times, while *b1500-iter-Rigid* and *b1500-iter-Similarity* showed improved lesion characterization. Further, the *b1500-iter-SyNCC* method optimized the Std and CoV metrics, which are critical for data consistency, and reduced outliers significantly affecting individual diagnosis. Nevertheless, when evaluating the translation into diagnostic performance *b1500-iter-SyNCC* lagged behind all approaches. To understand the underlying causes of this difference in technical and diagnostic evaluation, conducting a clinical trial incorporating manual segmentations performed directly on the ADC maps after co-registration may provide valuable insights.

In addition, cutoff analysis from the A6702 study showed comparable accuracy performance for most methods yet a considerable influence on sensitivity with a range from 0.75–0.97 with *b1500-iter-SyNCC* performing worst and *b1500-iter-Rigid* performing best, marginally improving the sensitivity of *no-coreg* assessments. Whilst none of the co-registration methods statistically outperformed *no-coreg* in both technical and lesion characterization for our cohort, the influence on sensitivity in fixed threshold assessments indicated the translational relevance of the study.

Even though co-registration did not significantly influence lesion differentiation in ADC assessments in our study, this should not be interpreted as to co-registration not providing value for image assessments. As our study only focused on mean ADC values of lesions, a detailed analysis of radiomic features could provide a deeper understanding of the impact of co-registration on quantification of lesion characteristics which might reveal the impact of co-registration in more details.

In conclusion, our study highlights the influence of co-registration techniques on ADC assessments in breast DWI aiming to enhance diagnostic accuracy in distinguishing benign from malignant lesions. Further research is needed to understand better how different co-registration methods impact clinical outcomes.

Acknowledgments. This project is partially funded by the Bavarian Ministry of Economic Affairs, Regional Development and Energy.

Disclosure of Interests. The authors have no competing interests.

References

1. Mann, R.M., et al., Breast MRI: EUSOBI recommendations for women's information. Eur Radiol, 2015. 25(12): p. 3669-78.
2. Luczynska, E., et al., The Utility of Diffusion Weighted Imaging (DWI) Using Apparent Diffusion Coefficient (ADC) Values in Discriminating Between Prostate Cancer and Normal Tissue. Pol J Radiol, 2014. 79: p. 450-5.
3. Hausmann, D., et al.: Advanced diffusion-weighted imaging sequences for breast MRI: comprehensive comparison of improved sequences and ultra-high B-values to identify the optimal combination. Diagnostics (Basel), **13**(4) (2023)

4. Peters, N.H., et al., Meta-analysis of MR imaging in the diagnosis of breast lesions. Radiology, 2008. 246(1): p. 116-24.
5. Youn, I., et al., Diagnostic Performance of Point-of-Care Apparent Diffusion Coefficient Measures to Reduce Biopsy in Breast Lesions at MRI: Clinical Validation. Radiology, 2024. 310(2): p. e232313.
6. Partridge, S.C., et al., Impact of Alternate b-Value Combinations and Metrics on the Predictive Performance and Repeatability of Diffusion-Weighted MRI in Breast Cancer Treatment: Results from the ECOG-ACRIN A6698 Trial. Tomography, 2022. 8(2): p. 701-717.
7. Shahbazi-Gahrouei, D., et al.: Advanced magnetic resonance imaging modalities for breast cancer diagnosis: an overview of recent findings and perspectives. Diagnostics (Basel), **12**(11) (2022)
8. Partridge, S.C., et al., Diffusion-weighted breast MRI: Clinical applications and emerging techniques. J Magn Reson Imaging, 2017. 45(2): p. 337-355.
9. Le Bihan, D., et al., Artifacts and pitfalls in diffusion MRI. J Magn Reson Imaging, 2006. 24(3): p. 478-88.
10. Kapsner, L.A., et al., Image quality assessment using deep learning in high b-value diffusion-weighted breast MRI. Sci Rep, 2023. 13(1): p. 10549.
11. Avants, B.B., et al., The Insight ToolKit image registration framework. Front Neuroinform, 2014. 8: p. 44.
12. Avants, B.B., et al., Symmetric diffeomorphic image registration with cross-correlation: evaluating automated labeling of elderly and neurodegenerative brain. Med Image Anal, 2008. 12(1): p. 26-41.
13. Park, S.Y., et al., Comparison of apparent diffusion coefficient calculation between two-point and multipoint B value analyses in prostate cancer and benign prostate tissue at 3 T: preliminary experience. AJR Am J Roentgenol, 2014. 203(3): p. W287-94.
14. Higaki, T., et al., Introduction to the Technical Aspects of Computed Diffusion-weighted Imaging for Radiologists. Radiographics, 2018. 38(4): p. 1131-1144.
15. Lee, S. and D.K. Lee, What is the proper way to apply the multiple comparison test? Korean J Anesthesiol, 2018. 71(5): p. 353-360.
16. Dash, C.S.K., et al.: An outliers detection and elimination framework in classification task of data mining. Decis. Analytics J. **6** 2023
17. McNemar, Q., Note on the sampling error of the difference between correlated proportions or percentages. Psychometrika, 1947. 12(2): p. 153-157.
18. Rahbar, H., et al., Utility of Diffusion-weighted Imaging to Decrease Unnecessary Biopsies Prompted by Breast MRI: A Trial of the ECOG-ACRIN Cancer Research Group (A6702). Clin Cancer Res, 2019. 25(6): p. 1756-1765.
19. Klein, S., et al., elastix: a toolbox for intensity-based medical image registration. IEEE Trans Med Imaging, 2010. 29(1): p. 196-205.

A Learning-Free Approach to Mitigate Abnormal Deformations in Medical Image Registration

Abdullah F. Al-Battal[1,2](✉), Soan T. M. Duong[3,4], Chanh D. Tr. Nguyen[3,5], Steven Q. H. Truong[3], Chien Phan[6], Truong Q. Nguyen[1], and Cheolhong An[1]

[1] Electrical and Computer Engineering Department, UC San Diego, La Jolla, CA, USA
aalbatta@ucsd.edu

[2] Electrical Engineering Department, King Fahd University of Petroleum and Minerals, Dhahran, Saudi Arabia

[3] Vinbrain JSC, Hanoi, Vietnam

[4] Department of Computer Science, Le Quy Don Technical University, Hanoi, Vietnam

[5] College of Engineering and Computer Science, VinUniversity, Hanoi, Vietnam

[6] Diagnostic Imaging Department, University Medical Center, Ho Chi Minh City, Vietnam

Abstract. Medical image registration is a critical process in several diagnostic and therapeutic procedures. While deep-learning deformable registration models have demonstrated reasonable accuracy, they can produce abnormal deformations that introduce substantial artifacts in medical images by unrealistically modifying the shape and position of anatomical structures. These abnormal deformations may not be effectively detected or mitigated during inference due to their similarity to the natural elastic deformation of soft tissue. Moreover, the limited generalizability of learning-based approaches restricts their ability to assess image registration beyond their training scope. In this paper, we propose a learning-free method for estimating and correcting abnormal deformations, which are responsible for the maximum deformation error. Our proposed model-agnostic approach introduces variations to both the input images of the registration model and the model weights at inference, making it adaptable to a wide range of deep-learning-based medical image registration models. Next, the proposed approach uses the variabilities in the estimated registration deformation fields to mitigate significant deformation errors. We evaluate our proposed approach on two datasets: a 3D abdominal computed tomography dataset (the LiTS dataset), and a 3D brain magnetic resonance imaging dataset (the OASIS dataset) using synthetically generated deformation fields that resembles patient and organ movement as well as changes in organ sizes; reducing the maximum registration error by up to 6.1% for the first and 5.7% for the second. These findings demonstrate that our approach can significantly

This work is supported by a grant from Vinbrain, JSC.

M. Modat et al. (Eds.): WBIR 2024, LNCS 15249, pp. 137–147, 2024.
https://doi.org/10.1007/978-3-031-73480-9_11

mitigate abnormal deformations in medical image registration, improving accuracy and reducing artifacts.

Keywords: Medical image registration · Deformable registration · Convolutional neural networks

1 Introduction

Image registration is the process of aligning one or more images to the space of a fixed (reference) image [8]. In medical imaging, registration is essential for many diagnostic and therapeutic procedures, where different images contain useful, but different information about the anatomical structure of interest [8,16,18]. By aligning multiple medical images acquired at different times or using different modalities, information from different images can be integrated, improving diagnostic accuracy. Medical image registration is also used to correct for motion artifacts and errors during cardiac, abdominal, and brain imaging, especially in magnetic resonance imaging due to lengthy acquisition times [12,14,17].

Image registration techniques can be categorized as rigid, affine, or deformable depending on the spatial transformation degrees of freedom and the scope at which the transformation varies spatially. Rigid and affine registration transformations are parameterized using a set of parameters that defines the rotation and translation operations for rigid registration in addition to scaling and shearing for affine registration [13]. Deformable registration, on the other hand, can be defined by a parametric correspondence [20] or by a non-parametric dense correspondence representing the shifts in all directions for each pixel in a 2D image, or each voxel in a 3D volume, which is usually called the deformation or registration field (ϕ) [7,8]. This allows the alignment of different anatomical structures using a more realistic local representation of their movement beyond the constrained global rigid and affine transformations [23]. Regardless of their category, registration methods seek to solve the problem defined as:

$$\widehat{\phi} = \underset{\phi}{\operatorname{argmin}}\, \mathcal{S}(F_I, T(M_I, \phi)), \tag{1}$$

where F_I represents the fixed image, M_I the moving image, ϕ the transformation parameter, which is a 4×4 transformation matrix for affine and rigid registration or the displacement field for deformable registration, and T the warping operation. $\mathcal{S}$ is the similarity measure, which is correlated with the level of alignment between the transformed M_I and F_I. For affine and rigid registration, once the transformation matrix is estimated, it is used to transform the voxels of the moving image to their respective location on the fixed image coordinate system. For deformable registration, once the deformation field (ϕ) is estimated, it defines the displacement of each of the voxels in M_I to align it with F_I.

Traditional deformable registration methods such as HAMMER [21], demons [25], ELASTIX [11], and ANTs [2] used iterative optimization algorithms to solve the registration problem defined in (1), These methods yield precise alignment, but converge at a significantly slow pace taking hours for a single alignment task [10,20,24]. To improve registration speeds while maintaining accuracy, deep-learning-based registration methods use a learning-based approach to train a model on predicting the transformation between two images; eliminating the need for the time-consuming iterative optimization at inference. These models are trained using different approaches, including unsupervised [1,26], semi-supervised [29], and fully supervised methods [6,22]. Unsupervised methods use image similarity metrics such as the mean square error (MSE) [26] or the normalized cross-correlation (NCC) [1]. Semi-supervised methods use a segmentation mask of an anatomical structure of interest as guidance through metrics such as the Dice Score [27,29]. Fully supervised models use the error between predicted and synthetically generated ground truth deformations [6,22].

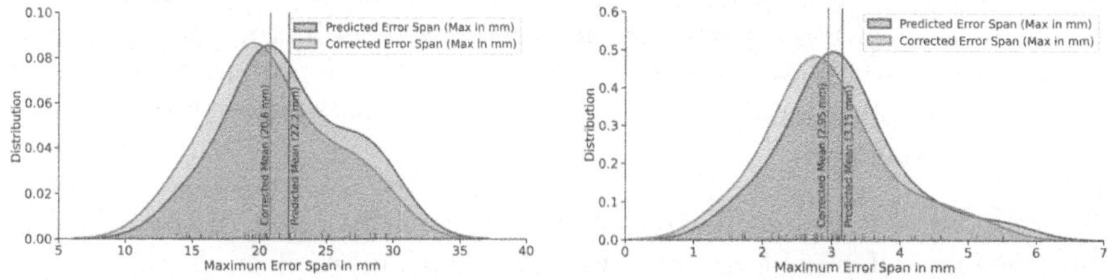

(a) LiTS dataset (b) OASIS dataset

Fig. 1. The deformation error span distribution for the predicted deformations using VoxelMorph and the corrected deformations using the proposed method. Synthetically generated deformation fields that resemble patient and organ movement, as well as changes in organ sizes, are used as the ground truth reference.

Early deep-learning methods, such as those proposed by Cao et al. and Yang et al., use key patch-wise alignment to improve computational efficiency [5,28]. Other methods use both whole image volumes and patches to predict the deformation that aligns images [22]. Patch-wise registration methods require pre-alignment of images as the range of predicted deformation is bounded by the patch size [5]. VoxelMorph [3] is among the most prominent deep-learning methods for medical image registration that uses an encoder-decoder convolutional neural network (CNN) architecture inspired by the U-Net model [19]. Despite performing well, VoxelMorph can produce large distortions where anatomical structures are deformed and aligned incorrectly by several millimeters [3], which may hinder the use of registered images for diagnostic and therapeutic procedures as these images would not represent the actual shape and location of anatomical structures. Other deep-learning methods such as SynthMorph [9], which is trained to learn contrast-invariant registration, have demonstrated promising registration performance in aligning anatomical structures of interest with high accuracy. To reduce deformation errors and distortions, which cause unrealistic shape alteration of anatomical structures, and incorrect alignment and placement, Mok et al. proposed cLapIRN [15]. cLapIRN learns the conditional features that are correlated with the regularization hyperparameters to select the optimal smoothness regularization during inference. Nevertheless, all of these models still introduce deformation errors, although with varying levels.

We propose a learning-free statistical framework based on randomized variation injection into inputs and model weights at inference to identify and reduce distortions. Figure 1 shows the deformation error span distribution with and without using our proposed approach on two different datasets. Our approach is tested on the VoxelMorph, SynthMorph, and cLapIRN models, as well as two datasets: a computed tomography (CT) abdomen dataset (the LiTS dataset) [4] and a magnetic resonance imaging (MRI) brain dataset (the OASIS dataset) [14]. We test our approach using the absolute error between the predicted and ground truth deformation field to evaluate its ability to reduce actual deviations from the true fields. By testing our approach using two different datasets of two different anatomical body parts and acquired via two different imaging modalities, we aim to validate its ability to reduce large distortions across a wide variety of registration applications on inter-modality and inter-subject registration. Our main contributions are:

- We propose a learning-free framework to identify large distortions during inference for deformable registration deep-learning models.
- We introduce a post-processing technique that selectively corrects identified large distortions.
- Our approach is model-agnostic, making it applicable to various CNN-based deformable registration models. We validate its effectiveness by testing it with three different models and two different datasets.

2 Methodology

The proposed approach aims to identify abnormal distortions in medical image registration and minimize their effect by injecting randomized variations into both the input images and model weights repeatedly. Algorithm 1 outlines the proposed approach. First, a set of variations are chosen randomly to be applied to the input intensity values and the model weights. Next, the registration model is used to estimate the deformation field. These two steps are repeated multiple times, and the estimated deformation field at each repetition is stored. The sample standard deviation is used to calculate the spread of estimated deformations at each voxel, and deformations with the largest spread are replaced with interpolated deformations using third-order spatial b-spline interpolation.

2.1 Deformable Registration Models

Our approach is implemented at inference without any training or learning. However, it uses a pre-trained deformable registration CNN to estimate the deformation fields at each of the stages outlined in the previous section. In our work, we first trained VoxelMorph [3] on registering 3D CT scans of the abdomen with the goal of aligning the liver. We also used 3 different pre-trained models (VoxelMorph, SynthMorph [9], and cLapIRN [15]) on the brain MRI registration dataset, OASIS [14]. We apply our proposed approach to these models at inference to estimate the distortions locations in the deformation fields they generate and reduce the magnitude of these distortions.

2.2 Deformation Abnormalities Estimation and Correction

During inference, we apply the deformable registration model several times on the two image volumes. Each time we inject a randomized set of variations into the input, the model weights, or both as outlined in Algorithm 1. For the input volumes, the first of these variations is noise, where we add Gaussian noise with mean 0 and a standard deviation equal to 5% of the maximum volume intensity value. The second is Gamma correction, which is used to vary the contrast and luminance of different parts of the volume to simulate different contrast-based imaging in CT and MRI. We randomly varied the Gamma coefficient at each iteration between 0.95 and 1.05. The third input variation we incorporate is the zeroing of intensity values at random. We randomly select a number of voxels between 2.5% and 7.5% and replaced their intensity values with zero. For model weights, the variations injected are noise and weights zeroing. For the first, we added Gaussian random noise with a standard deviation equal to 0.5% of the weight we are adding noise to. As for zeroing the model weights, we randomly select a subset of weights ranging in size from 2.5% to 7.5% of the total set of weights at each convolutional layer. The generated deformation fields from each iteration are then aggregated, and the standard deviation of the deformation at each voxel is computed. This is the metric we use to estimate the possibility a deformation is abnormal and not accurate. An average correlation coefficient of 0.61 is found to be present between the standard deviation and deformation error in the experiments we conducted.

Once the aggregate deformation fields are generated, and the standard deviation of the deformations at each voxel is computed, we use them to correct for abnormal deformations. At locations with high standard deviations, we replace the deformations with the interpolated deformation using third-order b-spline interpolation. The deformations selected for replacement are based on the distribution of standard deviations for that specific volume and are selected based on experimental testing of performance. We found that replacing the deformations with a standard deviation higher than the 75th percentile works best. This approach provides selective regularization on the value of deformation, smoothing the deformation field at specific locations rather than globally.

Algorithm 1: Abnormal deformation estimation & correction

Input : F_I, M_I, pre-trained registration model, number of repetitions (N)
Output: Deformation field spread (standard deviation), corrected deformation field

for $n \leftarrow 0$ **to** N **do**
  for *input variation* ***in*** *set of input variations* **do**
    `// set of input variations are: {random noise, gamma transformation, random zeroing of intensity values}`
    if *random number* ≥ 0.5 **then**
      `// random numbers in range [0, 1]`
      Apply input variation to a copy of F_I and M_I
    end
  end
  for *model variation* ***in*** *set of model variations* **do**
    `// set of model variations are: {random noise, random zeroing of weights (dropout at inference)}`
    if *random number* ≥ 0.5 **then**
      `// random numbers in range [0, 1]`
      Apply model variation to the model weights
    end
  end
  Generate deformation field (ϕ_n) from modified input and model
end
Replace the deformations with the largest set of standard deviations across ϕ_n with interpolated deformations using third order b-spline interpolation

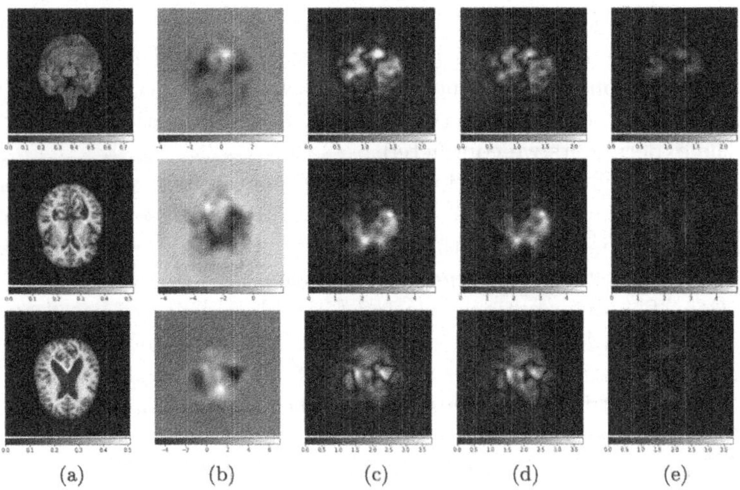

Fig. 2. Qualitative results of the proposed error reduction approach on three example slices from the OASIS dataset. (a) The example slices from the OASIS dataset, (b) the ground truth registration deformation field, (c) the predicted deformation field error using VoxelMorph, (d) the corrected deformation field error using our approach, and (e) the error difference between the corrected and predicted deformation error. Deformation fields and errors are in millimeters.

3 Experiments and Results

3.1 Dataset and Implementation

We tested the proposed approach using 3D CT scans from the Liver Tumor Segmentation (LiTS) dataset [4] and 3D MRI brain scans from the Open Access Series of Imaging Studies (OASIS) dataset [14]. The LiTS dataset contains scans of the abdomen from 131 subjects. The goal of this dataset is to benchmark liver and liver lesion segmentation models, and we adapted it to our registration work due to the extensive complexities present in abdominal scans of the body. We trained the VoxelMorph model on this dataset using randomly selected pairs of images from the dataset as inputs to the model; there are 2,162 such pairs. The 3D CT volumes were resized so that each voxel size is $1 \times 1 \times 1$ mm. They are then resized to $256 \times 256 \times 128$ voxels as inputs to the model. For the OASIS dataset, we used pre-trained models that were made available by the authors of the three registration models we evaluated (VoxelMorph, SynthMorph and cLapIRN). For the OASIS dataset, we used the original scan size of $176 \times 208 \times 176$ voxels.

For both datasets, we generated ground truth deformations and applied these deformations to image volumes. We then used the registration models to predict the deformation field. Using these ground truth deformations allows us to calculate the exact difference between the predicted and actual deformation fields in spatial units, which is millimeters in our analysis. It also allows us to quantify our approach's ability to estimate and reduce abnormal deformations. For the LiTS dataset, we used fifteen subjects, while for the OASIS dataset, we used twenty five subjects to evaluate our approach's performance. For each of the subjects, the predicted deformation field using the registration model and the corrected one using our approach are compared to the ground truth deformation field; using the absolute error in millimeters as the evaluation metric.

3.2 Evaluation and Results

We evaluated the proposed approach from two perspectives. The first is its ability to estimate where the predicted deformation fields would be abnormal and cause distortions by deviating significantly from the ground truth. The second is its ability to use this information to mitigate and correct these abnormal deformations. Using these two perspectives, we tested the approach's ability to estimate abnormal deformations based on a single form of variation injection first. However, we observed that using multiple types of randomized variations improved the correlation coefficient between the deformations' standard deviation and deformation error by approximately 20% from 0.5 to 0.61.

Using multiple types of randomized variations, our approach outlined in Algorithm 1 reduced the maximum deformation error by up to 6.1% as shown in Table 1. Our approach was also consistent in reducing the maximum deformation error and the errors at the 99th percentile across the different datasets and models. Figure 2 provides a qualitative representation of the proposed approach's ability to reduce deformation error for the OASIS dataset, while Fig. 3 provides

Table 1. Deformation error reduction performance of the proposed approach. The maximum deformation error and the 99th percentile error are outlined in millimeters (averaged across subjects). The improvement in reducing each of these errors is also outlined as a percentage of the original error.

Dataset	Approach	Avg. Max. Error	Avg. Error @ 99th Perc.	Improvement (%)	
				Max. Err.	99th% Err.
LiTS	VoxelMorph	22.2	6.8	–	–
	Proposed	20.8	6.6	6.1%	2.5%
OASIS	VoxelMorph	3.1	0.82	–	–
	Proposed	2.95	0.8	4.6%	2.3%
	SynthMorph	2.08	0.81	–	–
	Proposed	1.98	0.8	4.9%	1.2%
	cLapIRN	1.75	0.76	–	–
	Proposed	1.64	0.76	5.7%	0.3%

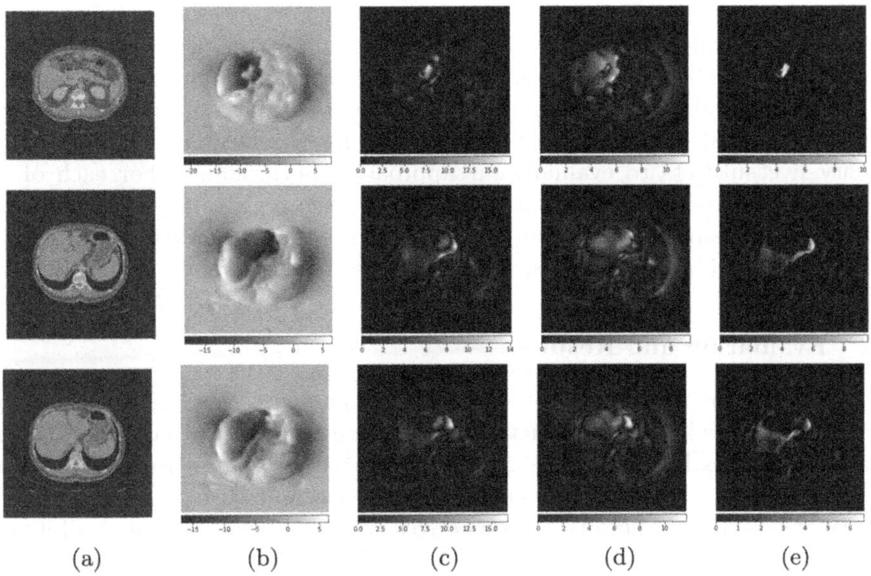

Fig. 3. Qualitative results of the proposed error reduction approach on six example slices from the LiTS dataset. (a) The example slices from the LiTS dataset, (b) the ground truth registration deformation field, (c) the predicted deformation field error using VoxelMorph, (d) the corrected deformation field error using our approach, and (e) the error difference between the corrected and predicted deformation error. Deformation fields and errors colormap is in millimeters. The colormap scale for the deformations and errors is different for each image, and depends on the maximum value within that image.

that representation for the LiTS dataset. As the number of repetitions increases, the performance of the proposed approach tends to saturate (see Fig. 4). 12 to 15 repetitions were sufficient in our tests to achieve the results outlined in Table 1. Although the proposed approach is successful at consistently reducing the predicted deformation error without the need for training, there are still future prospects for improvement as the deformation error extends beyond 20 mm for some volumes, even after correction.

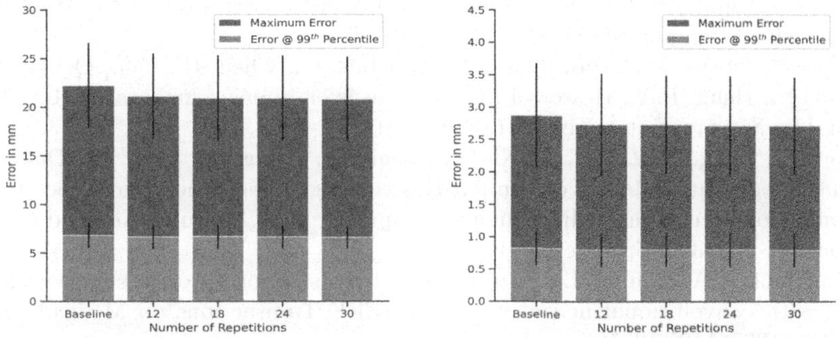

(a) LiTS dataset (b) OASIS dataset

Fig. 4. The number of repetitions effect (of model and input variation injection) on the correction algorithm performance.

4 Conclusion

In this paper, we presented a learning-free statistical-based framework to estimate abnormal deformations and correct them for deformable image registration models. The framework randomly and repeatedly injects variations into the input images and model weights to generate multiple deformation Fields. It uses the spread of these deformation fields at each voxel to estimate if the predicted deformation is abnormal. When a deformation is deemed abnormal, it is replaced by an interpolated deformation using neighboring deformations. This approach provides selective regularization on the value of deformation, smoothing the deformation field at specific locations rather than globally. We tested the framework by estimating and mitigating abnormal deformations generated by three different deformable registration models for 3D abdominal CT and 3D brain MRI scan registration. With its initial success in estimating and mitigating abnormal deformations, we hope that the proposed framework would encourage more research in this area of learning-free registration evaluation and correction.

Acknowledgements. This work is supported by a grant from Vinbrain, JSC.

References

1. Avants, B.B., Epstein, C.L., Grossman, M., Gee, J.C.: Symmetric diffeomorphic image registration with cross-correlation: evaluating automated labeling of elderly and neurodegenerative brain. Medical Image Analysis **12**(1), 26–41 (2008)
2. Avants, B.B., Tustison, N.J., Song, G., Cook, P.A., Klein, A., Gee, J.C.: A reproducible evaluation of ants similarity metric performance in brain image registration. NeuroImage **54**(3), 2033–2044 (2011)
3. Balakrishnan, G., Zhao, A., Sabuncu, M.R., Guttag, J., Dalca, A.V.: Voxelmorph: a learning framework for deformable medical image registration. IEEE Transactions on Medical Imaging **38**(8), 1788–1800 (2019)
4. Bilic, P., Christ, P.F., Vorontsov, E., Chlebus, G., Chen, H., Dou, Q., Fu, C.W., Han, X., Heng, P.A., Hesser, J., et al.: The liver tumor segmentation benchmark (lits). arXiv preprint arXiv:1901.04056 (2019)
5. Cao, X., Yang, J., Zhang, J., Nie, D., Kim, M., Wang, Q., Shen, D.: Deformable image registration based on similarity-steered cnn regression. In: Proc. International Conference on Medical Image Computing and Computer-Assisted Intervention. pp. 300–308. Springer (2017)
6. Eppenhof, K.A., Pluim, J.P.: Pulmonary ct registration through supervised learning with convolutional neural networks. IEEE Transactions on Medical Imaging **38**(5), 1097–1105 (2018)
7. Fischer, B., Modersitzki, J.: Ill-posed medicine—an introduction to image registration. Inverse problems **24**(3), 034008 (2008)
8. Hill, D.L., Batchelor, P.G., Holden, M., Hawkes, D.J.: Medical image registration. Physics in Medicine & Biology **46**(3), R1 (2001)
9. Hoffmann, M., Billot, B., Greve, D.N., Iglesias, J.E., Fischl, B., Dalca, A.V.: Synthmorph: learning contrast-invariant registration without acquired images. IEEE transactions on medical imaging **41**(3), 543–558 (2021)
10. Klein, A., Andersson, J., Ardekani, B.A., Ashburner, J., Avants, B., Chiang, M.C., Christensen, G.E., Collins, D.L., Gee, J., Hellier, P., et al.: Evaluation of 14 nonlinear deformation algorithms applied to human brain mri registration. Neuroimage **46**(3), 786–802 (2009)
11. Klein, S., Staring, M., Murphy, K., Viergever, M.A., Pluim, J.P.: Elastix: a toolbox for intensity-based medical image registration. IEEE Transactions on Medical Imaging **29**(1), 196–205 (2009)
12. Küstner, T., Pan, J., Qi, H., Cruz, G., Gilliam, C., Blu, T., Yang, B., Gatidis, S., Botnar, R., Prieto, C.: Lapnet: non-rigid registration derived in k-space for magnetic resonance imaging. IEEE transactions on medical imaging **40**(12), 3686–3697 (2021)
13. Livyatan, H., Yaniv, Z., Joskowicz, L.: Gradient-based 2-d/3-d rigid registration of fluoroscopic x-ray to ct. IEEE transactions on medical imaging **22**(11), 1395–1406 (2003)
14. Marcus, D.S., Fotenos, A.F., Csernansky, J.G., Morris, J.C., Buckner, R.L.: Open access series of imaging studies: longitudinal mri data in nondemented and demented older adults. Journal of cognitive neuroscience **22**(12), 2677–2684 (2010)
15. Mok, T.C., Chung, A.C.: Conditional deformable image registration with convolutional neural network. In: Medical Image Computing and Computer Assisted Intervention–MICCAI 2021: 24th International Conference, Strasbourg, France, September 27–October 1, 2021, Proceedings, Part IV 24. pp. 35–45. Springer (2021)

16. Oh, S., Kim, S.: Deformable image registration in radiation therapy. Radiation oncology journal **35**(2), 101 (2017)
17. Pan, J., Rueckert, D., Küstner, T., Hammernik, K.: Efficient image registration network for non-rigid cardiac motion estimation. In: Machine Learning for Medical Image Reconstruction: 4th International Workshop, MLMIR 2021, Held in Conjunction with MICCAI 2021, Strasbourg, France, October 1, 2021, Proceedings 4. pp. 14–24. Springer (2021)
18. Pluim, J.P., Fitzpatrick, J.M.: Image registration. IEEE Transactions on Medical Imaging **22**(11), 1341–1343 (2003)
19. Ronneberger, O., Fischer, P., Brox, T.: U-net: Convolutional networks for biomedical image segmentation. In: Proc. International Conference on Medical Image Computing and Computer-Assisted Intervention. pp. 234–241. Springer (2015)
20. Rueckert, D., Sonoda, L.I., Hayes, C., Hill, D.L., Leach, M.O., Hawkes, D.J.: Nonrigid registration using free-form deformations: application to breast mr images. IEEE transactions on medical imaging **18**(8), 712–721 (1999)
21. Shen, D.: Image registration by local histogram matching. Pattern Recognition **40**(4), 1161–1172 (2007)
22. Sokooti, H., de Vos, B., Berendsen, F., Ghafoorian, M., Yousefi, S., Lelieveldt, B.P., Isgum, I., Staring, M.: 3d convolutional neural networks image registration based on efficient supervised learning from artificial deformations. arXiv preprint arXiv:1908.10235 (2019)
23. Sotiras, A., Davatzikos, C., Paragios, N.: Deformable medical image registration: A survey. IEEE transactions on medical imaging **32**(7), 1153–1190 (2013)
24. Thirion, J.P.: Image matching as a diffusion process: an analogy with maxwell's demons. Medical image analysis **2**(3), 243–260 (1998)
25. Vercauteren, T., Pennec, X., Perchant, A., Ayache, N.: Diffeomorphic demons: Efficient non-parametric image registration. NeuroImage **45**(1), S61–S72 (2009)
26. Viola, P., Wells III, W.M.: Alignment by maximization of mutual information. International Journal of Computer Vision **24**(2), 137–154 (1997)
27. Xu, Z., Niethammer, M.: Deepatlas: Joint semi-supervised learning of image registration and segmentation. In: Medical Image Computing and Computer Assisted Intervention–MICCAI 2019: 22nd International Conference, Shenzhen, China, October 13–17, 2019, Proceedings, Part II 22. pp. 420–429. Springer (2019)
28. Yang, X., Kwitt, R., Styner, M., Niethammer, M.: Quicksilver: Fast predictive image registration–a deep learning approach. NeuroImage **158**, 378–396 (2017)
29. Zijdenbos, A.P., Dawant, B.M., Margolin, R.A., Palmer, A.C.: Morphometric analysis of white matter lesions in mr images: method and validation. IEEE Transactions on Medical Imaging **13**(4), 716–724 (1994). https://doi.org/10.1109/42.363096

Deformable MRI Sequence Registration for AI-Based Prostate Cancer Diagnosis

Alessa Hering[1,2(✉)], Sarah de Boer[1], Anindo Saha[1], Jasper J. Twilt[1], Mattias P. Heinrich[3], Derya Yakar[4,5], Maarten de Rooij[1], Henkjan Huisman[1,6], and Joeran S. Bosma[1,3]

[1] Department of Medical Imaging, Radboudumc, Nijmegen, The Netherlands
alessa.hering@radboudumc.nl
[2] Fraunhofer MEVIS, Lübeck, Germany
[3] Institut für Medizinische Informatik, Universität zu Lübeck, Lübeck, Germany
[4] Department of Radiology, University Medical Center, Groningen, The Netherlands
[5] Department of Radiology, Netherlands Cancer Institute, Amsterdam, The Netherlands
[6] Department of Circulation and Medical Imaging, Norwegian University of Science and Technology, Trondheim, Norway

Abstract. The PI-CAI (Prostate Imaging: Cancer AI) challenge led to expert-level diagnostic algorithms for clinically significant prostate cancer detection. The algorithms receive biparametric MRI scans as input, which consist of T2-weighted and diffusion-weighted scans. These scans can be misaligned due to multiple factors in the scanning process. Image registration can alleviate this issue by predicting the deformation between the sequences. We investigate the effect of image registration on the diagnostic performance of AI-based prostate cancer diagnosis. First, the image registration algorithm, developed in MeVisLab, is analyzed using a dataset with paired lesion annotations. Second, the effect on diagnosis is evaluated by comparing case-level cancer diagnosis performance between using the original dataset, rigidly aligned diffusion-weighted scans, or deformably aligned diffusion-weighted scans. Rigid registration showed no improvement. Deformable registration demonstrated a substantial improvement in lesion overlap (+10% median Dice score) and a positive yet non-significant improvement in diagnostic performance (+0.3% AUROC, $p = 0.18$). Our investigation shows that a substantial improvement in lesion alignment does not directly lead to a significant improvement in diagnostic performance. Qualitative analysis indicated that jointly developing image registration methods and diagnostic AI algorithms could enhance diagnostic accuracy and patient outcomes.

Keywords: Image Registration · Prostate Cancer · Artificial Intelligence · MRI

Supplementary Information The online version contains supplementary material available at https://doi.org/10.1007/978-3-031-73480-9_12.

M. Modat et al. (Eds.): WBIR 2024, LNCS 15249, pp. 148–162, 2024.
https://doi.org/10.1007/978-3-031-73480-9_12

1 Introduction

Prostate cancer (PCa) has 1.4 million new cases each year [21], a high incidence-to-mortality ratio and risks associated with treatment and biopsy; making non-invasive diagnosis of clinically significant prostate cancer (csPCa) crucial to reduce both overtreatment and unnecessary (confirmatory) biopsies [20]. MRI scans provide the best non-invasive diagnosis for prostate cancer [5], for which a 47% increase in demand is expected by 2040 [21]. Due to the world-wide shortage of diagnostic personnel [10], workload efficiency optimization is necessary to maintain healthcare accessibility in high-income countries and improve accessibility in low and middle-income countries.

Computer-aided diagnosis (CAD) can assist radiologists to diagnose csPCa and reduce the radiology workload [23], but the observed workload reduction is limited. Larger workload reduction can be achieved through autonomous operation of diagnostic algorithms. Recent advances resulted in expert-level diagnostic performance for csPCa detection algorithms using biparametric MRI [18].

Biparametric MRI (bpMRI) consists of T2-weighted (T2W) and diffusion-weighted imaging (DWI), and the DWI is used to calculate the apparent diffusion coefficient (ADC) and typically also the high b-value (HBV) map. T2W and DWI scans are usually acquired in immediate succession in about 15–30 min, but slight patient movement and processes like bladder filling can lead to misalignment between sequences [12]. This misalignment results in lesion image features being misaligned between the sequences. For an accurate csPCa diagnosis, the information of both sequences are necessary to consider [22], meaning that csPCa detection algorithms have to combine information from different spatial locations when misalignment occurs. Current state-of-the-art csPCa algorithms use an early fusion strategy for the combination of the different sequences, which may lead to challenges in accurate lesion detection and characterization when the lesion image features are not well aligned [18].

To address this, misalignment in the Prostate Imaging - Cancer Artificial Intelligence (PI-CAI) dataset was manually corrected (85/1000 (8.5%) of the test cases and 54/9107 (0.6%) of training cases), and algorithms were subsequently trained and evaluated on these manually aligned MRI studies [18]. However, manual alignment is labor-intensive, potentially undermining the efficiency gains offered by automated csPCa diagnostic methods when required during inference. Consequently, the efficacy of these algorithms in scenarios where sequences are not manually aligned remains uncertain and might be limited.

During inference, image registration can address the issue of misaligned sequences, by providing a plausible estimation of the patient movement and deformation and thus replacing the manual alignment step. Although the prostate cancer detection research field is vibrant, there has been limited focus on the registration of prostate MRI sequences. To address the issue of global misalignment, [19] proposed an affine registration approach based on prostate gland segmentation and [3] presented a rigid registration based on Mutual Information. For compensating local deformations, both [15] and [14], employed the SimpleITK non-rigid B-Spline registration using Mutual Information. However, the focus of these studies was not on the evaluation of registration performance, resulting in only [14] examining this using the Dice Score of automatically generated prostate segmentations. In contrast, the other studies have assessed registration performance through visual examination of registered ADC images. To the best of our knowledge, only recently [12] explored the impact of image registration on prostate cancer detection performance of algorithms using bpMRI. The results show that the B-Spline registration, which is based on [14], improves the overlap of manually annotated lesions as measured by the Dice score. Additionally, the performance of the downstream task of patient-level csPCa diagnosis measured by the AUROC showed a non-significant improvement from 0.76 to 0.79. These results suggest that registration is a useful preprocessing step in an automated prostate cancer diagnosis pipeline. However, due to limited sample size (only 46 positive cases in the test set) and the lack of external testing, the ability to draw definitive conclusions is hindered.

In this study, we conduct a comprehensive analysis of the impact of image registration on the clinical downstream task of case-level csPCa diagnosis, utilizing two extensive evaluation datasets. Registration accuracy is assessed through the measurement of lesion alignment across an independent dataset comprising 473 cases, each annotated with paired lesions per modality. Further, we evaluate the downstream diagnostic efficacy on an external testing set consisting of 546 cases.

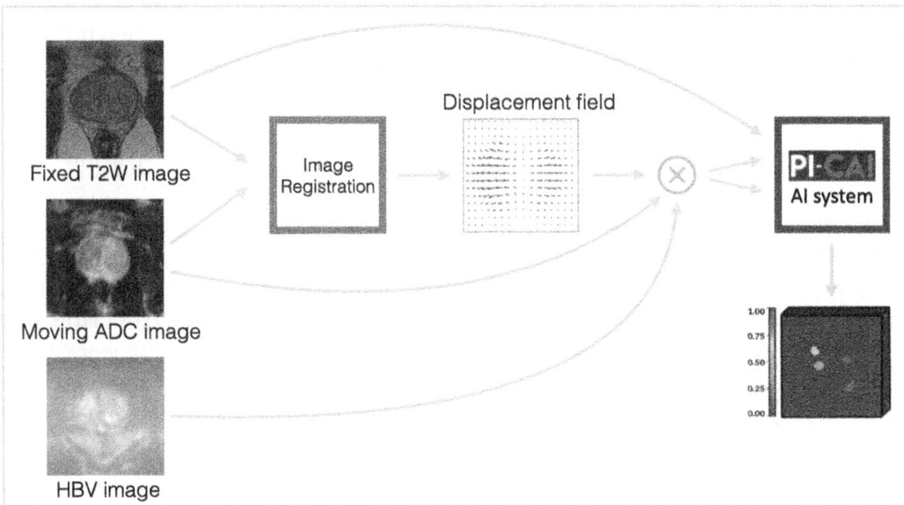

Fig. 1. Overview of our method. The T2W scan is used as fixed image and the ADC map as moving image to find the displacement field using the registration method. The displacement field is applied to the ADC and HBV maps. The registered and original scans are used as input for the PI-CAI AI system (see Sect. 2.3) to detect clinically significant prostate cancer. The case-level diagnosis performance of the end-to-end pipeline is evaluated and used as a measure of effectiveness.

2 Materials and Methods

2.1 Registration

The aim of the image registration approach is to align the DWI maps (ADC and HBV) with the T2W scan (see Fig. 1). Since the csPCa detection algorithms resample the DWI maps to the T2W scan, we chose the T2W scan as the fixed image and the ADC map as the moving image for the registration. The image registration algorithm is developed in the MeVisLab framework using the RegLib. We adopt a two-step approach which consists of a rigid registration and a deformable registration. Hereby, the registration pipeline starts with robust methods with fewer degrees of freedom and moves on to more precise, but less robust methods, which require better starting points due to their higher degrees of freedom. The calculated rigid and deformable transformation are applied to both DWI maps.

Let $\mathcal{F}, \mathcal{M} : \mathbb{R}^3 \to \mathbb{R}$ denote the fixed image and moving image, respectively, and let $\Omega \subset \mathbb{R}^3$ be a domain modeling the field of view of $\mathcal{F}$. The registration method aims to compute a deformation $y : \Omega \to \mathbb{R}^3$ that aligns the fixed image $\mathcal{F}$ and the moving image $\mathcal{M}$ on the field of view Ω such that $\mathcal{F}(x)$ and $\mathcal{M}(y(x))$ are similar for $x \in \Omega$.

Rigid Registration. The rigid registration adopts the method of [17]. We use the normalized gradient field distance measure [8],

$$\mathcal{D}^{NGF}(\mathcal{F}, \mathcal{M}(\mathrm{y})) = \int_{\Omega} 1 - \frac{\langle \nabla \mathcal{M}(\mathrm{y}(x)), \nabla \mathcal{F}(x) \rangle_{\epsilon}^2}{\|\nabla \mathcal{M}(\mathrm{y}(x))\|_{\epsilon}^2 \|\nabla \mathcal{F}(x)\|_{\epsilon}^2} \, \mathrm{d}x,$$

with $\langle f, g \rangle_{\epsilon} := \sum_{j=1}^{3}(f_j g_j + \epsilon^2)$, $\|f\|_{\epsilon} := \sqrt{\langle f, f \rangle_{\epsilon}}$, that focuses on the alignment of image gradients of the fixed image $\mathcal{F}$ and the deformed moving image $\mathcal{M}(y)$. The edge hyper-parameter $\epsilon > 0$ is used to suppress small image noise, without affecting image edges. The optimization problem is solved using a Gauss-Newton optimization scheme and is embedded into a multi-level scheme with two levels.

Deformable Registration. We deploy the matrix-free deformable registration of [11]. The deformation is defined as a minimizer of the cost function

$$\min_{y} \mathcal{D}^{NGF}(\mathcal{F}, \mathcal{M}(y)) + \alpha \mathcal{R}^{curv}(y),$$

with the normalized gradient field distance measure $\mathcal{D}^{NGF}$. To focus the registration to inside the prostate, we restrict Ω to the support of the prostate segmentation of the fixed image, which is automatically generated with the prostate segmentation algorithm provided by [18]. The second-order curvature regularizer $\mathcal{R}^{curv}$ [7] enforces smooth deformation by penalizing spatial derivatives. The parameter α is a weighting factor of the regularizer. The optimization problem is solved using the limited-memory Broyden-Fletcher-Goldfarb-Shannon (L-BFGS) optimization scheme [13]. Optimization was performed in a multi-level scheme with two levels on images with successively declining levels of smoothing to guide registration from larger structures to smaller refinements. During each registration level a grid size of the displacement field of $31 \times 31 \times 31$ is used to warp the moving image using trilinear interpolation. The deformable registration uses the output of the rigid registration as an initial starting point. Hyperparameters of the registration method were experimentally set using the first ten cases of the PI-CAI public training dataset.

2.2 Data

Three datasets with bpMRI scans (axial T2W, ADC and HBV ($b \geq 1000$) imaging) for prostate cancer detection were used. For each dataset, the reference standard was set by histopathology, with clinically significant prostate cancer defined as ISUP 2–5 (intermediate to very high risk) [6]. Informed consent was waived, given the retrospective scientific use of deidentified patient data. Scan characteristics are given in the supplementary material.

PI-CAI: For csPCa detection model development, 10,207 cases of 9129 patients from 10 Dutch hospitals and 1 Norwegian hospital were used [18]. Cases were acquired using 1.5 or 3-Tesla MRI scanners between 2012 and 2021 from patients with suspicion of harboring prostate cancer. Exclusion criteria included prior prostate-specific treatment, prior ISUP ≥ 2 findings, incomplete studies, and diagnostically insufficient image quality. Manual voxel-level annotations were available for 1175 positive training cases (1323 lesions) and for an additional 892 positive training cases (1037 lesions) AI-derived voxel-level annotations were provided.

PCNN: For testing of the registration algorithm, cases from the PI-CAI training set with manual voxel-level annotations per modality (T2W and ADC) were included. These annotations were not created independently, but the radiologists followed the lesion characteristics for each modality. In total, 473 cases of 438 patients from Prostaat Centrum Noord-Nederland (PCNN) (8 hospitals) were included.

PROMIS: For external testing, 546 cases of 546 patients from 11 United Kingdom hospitals were included [2]. Cases were acquired using 1.5-Tesla MRI scanners between 2012 and 2015 from patients with suspicion of harboring prostate cancer. Exclusion criteria included prior prostate treatment, prior biopsies, incomplete studies, and diagnostically insufficient image quality. No manual voxel-level annotations were available.

2.3 PI-CAI AI System

The PI-CAI AI system was developed in the PI-CAI challenge. The algorithm is the ensemble of the top 5 submissions, selected based on testing with 1000 cases. The models were trained using a dataset of 9107 cases. The algorithm uses the axial T2W, ADC and HBV scans and clinical variables (e.g. PSA density). The U-Net backbone was predominantly used, with early fusion of the scans. Early fusion allows networks to easily correlate information from each sequence, but also requires sequences to be aligned for optimal model performance. For details on the data and each of the top 5 submissions, see [18] and (pi-cai.grand-challenge.org/algorithms). No retraining of the AI system was performed in this study.

3 Experiments

The aim of this study is to evaluate the effect of image registration on the clinical downstream task of case-level csPCa diagnosis. To quantify the algorithm's performance degradation under severe and extreme misalignment conditions, we artificially misaligned the T2W and DWI images in two severity steps.

On the original data, we evaluated the registration performance by measuring lesion alignment and the plausibility of the displacement field. Subsequently, we employed the csPCa detection algorithms developed in the PI-CAI challenge for the diagnostic evaluation. For both experiments, we compare the results on three dataset variants: the original dataset, the dataset with rigidly aligned T2W and DWI scans, and the dataset with deformably aligned T2W and DWI scans.

3.1 Impact of Synthetic Misalignment

To investigate the impact of MRI sequence misalignment on the performance of a clinically significant prostate cancer detection algorithm, we conducted two synthetic misalignment tests (parameters were chosen by qualitatively analyzing original data):

Severe Misalignment: DWI scans were translated in the z-direction by a random selection from $\{-2, -1, 0, 1, 2\}$ slices and in-plane by $\{-5, -4, \ldots, 5\}$ voxels in both the x and y directions.

Extreme Misalignment: DWI scans were translated by a random selection from -5 or $+5$ slices in the z-direction and by -10 or $+10$ voxels in both the x and y directions.

3.2 Registration Performance

The evaluation of registration performance was conducted using the PCNN validation dataset, chosen for its availability of lesion annotations across both T2W and ADC scans. The hyperparameters for the registration method were manually fine-tuned using only the first 10 cases from the PI-CAI Public Training and Development dataset, which did not overlap with this PCNN dataset. Therefore, the PCNN dataset serves as an independent evaluation set for assessing the registration performance.

To quantitatively assess the quality of image registration in the absence of reference displacement fields, we employed two surrogate metrics. The Dice coefficient was utilized to quantify the overlap of lesion segmentations between T2W scans and ADC maps. Although we recognize that the Dice coefficient may not be the perfect metric for assessing the registration performance [16], its usage is justified in this context given the critical importance of accurate lesion alignment in T2W scans and ADC maps for the reliable performance of csPCa detection algorithms. The choice of the Dice score, therefore, aligns with our objective to prioritize lesion alignment in the evaluation of registration effectiveness. Smooth deformations within the prostate are important to preserve diagnostic features, therefore we evaluated the plausibility of the displacement field by examining the percentage of voxels exhibiting folding within the prostate region of the predicted deformation field.

csPCa Detection Performance. Diagnostic performance is assessed using the area under the receiver operator characteristic curve (AUROC). For case-level risk estimation of significant cancer, we utilized voxel-level detection maps generated by the PI-CAI AI system on the external PROMIS test dataset.

Additionally, we evaluated diagnostic performance using the PCNN dataset, to facilitate an evaluation of diagnostic performance in relation to the registration accuracy. We note that this dataset is not independent for the diagnostic algorithms, since this is a subset of the training data of the algorithms.

Since the PROMIS dataset contains scans with very large field-of-views with anatomical structures not present in the PI-CAI training dataset, we filtered out lesion predictions further than 3 mm away from the prostate segmentation. For this, the prostate segmentation mask was dilated by 3 mm and all predictions outside of that dilated mask were removed. Following the approach used in the PI-CAI challenge, each algorithm's case-level prediction was the maximum lesion-level prediction, and the AI system's case-level prediction was the equally-weighted prediction of each algorithm.

Statistical Analysis. The diagnostic performance differences on the external testing set were statistically analyzed. The performance with the deformably and rigidly aligned images are compared against the performance with the original dataset. To determine the probability of one configuration outperforming another, we performed DeLong's test [4]. Multiplicity was corrected for using the Holm-Bonferroni method, with a base alpha value of 0.05. For details, see the pre-defined statistical analysis plan online [9].

4 Results

4.1 Impact of Misalignment

In Fig. 2, the impact of misalignment on the performance of the clinically significant prostate cancer detection algorithm is illustrated. When a severe misalignment is introduced, the AUC decreases from 0.793 to 0.720. In the case of extreme misalignment, the AUC further drops to 0.487.

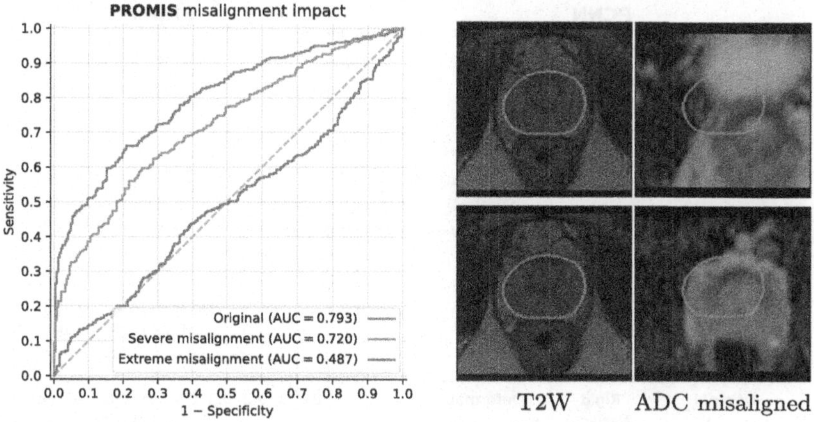

Fig. 2. This figure demonstrates the impact of synthetic misalignment. (left) the diagnostic performance of the PI-CAI AI system is shown on the PROMIS dataset. When a severe misalignment is introduced, the AUC decreases from 0.793 to 0.720. In the case of extreme misalignment, the AUC further drops to 0.487. (right) shows the T2W and misaligned ADC (top: severe misalignment, bottom: extreme misalignment) images, with ■ prostate gland contour of the T2W scan. (Color figure online)

4.2 Quantitative Results

Registration Performance. The median Dice score improved to 0.58 with deformable registration, compared to 0.48 for the original dataset and 0.47 with rigid registration (Fig. 3).

For one case, 1% of voxels in the prostate were folded. For all other cases, no foldings occurred in the deformation field.

csPCa Diagnosis Performance. For the PROMIS external testing dataset, the PI-CAI AI system showed a positive yet non-significant improvement in diagnostic performance (+0.3% AUROC, $p = 0.18$) with deformably aligned scans compared to the original dataset. A comprehensive qualitative analysis of representative cases is given in Sect. 4.3.

4.3 Qualitative Results

In this section, we present qualitative results of the image registration and subsequent csPCa detection algorithm. Results for the PCNN dataset are shown in Fig. 4, showing the case with the largest improvement in Dice score (first row) and the largest decrease in Dice score (second row) for the deformably aligned dataset, compared to the original dataset alongside the clinical interpretation for each case.

The first row shows the images with mild benign prostatic hyperplasia (BPH) in the transition zone. BPH is a benign condition, which grows over time. A typical transition zone with BPH shows so-called 'organized chaos', with multiple

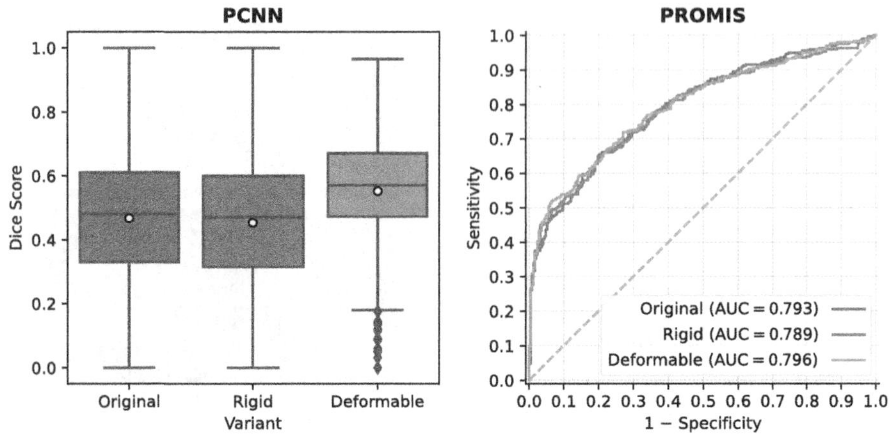

Fig. 3. Quantitative registration results. *(left)* Distribution of Dice scores between the lesion annotation on the T2W and ADC scans for the original, rigidly aligned, and deformably aligned PCNN datasets. *(right)* Model performance for the PI-CAI AI system with the original, rigidly aligned and deformably aligned PROMIS datasets. (Color figure online)

nodules with variable imaging appearance, often with diffusion restriction and enhancement. In transition zone tumors, the typical encapsulation is lost, and the organized aspect changes to a homogeneous low T2W signal with marked diffusion restriction and vivid enhancement. In the left transition zone of this patient, an encapsulated BPH nodule is annotated in yellow on T2W, with low T2W signal intensity. On the ADC map an area with marked diffusion restriction is annotated in red. A notable discrepancy is observed in the alignment between the T2W and ADC imaging, leading to misalignment between the lesion's features on the T2W and ADC scans. The encapsulated nature on T2W of this nodule is a non-suspicious sign in the transition zone. Consequently, the PI-CAI AI system with the original scans classifies the lesion in the middle of the transition zone instead of within an encapsulated nodule more laterally due to the misalignment, which suggests a higher risk level (prediction = 0.63). The deformable image registration method aligned the two modalities, and the PI-CAI AI system with the deformably aligned scans assigned a lower risk level (prediction = 0.47). Targeted biopsies revealed ISUP 1 in the left transition zone, which is an indolent prostate cancer that is often invisible on prostate MRI.

The second row shows the images for a 66-year-old man with a PSA level of 13 ng/mL and a PSA density of 0.11 ng/mL/cc. The images show a tumor suspicious area ventral in the apex of the prostate close to the anterior fibromuscular stroma (AFMS) ventral to the transition zone of the prostate. The delineation of the lesion mask was guided by the image features observed in the ADC scan and was subsequently adopted for the T2W scan as well. Upon reconsideration of the lesion segmentation with two radiologists, it appears that the extension into the AFMS is due to oversegmentation, rather than the lesion infiltrating the

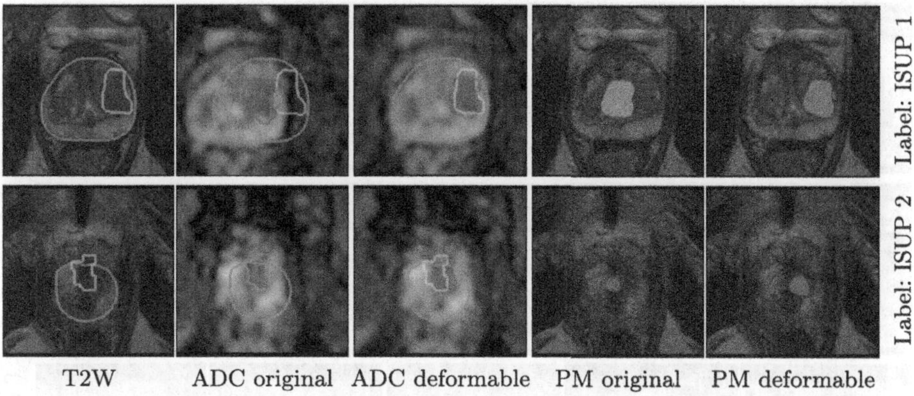

Fig. 4. Qualitative registration results showing two exemplary cases with ■ prostate gland, ■ lesion annotated on T2, ■ lesion annotated on ADC. In the last two columns, the prediction maps (PM) generated with the original dataset and the deformably aligned dataset are overlayed on the T2W scan. The color of the PMs indicates the predicted suspicion level, with red indicating high likelihood of csPCa and blue indicating a low likelihood. (Color figure online)

AFMS. As such, the model predictions capture the lesion extent very well. The prediction with the original dataset had a bit higher confidence (0.62 vs 0.56) for this positive case. Targeted biopsy of this area revealed ISUP 2 prostate cancer.

In Fig. 5, qualitative results on the PROMIS dataset are shown, which are explained in the following in more detail.

The first row shows images with a well-defined lesion in the left peripheral zone midprostate, with low signal intensity on T2W, and low signal intensity on the ADC map, consistent with a suspicious lesion (PI-RADS 4). The T2W and ADC map are misaligned, both in-plane and through-plane, resulting in the diffusion restriction on the original ADC map to be misaligned with the lesion features on the T2W sequence. The PI-CAI AI system identified the lesion with both variants of the dataset. With the deformably aligned dataset the algorithm confidence increased to 0.51, from a prediction of 0.36 before. Histopathological evaluation confirmed the aggressive nature of this lesion (ISUP 2).

The second row shows a T2W and ADC map that are misaligned in-plane. Consequently, a substantial part of the prostate on the ADC map appears outside of the prostate region of the T2W sequence. The ADC map shows diffusion restriction (low signal intensity; darker appearance) in the right transition zone midprostate. Due to the misalignment, this darker area on the original ADC map appears to be in the right peripheral zone on the T2W scan, and therefore misclassification can occur. After deformable alignment, the darker area on the ADC map aligns with the transition zone instead of peripheral zone. This is reflected in the lesion detection of the PI-CAI AI system, which predicts a lesion with confidence of 0.44 with the original scans and with a confidence of 0.18 with the deformably aligned scans. Targeted biopsies revealed ISUP 1 in the

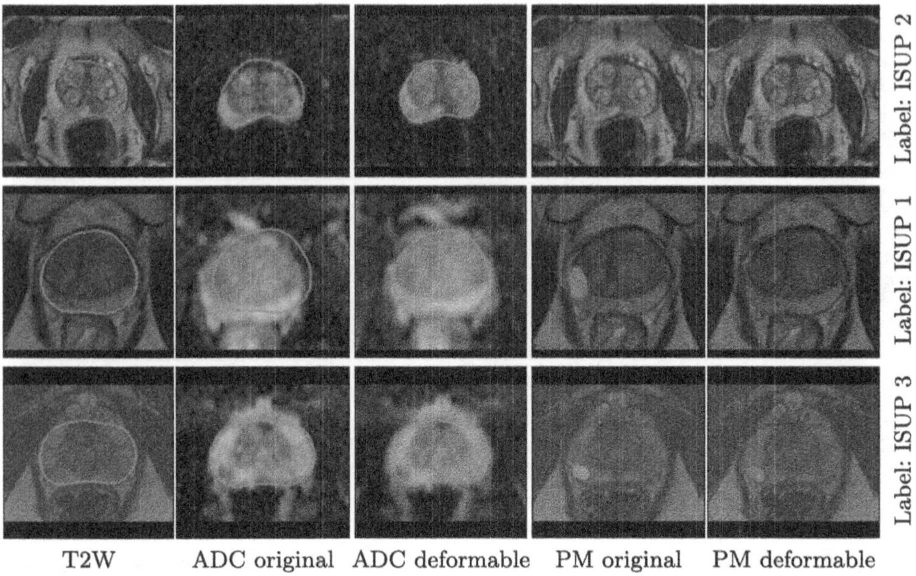

Fig. 5. Qualitative results on the PROMIS data set. The T2, ADC, and deformably aligned ADC are shown with ■ prostate gland. In the last two columns, the prediction maps (PM) generated with the original dataset and the deformbly aligned dataset are overlayed on the T2W image. The label shows the ISUP grade, where 1 is indolent cancer (negative), and ≥ 2 is intermediate to high-risk cancer (positive). The first two cases were selected to have the largest prediction increase and decrease for the deformably aligned dataset, compared to the original datasets, for cases with a case-level prediction above 0.3, respectively. The third case was a failure case with the deformably aligned scans. The color of the PMs indicates the predicted suspicion level, with red indicating high likelihood of csPCa and blue indicating a low likelihood. (Color figure online)

right transition zone, which is an indolent prostate cancer that is often invisible on prostate MRI. No aggressive PCa was detected.

The third row shows images with a small lesion in the right peripheral zone midprostate. The lesion appears as a well circumscribed area with low signal intensity (dark) on T2W images and the ADC map, suspicious for clinically significant cancer (PI-RADS 4). For this case, the deformable image registration slightly misaligned the T2W and ADC image features of the lesion, which resulted in the detection algorithm to decrease its lesion prediction from 0.49 to 0.39. Histopathological evaluation confirmed the aggressive nature of this lesion (ISUP 3).

5 Discussion and Conclusion

In this study, we investigated the effect of image registration on the clinical downstream task of case-level csPCa diagnosis when integrated at the inference stage. Deformable registration demonstrated a substantial improvement in lesion overlap on the validation dataset (+9% average Dice score) which is even slightly more than the one reported in [12] (+6% average Dice score). However, since different datasets were used, a direct comparison is not possible. Moreover, the Dice score achieved with deformable registration on the validation dataset was 0.58. This performance aligns closely with the inter-rater agreement typically observed in prostate tumor segmentation. Specifically, between two radiologists independently segmenting csPCa lesions using the same modalities, the observed Dice score was 0.60 [1].

Additionally, we showed a positive yet non-significant improvement in diagnostic performance on the PROMIS test dataset (+0.3% AUROC, $p = 0.18$) with deformably aligned scans. Our investigation shows that a substantial improvement in lesion alignment does not directly equal a significant improvement in diagnostic performance. To illustrate the impact of misalignment on the algorithmic results, we present detailed visualizations and analyses of several PCNN and PROMIS cases in Sect. 4.3. These results showed that the PI-CAI AI system demonstrated robustness to minor misalignments, particularly when these misalignments did not result in lesions being misrepresented in incorrect zones. Additionally, we anticipate a comparable number of misaligned cases in the PROMIS dataset as observed in the PI-CAI dataset, where the incidence was low. Therefore, the expected improvement in AUROC is limited. The positive yet non-significant improvement in diagnostic performance might be the result from those cases.

Rigid registration, as configured in this project, did not improve lesion overlap in the PCNN dataset and resulted in slightly lower diagnostic performance in the PROMIS dataset. We hypothesize that the lack of lesion overlap improvement is a result of using the full scan for this registration step, rather than focusing on the prostate. The slightly reduced diagnostic performance may be a result of losing information due to interpolation. Future research could focus on integrating the prostate mask into the rigid registration.

Our method had limitations. The deformable registration method potentially introduced unrecognized artifacts into the images which might result in worse diagnostic performance. Addressing this through retraining the csPCa algorithms to adapt to registration-induced image variations represents a promising strategy. It is crucial to note that the registration method avoided the generation of physiological unrealistic deformations. This is achieved by applying a high regularization weight to obtain smooth and plausible displacement fields. Another critical aspect is the choice of resampling strategy. This factor considerably impacts the smoothing of ADC values, especially for small lesions, and influences the diagnostic quality of images through the effects of multiple resamples. Merging all resampling steps into one would visibly increase the quality, but is only possible in an end-to-end approach.

The relevance of the PROMIS dataset in present-day analyses has been a subject of debate, particularly among radiologists. The diagnostic quality of MRI scans has markedly improved since the trial finished in 2015. Additionally, the PROMIS dataset contains acquired high b-value scans, while contemporary protocols calculate this based on acquired lower b-value scans, which results in less noise and better diagnostic quality. Despite these limitations, the relevance of the PROMIS dataset should not be understated. This dataset can serve as a benchmark for scenarios where access to high-end, expensive scanners is limited. This situation is a common reality for many institutions, highlighting the importance of developing algorithms that can perform well across a range of image acquisition methods.

In conclusion, our study shows that while image registration can substantially improve lesion overlap in csPCa diagnosis, it does not directly lead to a significant improvement in diagnostic performance. However, the qualitative analysis showed promising results and indicate that joint development of image registration methods and diagnostic AI algorithms could enhance diagnostic accuracy and patient outcome.

Acknowledgments. This research is funded by the European Union under HORIZON-HLTH-2022: COMFORT (101079894), Health Holland (LSHM20103), European Union H2020: ProCAncer-I project (952159), European Union H2020: PANCAIM project (101016851), and Siemens Healthineers (CID: C00225450). Views and opinions expressed are however those of the author(s) only and do not necessarily reflect those of the European Union or European Health and Digital Executive Agency (HADEA). Neither the European Union nor the granting authority can be held responsible for them.

The PROMIS data used in the analysis for this manuscript were provided from the PROMIS study, led by University College London (UCL). PROMIS was funded by the UK Government Department of Health, National Institute of Health Research-Health Technology Assessment Programme, (Project number 09/22/67). Support was also provided by National Institute for Health Research (NIHR) UCLH/UCL Biomedical Research Centre, National Institute for Health Research (NIHR) The Royal Marsden and Institute for Cancer Research Biomedical Research Centre and National Institute for Health Research (NIHR) Imperial Biomedical Research Centre. The original PROMIS study was coordinated by the Medical Research Council Clinical Trials Unit (MRC CTU) at UCL and sponsored by UCL. The PROMIS Biobank was funded by Prostate Cancer UK (PG10-17). The PROMIS dataset and the biobank is under the research governance of the ReIMAGINE Risk Trial Management Group (funded by Medical Research Council (UKRI) and Cancer Research UK: MR/R014043/1).

References

1. Adams, L.C., Makowski, M.R., Engel, G., Rattunde, M., Busch, F., Asbach, P., Niehues, S.M., Vinayahalingam, S., van Ginneken, B., Litjens, G., et al.: Prostate158-an expert-annotated 3t mri dataset and algorithm for prostate cancer detection. Computers in Biology and Medicine **148**, 105817 (2022)
2. Ahmed, H.U., Bosaily, A.E.S., Brown, L.C., Gabe, R., Kaplan, R., Parmar, M.K., Collaco-Moraes, Y., Ward, K., Hindley, R.G., Freeman, A., et al.: Diagnostic accuracy of multi-parametric MRI and TRUS biopsy in prostate cancer (PROMIS): a paired validating confirmatory study. The Lancet **389**(10071), 815–822 (2017)
3. De Vente, C., Vos, P., Hosseinzadeh, M., Pluim, J., Veta, M.: Deep learning regression for prostate cancer detection and grading in bi-parametric MRI. IEEE Transactions on Biomedical Engineering **68**(2), 374–383 (2020)

4. DeLong, E.R., DeLong, D.M., Clarke-Pearson, D.L.: Comparing the areas under two or more correlated receiver operating characteristic curves: a nonparametric approach. Biometrics pp. 837–845 (1988)
5. Eldred-Evans, D., Burak, P., Connor, M.J., Day, E., Evans, M., Fiorentino, F., Gammon, M., Hosking-Jervis, F., Klimowska-Nassar, N., McGuire, W., Padhani, A.R., Prevost, A.T., Price, D., Sokhi, H., Tam, H.H., Winkler, M., Ahmed, H.U.: Population-based prostate cancer screening with magnetic resonance imaging or ultrasonography. JAMA Oncology **7**, 395 – 402 (2021)
6. Epstein, J.I., Egevad, L., Amin, M.B., Delahunt, B., Srigley, J.R., Humphrey, P.A.: The 2014 international society of urological pathology (ISUP) consensus conference on gleason grading of prostatic carcinoma: Definition of grading patterns and proposal for a new grading system. The American Journal of Surgical Pathology **40**, 244-252 (2016)
7. Fischer, B., Modersitzki, J.: Curvature based image registration. Journal of Mathematical Imaging and Vision **18**(1), 81–85 (2003)
8. Haber, E., Modersitzki, J.: Intensity gradient based registration and fusion of multimodal images. In: Medical Image Computing and Computer-Assisted Intervention – MICCAI 2006. vol. 3216, pp. 591–598 (2006)
9. Hering, A., de Boer, S., Saha, A., Twilt, J.J., Heinrich, M.P., Yaker, D., de Rooij, M., Huisman, H., Bosma, J.S.: Statistical Analysis Plan - Deformable MRI Sequence Registration for AI-based Prostate Cancer Diagnosis (2024). https://doi.org/10.5281/zenodo.12170878
10. Hricak, H., Abdel-Wahab, M., Atun, R., Lette, M.M., Paez, D., Brink, J.A., Donoso-Bach, L., Frija, G., Hierath, M., Holmberg, O., et al.: Medical imaging and nuclear medicine: a lancet oncology commission. The Lancet Oncology **22**(4), e136–e172 (2021)
11. König, L., Rühaak, J., Derksen, A., Lellmann, J.: A matrix-free approach to parallel and memory-efficient deformable image registration. SIAM Journal on Scientific Computing **40**(3), B858–B888 (2018)
12. Kovacs, B., Netzer, N., Baumgartner, M., Schrader, A., Isensee, F., Weißer, C., Wolf, I., Görtz, M., Jaeger, P.F., Schütz, V., et al.: Addressing image misalignments in multi-parametric prostate MRI for enhanced computer-aided diagnosis of prostate cancer. Scientific Reports **13**(1), 19805 (2023)
13. Liu, D.C., Nocedal, J.: On the limited memory BFGS method for large scale optimization. Mathematical programming **45**(1-3), 503–528 (1989)
14. Netzer, N., Weißer, C., Schelb, P., Wang, X., Qin, X., Görtz, M., Schütz, V., Radtke, J.P., Hielscher, T., Schwab, C., et al.: Fully automatic deep learning in bi-institutional prostate magnetic resonance imaging: effects of cohort size and heterogeneity. Investigative radiology **56**(12), 799–808 (2021)
15. Pellicer-Valero, O.J., Marenco Jimenez, J.L., Gonzalez-Perez, V., Casanova Ramon-Borja, J.L., Martin Garcia, I., Barrios Benito, M., Pelechano Gomez, P., Rubio-Briones, J., Rupérez, M.J., Martín-Guerrero, J.D.: Deep learning for fully automatic detection, segmentation, and gleason grade estimation of prostate cancer in multiparametric magnetic resonance images. Scientific reports **12**(1), 2975 (2022)
16. Rohlfing, T.: Image similarity and tissue overlaps as surrogates for image registration accuracy: widely used but unreliable. IEEE transactions on medical imaging **31**(2), 153–163 (2011)
17. Rühaak, J., König, L., Tramnitzke, F., Köstler, H., Modersitzki, J.: A matrix-free approach to efficient affine-linear image registration on CPU and GPU. Journal of Real-Time Image Processing **13**, 205–225 (2017)

18. Saha, A., Bosma, J.S., Twilt, J.J., van Ginneken, B., Bjartell, A., Padhani, A.R., Bonekamp, D., Villeirs, G., Salomon, G., Giannarini, G., Kalpathy-Cramer, J., Barentsz, J., Maier-Hein, K.H., Rusu, M., Rouvière, O., van den Bergh, R., Panebianco, V., Kasivisvanathan, V., Obuchowski, N.A., Yakar, D., Elschot, M., Veltman, J., Fütterer, J.J., de Rooij, M., Huisman, H., the PI-CAI consortium: Artificial intelligence and radiologists in prostate cancer detection on mri (pi-cai): an international, paired, non-inferiority, confirmatory study. The Lancet Oncology (2024).https://doi.org/10.1016/S1470-2045(24)00220-1
19. Sanyal, J., Banerjee, I., Hahn, L., Rubin, D.: An automated two-step pipeline for aggressive prostate lesion detection from multi-parametric MR sequence. AMIA Summits on Translational Science Proceedings **2020**, 552 (2020)
20. Stavrinides, V., Giganti, F., Emberton, M., Moore, C.M.: MRI in active surveillance: a critical review. Prostate Cancer and Prostatic Diseases **22**(1), 5–15 (2019)
21. Sung, H., Ferlay, J., Siegel, R.L., Laversanne, M., Soerjomataram, I., Jemal, A., Bray, F.: Global cancer statistics 2020: GLOBOCAN estimates of incidence and mortality worldwide for 36 cancers in 185 countries. CA: A Cancer Journal for Clinicians **71**(3), 209–249 (2021)
22. Weinreb, J.C., Barentsz, J.O., Choyke, P.L., Cornud, F., Haider, M.A., Macura, K.J., Margolis, D., Schnall, M.D., Shtern, F., Tempany, C.M., et al.: Pi-rads prostate imaging–reporting and data system: 2015, version 2. European urology **69**(1), 16–40 (2016)
23. Winkel, D.J., Tong, A., Lou, B., Kamen, A., Comaniciu, D., Disselhorst, J.A., Rodríguez-Ruiz, A., Huisman, H., Szolar, D., Shabunin, I., et al.: A novel deep learning based computer-aided diagnosis system improves the accuracy and efficiency of radiologists in reading biparametric magnetic resonance images of the prostate: results of a multireader, multicase study. Investigative radiology **56**(10), 605–613 (2021)

Atlas/Fusion

SINA: Sharp Implicit Neural Atlases by Joint Optimisation of Representation and Deformation

Christoph Großbröhmer(✉), Ziad Al-Haj Hemidi, Fenja Falta, and Mattias P. Heinrich

Institute of Medical Informatics, University of Lübeck, Lübeck, Germany
{c.grossbroehmer,heinrich}@uni-luebeck.de

Abstract. Medical image atlases simplify the analysis of large-scale datasets and are used for various applications, such as population analysis or probabilistic segmentation. They are often synthesized by minimising geometric distances. This formulation allows for fuzzy atlases, resulting in reduced interpretability and flexibility. Furthermore, most atlas generation methods are limited in terms of large anatomical variations and non-aligned data. We propose SINA (Sharp Implicit Neural Atlases), a novel framework for medical image atlas synthesis, leveraging the joint optimisation of data representation and registration. By iteratively refining sample-to-atlas registrations and modelling the atlas as a continuous function in an Implicit Neural Representation, we demonstrate the possibility of achieving atlas sharpness while maintaining data fidelity. Our approach is evaluated on 2D and 3D datasets of varying complexity against popular atlas synthesis methods. Our method ranks first regarding the synthesis of a complex chest X-ray atlas, second in the easier task of pre-aligned MRI images, and is capable of handling high-resolution abdominal CTs while composing representative and sharp atlases. Source code and trained models are published on GitHub.

Keywords: Atlas Generation · Implicit Neural Representation · Image Registration

1 Introduction

The creation of medical image atlases (or templates) is a central task in the field of medical informatics to simplify the analysis of large amounts of data. In addition to their role as an educational tool, they form the basis for image-based population studies, enable probabilistic atlas-based segmentation, and serve as a data-driven reference model for different anatomies of various imaging modalities. An atlas combines information from both geometry and appearance across a large number of images of a population. In order to aggregate this information, it must be brought into spatial correspondence, linking the methodology for creating an atlas to the field of medical image registration [17]. An optimal

M. Modat et al. (Eds.): WBIR 2024, LNCS 15249, pp. 165–180, 2024.
https://doi.org/10.1007/978-3-031-73480-9_13

template is often regarded as a minimization of the geometric distance to all images in a dataset. This definition is implicitly accompanied by the demand, or at least acceptance, of a fuzzy atlas. On the other hand, image registration methods benefit from sharp structures providing correspondences and the delineation of objects. While the necessity of sharpness for template synthesis is still a subject of discussion, a blurry template comes with the disadvantage of reduced interpretability.

Technical and methodological innovations in the field of image registration, such as the use of convolutional networks, have simultaneously advanced methods for atlas generation. Recently, Implicit Neural Representations (INRs), which approximate data-inherent continuous functions of individual instances in network weights, have gained popularity. A key feature is the abandonment of regular grid structures, providing flexibility in querying the network on arbitrary sizes. Especially in the field of computer vision, INRs demonstrate their ability to learn scenes as neural radiance fields (NeRF) from image and video data using known camera positions [24]. To mitigate the problem of inaccurate or missing poses, the joint optimisation of NeRF and camera positions was proposed with BARF [18]. In this method, inconsistent observations are correlated or registered by additionally learning corresponding camera positions. Since the goal of obtaining a coherent representation is achieved through registration, this idea can be linked to our approach of atlas synthesis. Furthermore, by leveraging the continuous properties of INRs, an atlas can be trained and inferred involving sub-pixel accuracy.

In this work, we propose the simple joint optimisation of INR and transformation parameters to synthesize a sharp and easily interpretable dataset representation.

1.1 Related Work

The classical approach for **Medical Image Atlas Synthesis** is to average the displacement vectors of registrations on a manually selected template to obtain a mean shape and intensity [12]. However, a disadvantage of this method is the manual bias that comes with choosing a suitable template, which can be tackled by employing groupwise registration [17]. For this step, deep learning-based approaches pose a clear advantage in terms of runtime and computational costs, especially for 3D data, and can be extended with pairwise registration to enhance image alignment [9] or include joint learning of registration and segmentation [31]. Dalca et al. propose a CNN-based model to synthesize atlases based on population attributes [7]. In contrast, Yu et al. learn conditioned deformation fields warping an unconditioned template, effectively encoding population attributes in deformations [39]. However, such approaches are limited by fixed coordinate systems, which makes them inflexible in terms of spatial resolution. Additionally, such models often require rigid pre-alignment and are restricted to small image variations, limiting their application. Determining **Atlas Quality** is not straightforward and depends on the intended use and the method of atlas synthesis. One pitfall is the comparison of forward- and backward-generated atlases that contain different information. The latter (e.g. [14,17]) is characterised by the fact that they warp to images in the synthesis templates and thus represent a mean of all warped images. Forward creation methods (e.g. [7]), on the other hand, warp the images onto the atlas and generate the mean of all warped images, weighted with the required deformation. Evaluating both with the same metrics can distort the comparison. This can be alleviated by employing the atlas as a bridge, i.e. calculating metrics on consecutive forward and backward warped images [9]. Another metric used to evaluate templates is **Atlas Sharpness** [20]. Although there is no consensus on whether sharper templates lead to better registration or segmentation results [9,38], the inverse relationship is unlikely. Our method can be loosely linked to the strategy presented by Wu et al., who improve groupwise registration results by explicitly enforcing a sharp mean image using an adaptive spatial weighting strategy [37]. Our approach can further be associated with the concept of **Metamorphosis**, which models images as a combination of intensity and geometric transforms of a prototype [1,27], often applied to image registration methods [23], and active appearance models [6].

INRs were introduced as a powerful tool for computer vision tasks [32]. They use multiple-layer perceptrons (MLPs) to represent complex data, such as shapes and images, as continuous functions. Traditional INRs with ReLUs as activation functions were limited by spectral bias, but incorporating Fourier features [34], periodic activations [32], or multi-grid features [26] significantly improved their ability to fit higher frequencies. One of the primary advantages of INRs is their ability to offer high levels of detail with relatively low storage requirements, as the information is stored in the weights of the neural network [28]. This efficiency is particularly beneficial in applications like 3D scene reconstruction, where high fidelity is necessary but memory is limited. Another significant benefit of INRs is their inherent smoothness and continuous nature, preventing aliasing artefacts common in discrete representations [24]. This characteristic makes them ideal for tasks requiring high-quality interpolation, such as super-resolution and novel view synthesis. Moreover, INRs can naturally handle variable levels of detail, providing fine details when needed and abstracting them when not. Additionally, INRs demonstrate versatility across various domains. They have been successfully applied to represent images, 3D shapes, and even entire scenes. Their ability to generalize across different data types and their flexibility in encoding spatial and temporal dimensions make them a powerful tool in machine learning and computer graphics [34]. **INRs for image registration** [36], which represent deformation fields in MLPs, came recently into focus. In the medical fields, multiple extensions of this idea have been explored, e.g. different activation functions, cycle consistency, cascades and the incorporation of prior information and implicit regularisation [2,13,30]. While these methods are typically tailored to individual instances, efforts have been made to generalize INRs for image registration [40]. **Generalized INRs** employed in computer vision are commonly trained to represent a population of images where the INR learns a common representation (template) of the presented images, hence sharing similarities to the atlas creation. The input of such functions usually comprises continuous encoded coordinates and instance-specific information, e.g. latent codes or features [28,33]. Generalization can also be achieved through meta-learning [4,19]. Such techniques enable optimisation of a global representation, e.g. template image or shape, that enables faster and improved fitting of unseen data points from the same distribution.

INR Atlas creation was previously explored in three works; Low et al. [21] create an atlas to represent shapes of natural objects, while [5] and [8] pioneered the use of INR atlases in the medical image analysis field. Chi et al. [5] use a complex encoder-decoder training strategy with sampling from three separate neural fields for registration and intensity estimation for 4D fetal MRI time series to accelerate the speed of atlas synthesis. In a paper not yet published under peer-review, Dannecker et al. [8] propose a conditional INR encoding both grey value and segmentation atlas based on joint optimisation of latent codes and network weights. During training, this code is infused by scaling and shifting in modulation layers of the INR, while for new subjects, only the patient-specific latent code is optimised. An atlas can then be obtained using a Gaussian regression of these codes, which encodes geometric and intensity variations in the training data.

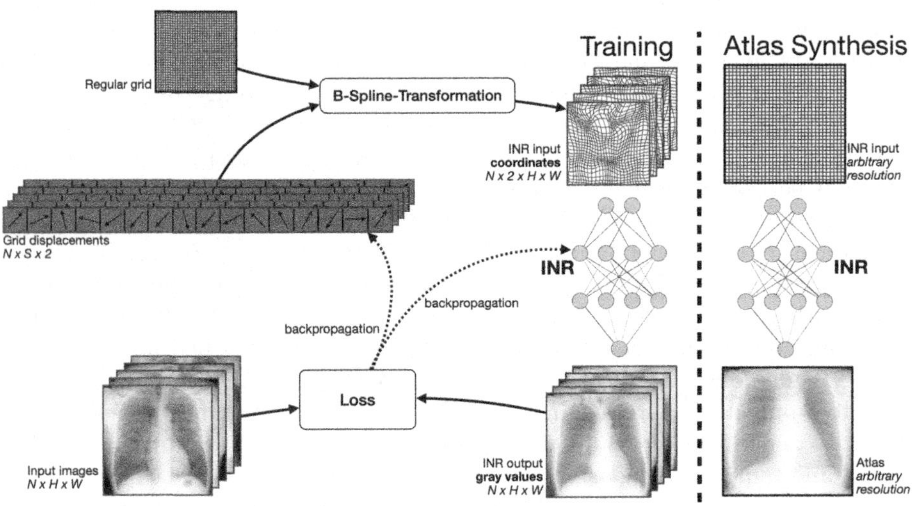

Fig. 1. Simple Implicit Neural Atlases (SINA) Framework. During training, N displacements for S grid points are subject to a B-spline transformation and are subsequently input into an Implicit Neural Network (INR). The calculated loss between INR outputs and input images is backpropagated to optimise grid displacements per instance and INR weights capturing the underlying intensity atlas. At inference, an implicit template can be obtained by sampling the INR on a regular grid of any resolution.

1.2 Contribution

In this paper, we propose Sharp Implicit Neural Atlases (SINA), a simple and comprehensive framework for unsupervised atlas generation using joint optimisation of transformation parameters and INR weights to simultaneously obtain a sharp and expressive medical image atlas and corresponding displacement fields. The resulting atlas is stored implicitly and can be queried on any resolution. The method does not require any pre-alignment and can handle various strengths of deformations. We compare our method against popular atlas synthesis frameworks on two 2D datasets of varying complexity (chest X-ray and brain MRI slices) and demonstrate its extension to 3D abdominal CT data. To our knowledge, this is the first work to combine explicit registration and implicit modelling for medical atlas generation. We rely on publicly available data and comparison methods and release code and trained models on GitHub[1].

2 Methods

2.1 SINA

A medical image atlas should contain both the geometric properties and intensities of the dataset. Another desired property is the presence of edges or high image frequencies that facilitate registration between the image and the atlas. We hypothesise that combining an implicit representation and explicit learning of displacements fulfils both requirements. This method is inspired by [18] from the field of computer vision, which demonstrated the simultaneous optimisation of homographies of image patches and the fitting of an INR. The main idea is to allow for transformations of the INR input to achieve a consistent implicit representation despite conflicting observations. We transfer this approach to the domain of medical images by replacing the homographic transformations of individual image patches with a deformable B-spline transformation model on a regular grid [25]. In contrast to natural photographs, medical images usually share a similar field of view due to their acquisition modalities, which holds true especially for the atlas generation. While patch-based training would have the advantage of incorporating individual image patches into the atlas, this scenario would be only necessary for image compounding and is not considered here. Nevertheless, a connection can be found between the two options with polyaffine transformations [35].

We aim to store an atlas or template of a medical image dataset in a single INR, which takes coordinates as an input and outputs grey values. One option to train the network would be to input regular grid coordinates and minimize the loss between each sample and the derived image intensities, leading to an undesirable blurry mean image at the expense of instance fidelity. We suggest to simultaneously optimise network weights and instance-specific grid displacements during training. While the INR captures the underlying intensities of the

[1] https://github.com/multimodallearning/SharpImplicitNeuralAtlas.

whole dataset, the displacements encode single dataset images. After training, data samples can be retrieved by querying the network with their respective optimised grids, while a sharp general representation of all data can be generated with a regular grid of arbitrary size. An overview of the proposed method is shown in Fig. 1.

Given a dataset $\mathcal{Y} = \{y_i\}_{i=1}^{N}$ of N discrete images $y_i : \Omega \subset \mathbb{R}^d \rightarrow \mathbb{R}$ with $d \in \{2, 3\}$ we aim to construct a continuous global atlas (or template) $\mathcal{A} : \mathbb{R}^d \rightarrow \mathbb{R}$. The atlas can be formulated as an INR

$$\mathcal{A}(\boldsymbol{x}) = f(\boldsymbol{x}; \varphi), \qquad \boldsymbol{x} \in \mathbb{R}^d, \tag{1}$$

where f is a multi-layer-perceptron, specifically a SIREN network, with trainable parameters φ that gets coordinates $\boldsymbol{x}$ as input. In addition to the atlas $\mathcal{A}$, we aim to find a deformation $\boldsymbol{u}_i : \mathbb{R}^d \rightarrow \mathbb{R}^d$ (defined by parameters ψ_i) for each image y_i such that the deformed atlas

$$\mathcal{A}(\boldsymbol{x} + \boldsymbol{u}_i(\boldsymbol{x}; \psi_i)), \qquad \boldsymbol{x} \in \Omega, \tag{2}$$

is similar to $y_i(\mathbf{x})$. Since $\mathcal{A}$ is formulated as an INR, we deform the atlas by deforming the input coordinates for the network f instead of deforming the output image itself. This removes the necessity to interpolate the deformed atlas image since we can work on the whole continuous definition area $\mathbb{R}^d$ instead of the smaller definition area Ω of the discretised images. The displacement field $\boldsymbol{u}_i$ is obtained via a B-spline transformation of learnable control point displacements ψ_i to ensure smoothness of the field. A regularisation loss can thus be omitted. In summary, the optimisation during training can be expressed as

$$\arg \min_{\varphi, \{\psi_i\}} \sum_{i=1}^{N} \mathcal{L}_{\text{sim}}(f(\boldsymbol{x} + \boldsymbol{u}_i(\boldsymbol{x}; \psi_i); \varphi), y_i(\boldsymbol{x})), \tag{3}$$

where $\mathcal{L}_{\text{sim}}$ denotes a similarity loss, for which we employ the structural similarity index measure (SSIM). By propagating the loss to both φ and $\psi_1, \ldots, \psi_N$, registration and representation are optimized simultaneously. To mitigate the problem of difficult (or impossible) registrations of images to an unfitted INR and thus to combat the early deterioration of displacements, the learning rate of the displacements optimiser is multiplied by a factor that increases linearly from 0 to 1 up to epoch e_{warm}.

2.2 Evaluation

We assess the fidelity of the atlas following [9] in three different domains: In the atlas space, the similarity between the warped atlas and the input image is determined, and in the image space, the warped image is compared with the synthesized atlas. Here, we apply the structural similarity index measure (SSIM) and the peak signal-to-noise ratio (PSNR) to calculate similarity after registration. To limit the influence of edge-related registration artefacts, we mask out a border frame in the size of 20 pixels on both datasets.

We also examine the results of a combined backward and forward transformation (atlas as a bridge) for a fair comparison between forward and backward-based methods. Here, target registration errors (TRE) and Dice scores (DSC) are computed on keypoints and labels, which have been consecutively propagated from source to atlas and from atlas to target images. Since we only obtain non-diffeomorphic displacement fields for one transformation direction per method, we follow Chen et al. [3] to approximate its inverse with an iterative fixed-point approach based on inverse consistency optimisation. We calculate the TRE for the JSRT dataset between all (over 60000) possible combinations of source and target images, while Dice scores are evaluated on $N - 1$ registration pairs.

We further evaluate the atlas synthesis methods by assessing the smoothness of deformation fields through the standard deviation of the logarithmic Jacobian determinant (logJacDet) and the sharpness of each template through the mean average gradient magnitude. Additionally, we visualize the template fidelity by assigning the most probable label to each pixel or voxel in an atlas through voting based on its label frequency in warped segmentations. Leveraging the properties of sharp atlases, the registration quality is indicated by the visualised alignment of structures and segmentations.

Analogous to 2D experiments, we investigate the extension of SINA to 3D by assessing its atlas-as-a-bridge performance through Dice scores between initial alignment and propagated labels. For qualitative evaluation, we show a plot of the different views (axial, coronal, sagittal) and a 3D render of majority-voted labels.

2.3 Implementation Details

Our methods and experiments are implemented in *PyTorch* 2.2.0 and conducted on a single GeForce RTX 2080. We select to implement the INR as a SIREN [32] with 7 layers of 64 channels each and initialise model parameters with $\omega = 20$. Deformations are based on a grid size of 28×28 ($S = 784$), corresponding to a total of 1568 trainable displacement parameters per image. We implement the B-spline transformation as two iterations of average pooling with kernel size 5 and padding length of 2, followed by bilinear interpolation. Both displacement and INR optimizers are updated in every minibatch using Adam with the respective learning rates of 10^{-5} and 10^{-3}, based on perceptual SSIM loss (window size of 11). The results shown are achieved with batch sizes of (2,2), epochs of (250,2000), and e_{warm} = (300,200) for JSRT and OASIS respectively, resulting in training times of 16 and 50 minutes. We find that small batch sizes drastically improve early convergence to a mean shape and thus can accelerate training. All image data was normalised to $[0, 1]$ prior to training.

The 3D version of our model employs a SIREN with 10 layers with 256 channels and initialises its weights with $\omega = 35$. We train the network with a batch size of 1 for 2000 epochs (approx. 4 h) using an NVIDIA A100 GPU using the mean squared error as a loss function. Displacements are mapped onto a grid scaled down to a quarter of the image's resolution ($S = 122880$).

Table 1. JSRT Results. We report SSIM, PSNR, and standard deviation of logarithmized Jacobian determinant for Atlas (comparison of warped atlas and image) and Image space (comparison of warped image with the atlas). Target Registration Error (TRE in mm) and Dice scores (DSC) are computed with propagated annotations through source-to-atlas and atlas-to-target transformations. Atlas synthesis methods are divided into forward (f) or backward (b) models. The row "zerofield" describes results without any deformation and utilises the mean shape between data samples as an atlas.

Method	f/b	Atlas space			Image space			Atlas as a bridge	
		SSIM ↑	PSNR ↑	logJacDet ↓	SSIM ↑	PSNR ↑	logJacDet ↓	TRE ↓	DSC ↑
zerofield	–	0.70 ± 0.07	24.46 ± 2.32	–	0.70 ± 0.07	24.46 ± 2.32	–	35.63 ± 14.52	45.10 ± 14.07
SimpleElastix	f	0.62 ± 0.04	26.09 ± 2.47	0.53 ± 0.44	0.67 ± 0.05	26.73 ± 2.53	0.62 ± 0.41	33.50 ± 12.40	49.14 ± 12.43
VoxelMorph	f	0.71 ± 0.07	25.00 ± 2.25	**0.02 ± 0.01**	0.81 ± 0.03	25.37 ± 2.12	**0.02 ± 0.01**	34.67 ± 14.32	46.38 ± 13.94
SINA (ours)	b	**0.85 ± 0.05**	**30.05 ± 2.40**	0.49 ± 0.48	**0.89 ± 0.02**	**28.43 ± 2.87**	0.11 ± 0.08	**21.03 ± 8.61**	**64.60 ± 11.72**

3 Experiments

3.1 Datasets

The **JSRT** chest X-ray dataset consists of a total of 247 X-ray images with (154) and without (93) pulmonary nodules [29]. We apply z-normalization and downsample the images to 256 × 256 pixels during preprocessing. Since the data is already manually cropped to compensate for translation misalignment, a random section of size 224 × 224 was taken from each image to restore the original difficulty of the registration task. In addition, some images have black rectangles at the edges for anonymity purposes, which introduces further challenges for atlas synthesis. For each image, a total of 166 manually created expert landmarks delineating 5 anatomical structures (left (44) and right lung (50), left (23) and right clavicle (23), heart (26)) are provided [10,11]. All cropped landmarks are excluded from evaluation. We further leverage landmarks to generate label masks for all 5 structures for Dice score calculation. Recognising the difficulty of lung registration, we also investigate the effectiveness of our method on the simpler 2D **OASIS** dataset [16,22], which contains 414 skull-stripped and pre-aligned MRI images with corresponding segmentations for a total of 24 anatomical structures. We evaluate the entire dataset and refrain from further preprocessing. For 3D, we employ the publicly available Learn2Reg **AbdomenCTCT**[2] dataset [15] comprised of 30 pre-aligned abdominal CT images of shape 192 × 160 × 256 and a total of 13 anatomical labels.

3.2 Baselines

In the interests of reproducibility, we have decided to rely on established and publicly available registration methods with existing instructions. First, we compare SINA against the popular registration framework **SimpleElastix** and follow their tutorial for groupwise B-spline registration[3]. As the initial results were

[2] https://learn2reg.grand-challenge.org/Datasets/.
[3] https://simpleelastix.readthedocs.io/GroupwiseRegistration.html.

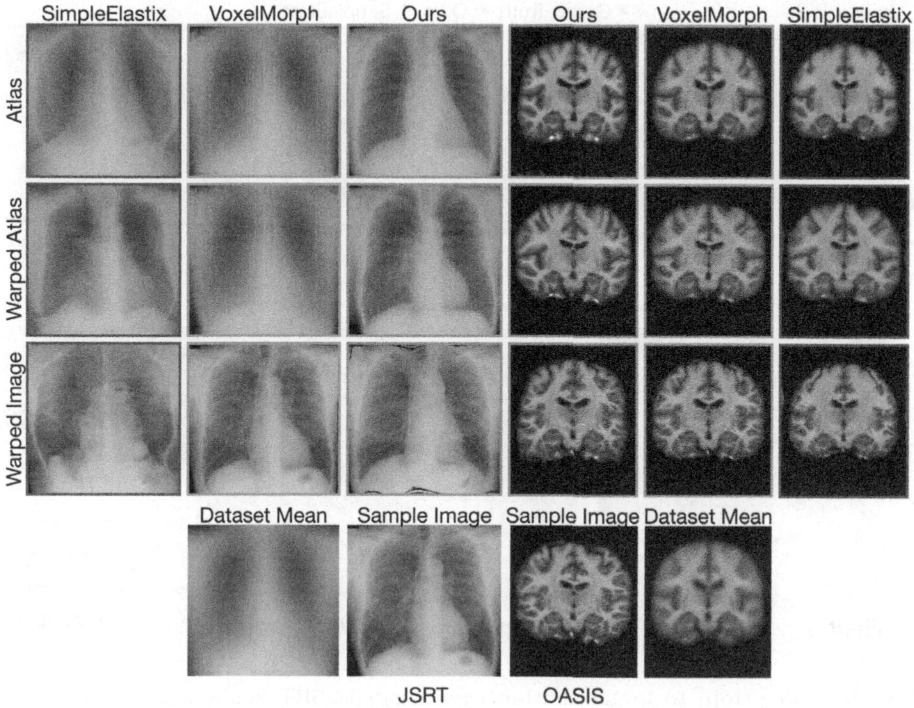

Fig. 2. Examples of forward and backward atlas registrations for JSRT (left) and OASIS (right) datasets. Learned deformations are applied to the sample image (mid-bottom) to the atlas (top row) and vice versa. Transformation results for warped atlases and warped images are shown in rows two and three, respectively. Mean dataset images are depicted at the bottom outside edges of the figure. Atlases and results are shown for • SINA (ours), • VoxelMorph and • SimpleElastix. (Color figure online)

not satisfactory, we performed a local histogram equalisation using images in the JSRT dataset before registration. Further adjustments did not lead to any improvement. In contrast to the method presented above, SimpleITK constructs a forward atlas in which the registrations of samples are averaged onto the atlas. Analogous to our method, we approximate the inverse displacement field for evaluation purposes. We further employ the unconditioned atlas generation variant of **VoxelMorph** [7] based on the provided script[4].

4 Results and Discussion

Table 1 shows the quantitative results for atlas fidelity on the JSRT dataset. Our proposed method outperforms all baselines (except for the smoothness of the deformation metric) by a large margin. While the SSIM and PSNR metrics

[4] https://github.com/voxelmorph/voxelmorph?#tutorial.

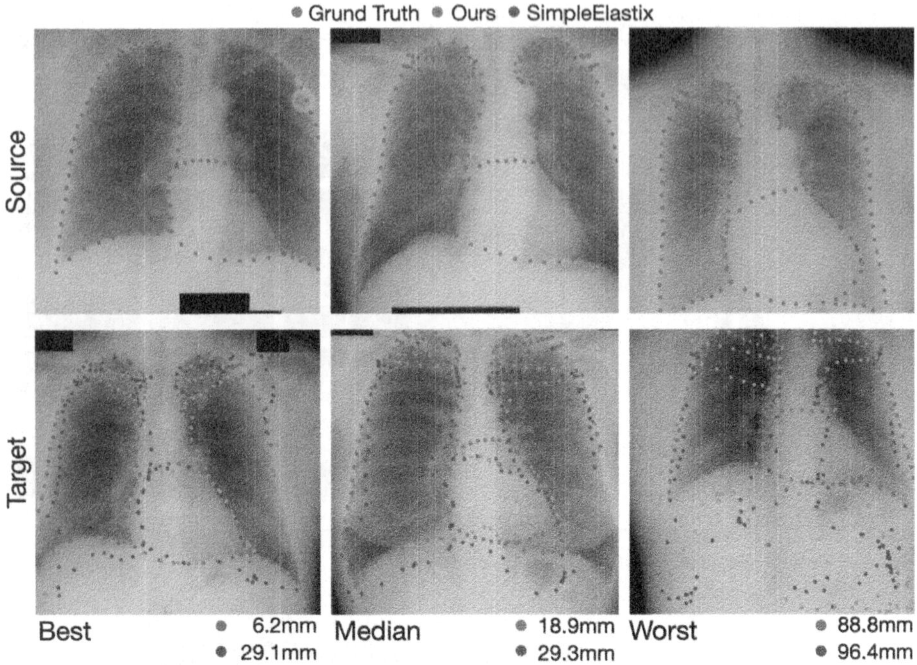

Fig. 3. Source (top) to target (bottom) propagated JSRT landmarks. Point colours represent • ground truth landmarks and landmarks warped with • SINA (ours) and • SimpleITK as an atlas-as-a-brigde evaluation. The resulting TREs are denoted at the bottom. (Color figure online)

already indicate some differences, the bridge metrics reveal a significant disparity between SINA and the other methods, with a difference in TRE of over 12 mm and approximately 15% points in Dice score. Both SimpleElastix and VoxelMorph were not able to provide the combination of a high-fidelity atlas and accurate transformations required for this transfer task. Dice and TRE only improved minimally compared to the initial scores. This impression is confirmed by qualitative results in Fig. 2. SINA was the only method able to produce reproducible results for both the warped image and the warped atlas. While the application of SimpleITK leads to unrealistic warped images, the deformations produced by VoxelMorph are extremely small, devaluing its low smoothness metrics. The visualisations of the contour point transformations in Fig. 3 demonstrate the superiority of SINA. In the best case, the SINA and ground truth align nearly perfectly, while in the median case, the transformation appears to capture lung movement but struggles with regard to clavicles and heart. The alignment for SimpleElastix is significantly worse. With very different source and target images and, accordingly, complex displacements, the registration also fails with SINA.

Table 2. OASIS Results. We report SSIM, PSNR, and standard deviation of logarithmized Jacobian determinant for Atlas (comparison of warped atlas and image) and Image space (comparison of warped image with the atlas). Dice scores (DSC) are computed with propagated annotations through source-to-atlas and atlas-to-target transformations. Atlas synthesis methods are divided into forward (f) or backward (b) models. The row "zerofield" describes results without any deformation and utilises the mean shape between data samples as an atlas.

Method	f/b	Atlas space			Image space			Atlas as a bridge
		SSIM ↑	PSNR ↑	logJacDet ↓	SSIM ↑	PSNR ↑	logJacDet ↓	Dice ↑
zerofield	–	0.51 ± 0.04	22.88 ± 1.03	–	0.51 ± 0.04	22.88 ± 1.03	–	53.09 ± 8.57
SimpleElastix	f	0.69 ± 0.04	26.26 ± 1.02	0.12 ± 0.07	**0.79 ± 0.04**	**29.80 ± 0.75**	0.09 ± 0.05	65.68 ± 4.56
VoxelMorph	f	0.66 ± 0.05	**33.67 ± 0.90**	**0.05 ± 0.00**	0.61 ± 0.06	27.29 ± 0.92	**0.06 ± 0.01**	**69.54 ± 5.13**
ours	b	**0.70 ± 0.04**	32.77 ± 0.98	0.14 ± 0.15	0.58 ± 0.03	25.90 ± 0.97	0.13 ± 0.10	67.38 ± 4.49

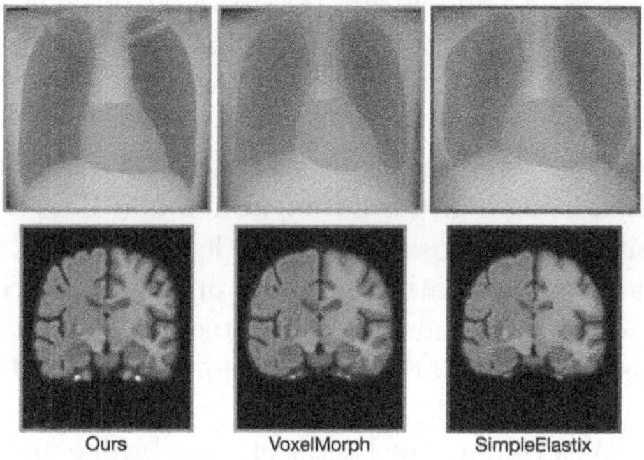

Fig. 4. Atlases and corresponding majority vote segmentations for • SINA (ours), • VoxelMorph and • SimpleElastix on JSRT (top) and OASIS (bottom). (Color figure online)

Table 3. Sharpness of synthesised templates for all methods regarding averaged gradient magnitude amplified by a factor of 100.

	JSRT	OASIS
mean	2.43	7.67
SimpleElastix	2.68	9.56
VoxelMorph	4.27	10.35
SINA (ours)	4.12	13.04

The results in the OASIS dataset differ to some degree: Here, VoxelMorph performs best in the Dice bridge metric with 69 %, while SINA follows with a gap of around 2% points, ranking second in most metrics and best in atlas-space SSIM. SimpleElastix also improved by over 12% points compared to the initial Dice. Metrics in the atlas and image space show no clear trend. The visual results from Fig. 2 underline the quantitative results. The VoxelMorph atlas is best able to map both transformations. Reported Jacobian determinants highlight that SINA tends to learn slightly stronger and more complex transformations than the other methods.

These findings align closely with the majority vote atlas segmentations shown in Fig. 4. Although the differences for OASIS are minimal, SINA stands out for the JSRT data as the only method that produces realistic segmentation maps for all five structures, including the relatively small clavicles. This is particularly impressive, considering no label information was utilized during the atlas synthesis.

As expected, SINA generates visually sharp templates in all experiments. The level of detail is particularly impressive with the JSRT atlas, as bone structures such as ribs or collarbones, as well as the heart, are well delineated. Quantitative results in Table 3 support these observations. The high gradient amplitude in VoxelMorph does not match the visual results and is probably due to the artifact-like vertical and horizontal lines. The results emphasise the broad applicability of the method presented. While results on the OASIS dataset are comparable with other methods, SINA is the only one that successfully creates a meaningful and representative atlas for chest X-rays. Compared to the other datasets, JSRT has a higher variance in terms of deformation and geometry, which limits or excludes the use of both SimpleElastix and VoxelMorph. Training times for optimal results differ between the 2D datasets at 16 (JSRT) and 50 min (OASIS), with the latter achieving 95% of the final training losses and 98% Dice scores (-1.62%) after just a quarter of the iterations. An elaborated optimisation strategy could further accelerate convergence.

Our approach can be seamlessly transferred to the generation of 3D templates and produces a sharp atlas even when using high-resolution abdominal CTs. The visual results in Fig. 5 show plausible structures with a high level of detail. In addition to bones, whose registration is comparatively easy to optimise, internal organs such as the liver, kidneys, and spleen can also be delineated. The depiction of fine structures, such as lung fissures, is remarkable. The good alignment of the majority vote segmentations with the depicted structures is an indication of accurately learned transformations. We evaluate the Dice score as a bridge metric and achieve an increase of over 17% points compared to the initial alignment regarding large structures (liver, spleen, and kidneys).

Although the results of our method produce state-of-the-art results for large and complex deformations, OASIS still trails slightly behind VoxelMorph. The gap might be closed by employing a more complex transformation model and explicit regularisation with regard to accuracy and smoothness. A diffeomorphic adaptation could further improve its image-space fidelity. Other atlas synthesis models are able to construct sub-population templates through model conditioning. This concept could be applied by optimisation of additional appearance latent codes during training.

The working principle of the proposed method can best be illustrated by observing the transformation of image-specific features, such as redacted patient info replaced by a black box or, in AbdomenCTCT, scanner-specific properties outside of the body. While conventional methods would probably lose this information by reduction to a common mean, SINA possesses the freedom to recover or hide these elements through deformations, if necessary, beyond image dimensions. We argue that SINA is thus more expressive than common dataset templates, effectively encoding image features through geometric transformations. Our approach could be reformulated as a generalised conditioned INR where the conditions are restricted to spatial deformations, consequently com-

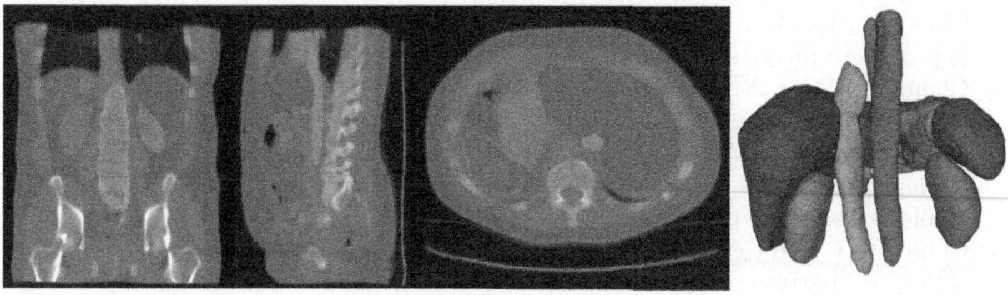

Fig. 5. Generated 3D atlas for the AbdomenCTCT dataset by our method SINA.

bining registration and representation. We, therefore, challenge the requirements one usually places on medical image atlases: Instead of seeking to minimise geometric distances, we strive for fidelity to individual dataset instances. Following this paradigm, the synthesis of sharp, realistic images would be an aesthetic by-product.

5 Conclusion

In this work, we presented SINA, a novel technique to synthesize sharp and high-fidelity atlases by joint optimisation of representation and registration. The method does not rely on affinely pre-aligned data and is capable of handling complex deformations to produce realistic and sharp templates and corresponding transformations. Our method is not restricted to two dimensions or a specific choice of INR and can be employed in any image domain. We evaluated SINA with datasets of varying complexity and compared the results against popular atlas synthesis frameworks. Our proposed method outperformed the other approaches for complex chest X-ray data and produces comparable results to the best baseline method for the simpler MRI brain dataset while synthesising the sharpest templates. Finally, we demonstrated SINA's applicability to construct a 3D high-fidelity abdominal CT template with strong delineation of structures.

Disclosure of Interests. This work has been funded by the German Federal Ministry of Education and Research (BMBF, FKZ 13GW0578C).

References

1. Bône, A., Vernhet, P., Colliot, O., Durrleman, S.: Learning joint shape and appearance representations with metamorphic auto-encoders. In: Medical Image Computing and Computer Assisted Intervention. Springer (2020)
2. Byra, M., Poon, C., Rachmadi, M.F., Schlachter, M., Skibbe, H.: Exploring the performance of implicit neural representations for brain image registration. Scientific Reports **13** (2023)

3. Chen, M., Lu, W., Chen, Q., Ruchala, K.J., Olivera, G.H.: A simple fixed-point approach to invert a deformation field. Medical physics **35**, 81–88 (2008)
4. Chen, Y., Wang, X.: Transformers as meta-learners for implicit neural representations. In: European Conference on Computer Vision. Springer (2022)
5. Chi, Z., Cong, Z., Wang, C.J., Liu, Y., Turk, E.A., Grant, P.E., Abulnaga, S.M., Golland, P., Dey, N.: Dynamic Neural Fields for Learning Atlases of 4D Fetal MRI Time-series. arXiv preprint arXiv:2311.02874 (2023)
6. Cootes, T.F., Edwards, G.J., Taylor, C.J.: Active appearance models. IEEE Transactions on pattern analysis and machine intelligence **23**, 681–685 (2001)
7. Dalca, A., Rakic, M., Guttag, J., Sabuncu, M.: Learning conditional deformable templates with convolutional networks. Advances in neural information processing systems **32** (2019)
8. Dannecker, M., Kyriakopoulou, V., Cordero-Grande, L., Price, A.N., Hajnal, J.V., Rueckert, D.: CINA: Conditional Implicit Neural Atlas for Spatio-Temporal Representation of Fetal Brains. arXiv preprint arXiv:2403.08550 (2024)
9. Ding, Z., Niethammer, M.: ALADDIN: Joint atlas building and diffeomorphic registration learning with pairwise alignment. In: Proceedings of the IEEE/CVF conference on computer vision and pattern recognition (2022)
10. Gaggion, N., Mansilla, L., Milone, D.H., Ferrante, E.: Hybrid Graph Convolutional Neural Networks for Landmark-Based Anatomical Segmentation. In: Medical Image Computing and Computer Assisted Intervention. pp. 600–610. Springer (2021)
11. Gaggion, N., Mansilla, L., Mosquera, C., Milone, D.H., Ferrante, E.: Improving anatomical plausibility in medical image segmentation via hybrid graph neural networks: applications to chest x-ray analysis. IEEE Transactions on Medical Imaging (2022)
12. Guimond, A., Meunier, J., Thirion, J.P.: Average brain models: A convergence study. Computer vision and image understanding **77**(2), 192–210 (2000)
13. van Harten, L., Van Herten, R.L.M., Stoker, J., Isgum, I.: Deformable image registration with geometry-informed implicit neural representations. In: Medical Imaging with Deep Learning. PMLR (2024)
14. He, Z., Chung, A.C.S.: Unsupervised end-to-end groupwise registration framework without generating templates. IEEE International Conference on Image Processing (ICIP) pp. 375–379 (2020)
15. Hering, A., Hansen, L., Mok, T.C., Chung, A.C., Siebert, H., Häger, S., Lange, A., Kuckertz, S., Heldmann, S., Shao, W., et al.: Learn2Reg: comprehensive multitask medical image registration challenge, dataset and evaluation in the era of deep learning. IEEE Transactions on Medical Imaging **42**, 697–712 (2022)
16. Hoopes, A., Hoffmann, M., Greve, D.N., Fischl, B., Guttag, J., Dalca, A.V.: Learning the effect of registration hyperparameters with hypermorph. The journal of machine learning for biomedical imaging **1** (2022)
17. Joshi, S.C., Davis, B.C., Jomier, M., Gerig, G.: Unbiased diffeomorphic atlas construction for computational anatomy. NeuroImage **23**, S151–S160 (2004)
18. Lin, C.H., Ma, W.C., Torralba, A., Lucey, S.: BARF: Bundle-adjusting neural radiance fields. In: Proceedings of the IEEE/CVF International Conference on Computer Vision. pp. 5741–5751 (2021)
19. Liu, C., Wang, Q., Si, W., Ni, X.: NuTracker: a coordinate-based neural network representation of lung motion for intrafraction tumor tracking with various surrogates in radiotherapy. Physics in Medicine & Biology **68**(1), 015006 (2022)

20. Lorenzen, P., Davis, B.C., Joshi, S.: Unbiased atlas formation via large deformations metric mapping. In: Medical Image Computing and Computer-Assisted Intervention. Springer (2005)
21. Low, W.F., Lee, G.H.: Minimal Neural Atlas: Parameterizing Complex Surfaces with Minimal Charts and Distortion. In: European Conference on Computer Vision (ECCV) (2022)
22. Marcus, D.S., Wang, T.H., Parker, J., Csernansky, J.G., Morris, J.C., Buckner, R.L.: Open access series of imaging studies (OASIS): cross-sectional mri data in young, middle aged, nondemented, and demented older adults. Journal of cognitive neuroscience **19**(9), 1498–1507 (2007)
23. Meng, M., Bi, L., Fulham, M., Feng, D.D., Kim, J.: Enhancing medical image registration via appearance adjustment networks. NeuroImage **259**, 119444 (2022)
24. Mildenhall, B., Srinivasan, P.P., Tancik, M., Barron, J.T., Ramamoorthi, R., Ng, R.: NeRF: Representing scenes as neural radiance fields for view synthesis. In: ECCV (2020)
25. Modat, M., Ridgway, G.R., Taylor, Z.A., Lehmann, M., Barnes, J., Hawkes, D.J., Fox, N.C., Ourselin, S.: Fast free-form deformation using graphics processing units. Computer methods and programs in biomedicine **98**, 278–284 (2010)
26. Müller, T., Evans, A., Schied, C., Keller, A.: Instant neural graphics primitives with a multiresolution hash encoding. ACM transactions on graphics (TOG) **41**, 1–15 (2022)
27. Niethammer, M., Hart, G.L., Pace, D.F., Vespa, P.M., Irimia, A., Van Horn, J.D., Aylward, S.R.: Geometric metamorphosis. In: Medical Image Computing and Computer-Assisted Intervention. Springer (2011)
28. Park, J.J., Florence, P., Straub, J., Newcombe, R., Lovegrove, S.: Deepsdf: Learning continuous signed distance functions for shape representation. In: Proceedings of the IEEE/CVF conference on computer vision and pattern recognition (2019)
29. Shiraishi, J., Katsuragawa, S., Ikezoe, J., Matsumoto, T., Kobayashi, T., Komatsu, K.i., Matsui, M., Fujita, H., Kodera, Y., Doi, K.: Development of a digital image database for chest radiographs with and without a lung nodule: receiver operating characteristic analysis of radiologists' detection of pulmonary nodules. American journal of roentgenology **174**(1), 71–74 (2000)
30. Sideri-Lampretsa, V., McGinnis, J., Qiu, H., Paschali, M., Simson, W., Rueckert, D.: Sinr: Spline-enhanced implicit neural representation for multi-modal registration. In: Medical Imaging with Deep Learning (2024)
31. Sinclair, M., Schuh, A., Hahn, K., Petersen, K., Bai, Y., Batten, J., Schaap, M., Glocker, B.: Atlas-ISTN: joint segmentation, registration and atlas construction with image-and-spatial transformer networks. Medical Image Analysis **78**, 102383 (2022)
32. Sitzmann, V., Martel, J.N., Bergman, A.W., Lindell, D.B., Wetzstein, G.: Implicit Neural Representations with Periodic Activation Functions. In: Proc. NeurIPS (2020)
33. Stolt-Ansó, N., McGinnis, J., Pan, J., Hammernik, K., Rueckert, D.: Nisf: Neural implicit segmentation functions. In: International Conference on Medical Image Computing and Computer-Assisted Intervention. Springer (2023)
34. Tancik, M., Srinivasan, P., Mildenhall, B., Fridovich-Keil, S., Raghavan, N., Singhal, U., Ramamoorthi, R., Barron, J., Ng, R.: Fourier features let networks learn high frequency functions in low dimensional domains. Advances in neural information processing systems **33**, 7537–7547 (2020)

35. Taquet, M., Macq, B., Warfield, S.K.: Spatially adaptive log-euclidean polyaffine registration based on sparse matches. In: Medical Image Computing and Computer-Assisted Intervention. Springer (2011)
36. Wolterink, J.M., Zwienenberg, J.C., Brune, C.: Implicit Neural Representations for Deformable Image Registration. In: Proceedings of The 5th International Conference on Medical Imaging with Deep Learning. PMLR (2022)
37. Wu, G., Jia, H., Wang, Q., Shen, D.: SharpMean: groupwise registration guided by sharp mean image and tree-based registration. NeuroImage **56**, 1968–1981 (2011)
38. Yeo, B.T.T., Sabuncu, M.R., Desikan, R., Fischl, B., Golland, P.: Effects of Registration Regularization and Atlas Sharpness on Segmentation Accuracy. Medical image computing and computer-assisted intervention (2007)
39. Yu, E.M., Dalca, A.V., Sabuncu, M.R.: Learning conditional deformable shape templates for brain anatomy. In: Machine Learning in Medical Imaging: 11th International Workshop. Springer (2020)
40. Zimmer, V.A., Hammernik, K., Sideri-Lampretsa, V., Huang, W., Reithmeir, A., Rueckert, D., Schnabel, J.A.: Towards generalised neural implicit representations for image registration. In: International Conference on Medical Image Computing and Computer-Assisted Intervention. Springer (2023)

A Novel Fusion of CT/MRI and US Images Based on Depth Camera and Electromagnetic Tracking

Martina Paccini[1(✉)], Giacomo Paschina[2], Stefano De Beni[2], Andrei Stefanov[3], Velizar Kolev[3], and Giuseppe Patanè[1]

[1] CNR-IMATI 'E. Magenes', via de Marini, 6, Genova, Italy
martina.paccini@ge.imati.cnr.it
[2] Esaote S.p.a., via E. Melen, 77, Genova, Italy
[3] MedCom GmbH, Dolivostr., 11 mù, 64293 Darmstadt, Germany

Abstract. This paper presents an automatic fusion imaging system that combines 3D CT (or MR) images with real-time ultrasound (US) acquisition using depth cameras and electromagnetic (EM) trackers. The system eliminates the need for physical markers and complex training, making image fusion accessible for physicians with varying experience levels. The fusion algorithm consists of two main parts: skin segmentation and rigid co-registration, both integrated into the US machine. The co-registration process aligns the surface extracted from CT images with the 3D surface acquired by a depth camera, facilitating rapid and effective fusion. Experimental tests in different settings confirm the system's accuracy, computational efficiency, noise robustness, and operator independence. The co-registration error remains within acceptable ranges.

Keywords: Image registration · Medical image fusion · Computed tomography · Depth camera · Magnetic resonance imaging · Image processing · Skin segmentation · Ultrasound

1 Introduction

Fusion imaging plays a crucial role in various clinical procedures, such as diagnostic and minimally invasive surgical interventions [1,4]. It increases the patient's comfort and produces more reliable results in tissue sampling [3,5]. However, achieving a balance between data registration accuracy and system complexity remains a challenge, as resulting systems are often too complex to be practical for clinical use. The process of image fusion involves tracking the position, orientation, and displacements of the ultrasound (US) probe during image acquisition within a common reference system for the US images and other considered modalities [10]. This requires the use of probe trackers of different natures, such as optical technology, which offers high precision but has line-of-sight limitations,

Supplementary Information The online version contains supplementary material available at https://doi.org/10.1007/978-3-031-73480-9_14.

M. Modat et al. (Eds.): WBIR 2024, LNCS 15249, pp. 181–191, 2024.
https://doi.org/10.1007/978-3-031-73480-9_14

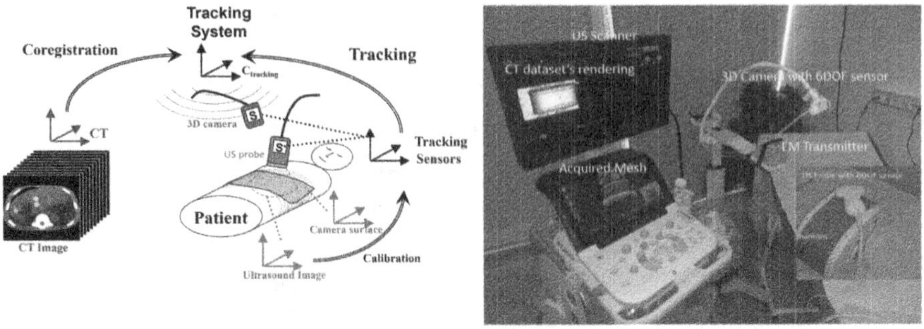

(a) (b)

Fig. 1. (a) System components and their mutual interaction for the image fusion. The tracking system comprehends: EM transmitter and 6DOF tracking sensors (one on the 3D camera and one on the US probe) (b) US system integrated and introduced in a set-up that mimics clinical environment.

and electromagnetic (EM) technology, which despite the problems linked to distortion and range, is often applied in clinical practice. However, EM tracking requires the placement of markers on the patient with a wand (or needle guide wire or catheter) and simultaneously selecting those markers on the preprocedural image (CT/MRI) or leveraging fiducial patches on the procedural (CT/MRI) image. These procedures are time-consuming and error-prone [12].

In [13], physical external markers are avoided by utilizing 3D US images to perform rigid registration through similarity metrics evaluation computed in volumetric patches. However, this approach needs further improvement for clinical application, as physicians typically use 2D US images in clinical practice. Another method based on similarity evaluation that considers 2D US and 3D MRI images [16] requires an initial coarse registration performed by an expert, enhancing the operator-dependency of the exam. A method based on the detection and matching of local features across modalities has been developed for the 2D/3D registration of US images with MRI [8]. The method discovers, extracts and matches key points from US and MR images by adapting a network method to the multi-modal and cross-dimensional setting while handling imprecise ground truth. However, in dealing with networks the method applies only to T1-weighted MRIs and would require training with an appropriate dataset to apply to other imaging modalities.

To address these challenges, a new image fusion system combining 3D CT/MRI images with real-time US acquisition has been introduced. This system does not rely on trained networks and works on both CT and MR images. The hardware components include a 3D depth camera, US system, and simplified EM tracking system, while the software components encompass tracking, surface co-registration, and visualisation. This innovative system can be applied in various clinical scenarios, providing aligned or superimposed visualisation of US and CT/MRI images to offer more information to the physician, especially when the volumetric image provides information not visible in the US image. Due to

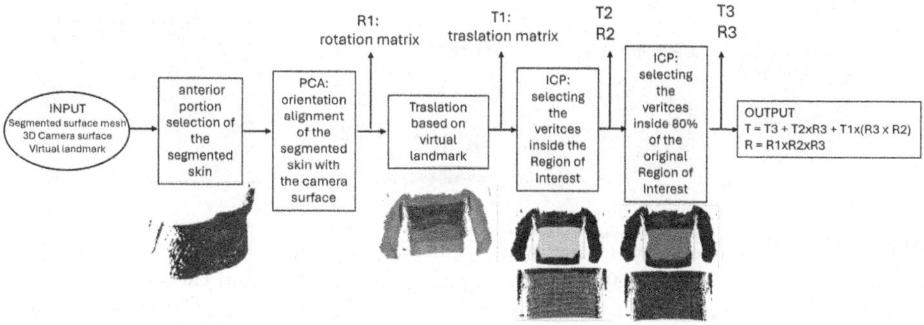

Fig. 2. Co-registration pipeline. From left to right the red surface represents the selected anterior portion of the segmented surface. PCA alignment and translation on the reference points (in green the surface acquired by the camera). Surface sub-regions for the first ICP run: the green vertices represent the first Region of Interest (ROI). Surface sub-regions tuning for the second ICP step, in red the 80% of the ROI of the first ICP run is considered in the second refinement run. (Color figure online)

its independence from the user's experience level and its capability to display image information in various formats (e.g., CT/MR parallel to US, CT/MR superimposed on US, and different co-registration error visualisations), the system is well-suited for both patient evaluations and educational activities for novice radiologists. For patient evaluations, the system's online fine-tuning feature enables image fusion accuracy that is appropriate for non-invasive interventions. In educational settings, the inherent accuracy of the system is adequate for instructional purposes.

The image fusion system has been tested for co-registration accuracy, computational cost for real-time applications, noise robustness, and independence from the operator. The main novelty of this image fusion system is the integration of a 3D camera in a state-of-the-art setup based on EM tracking eliminating the need for complicated calibrations. This integration has no impact on the EM system functioning principles but, conversely, it simplifies the acquisition procedure. The 3D camera allows for efficient fusion imaging, independently from the operator's experience. Furthermore, this paper presents a novel co-registration pipeline tailored for the system setup employing rigid registration techniques and eliminating external physical markers. Additionally, the developed skin segmentation algorithm is highly versatile and applicable to various anatomical regions and image types without extensive training datasets. Differently from state-of-the-art skin segmentation methods the proposed algorithm can be used directly on multiple types of volumetric images [2,9], it is not district-specific [6] and it is not AI-based [14,15].

2 MR and US Fusion System

Fusion Process. During the US examination, the volumetric image of the patient (CT or MRI) is loaded. The system segments the skin surface and generates a

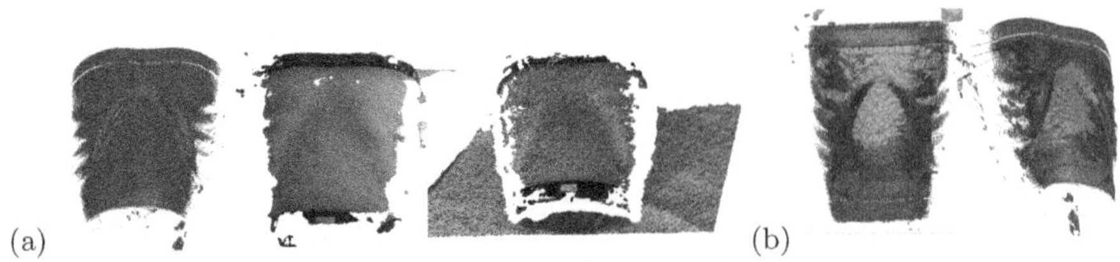

Fig. 3. (a) Difference between segmented surface extracted from the morphological phantom (Kyoto Kagaku dual modality human abdomen) CT and the surface acquired from the camera. The red surface is extracted from the phantom's CT (b) Coregistration result: the higher distance between the two surfaces corresponds to morphologically different regions. The red surface is the one extracted from the phantom's CT while the green surface is the phantom acquired by the 3D camera. (Color figure online)

3D surface overlay on the patient's 3D rendering. The clinician then acquires the 3D skin with the 3D camera. When capturing the surface with the 3D camera, the patient is in the same pose as in the volumetric image (e.g. if during the CT/MRI the patient positioned the arms above the head, the same should be done during the acquisition with the 3D camera). If possible, the patient is also requested to inhale or exhale according to the respiratory phase of the volumetric image. Due to the speed of surface acquisition with the 3D camera, this should be fine for the patient. Once the camera acquires the surface, the patient resumes normal breathing. The clinician selects a virtual landmark on the segmented surface and the corresponding point on the camera-acquired surface through the US system interface (i.e., the touch screen already on the US machine). Thus, the landmark is exclusively virtual and does not require any external physical placement. Then, the *co-registration software* automatically starts and enables image fusion in a few seconds, together with the visualisation of the registration error. Upon successful registration, the system presents MR or CT images alongside the US image in various visualisation modes (simultaneously or overlaid). This allows the clinician to access the corresponding anatomical information and real-time US data during the examination.

Fusion System Overview. The tracking system comprises an electronic unit, a mid-range transmitter, and two sensors for tracking traces of the position and coordinates of the 3D camera and the US probe. The transmitter generates an electromagnetic field and can simultaneously track up to four sensors at 70 times per second (Fig. 1). One sensor is connected to the US probe, and the other is associated with the 3D camera, allowing the representation of the US image and the 3D surface acquired by the camera in a unique coordinate system (i.e. the tracking coordinates). Software-wise, two main components integrate into the image fusion system. The first aims to segment the external skin surface of patients acquired by MR/CT. The second seeks to co-register the segmented skin surface with the 3D surface obtained by the camera, allowing the final fusion.

Fig. 4. Co-registration and error distribution on a phantom acquired by a 3D camera at different distances: (a,e) 20 cm, (b,f) 25 cm, (c,g) 30 cm, (d,h) 35 cm. The error visualisation scale in the colour map goes from 0 mm (blue) to 10 mm (red). (Color figure online)

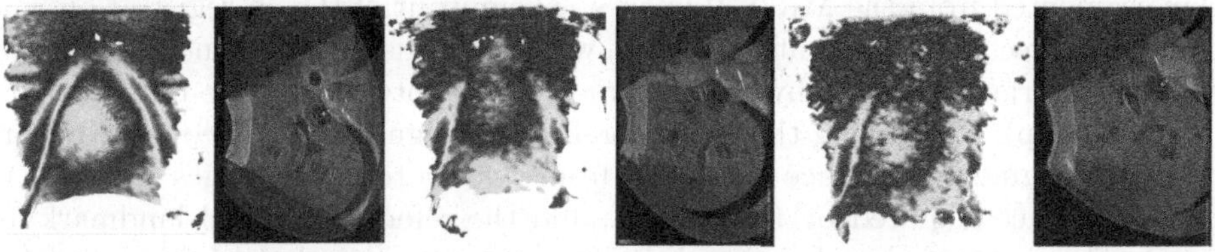

Fig. 5. Distance distribution and fusion error within the volume at different distance acquisition with the morphological phantom. From left to right: 30 cm distance with 6.2 mm error, 35 cm distance with 8 mm error, 40 cm distance with 4.7 mm error

Skin Segmentation. Extracting the body surface from volumetric imaging allows the seamless integration of heterogeneous data sources and facilitates subsequent analyses. Indeed, working with 3D surfaces allows us to have lighter data that represent the only region of the image that we leverage in the co-registration process (i.e., the skin). Given a CT/MRI image paired with the skin iso-value (retrievable according to the specification of the acquisition machine), the proposed segmentation identifies the skin surface. It is the first step for the co-registration. The segmentation proceeds one slice at a time and considers one of the corner pixels of the image as the first visited pixel to ensure that it starts from the background. Then, the growth of the background region proceeds iteratively based on the pixels' adjacency and stops when it encounters a pixel whose grey value is higher or equal to the iso-value of the skin. Through this region-growing algorithm, the evaluation expands only where the air is present; the body will be segmented as a whole object by exclusion. The segmented volume then undergoes the marching cube algorithm [7] to extract a 3D surface mesh of the segmented skin, representing the input for the co-registration phase to match the MR/CT and the 3D surface acquired by the camera.

Skin Co-registration. To align the MR/CT image with the US probe, the 3D surface acquired by the camera, which is in the same reference system as the US probe and of the magnetic tracking, is rigidly co-registered with the patient

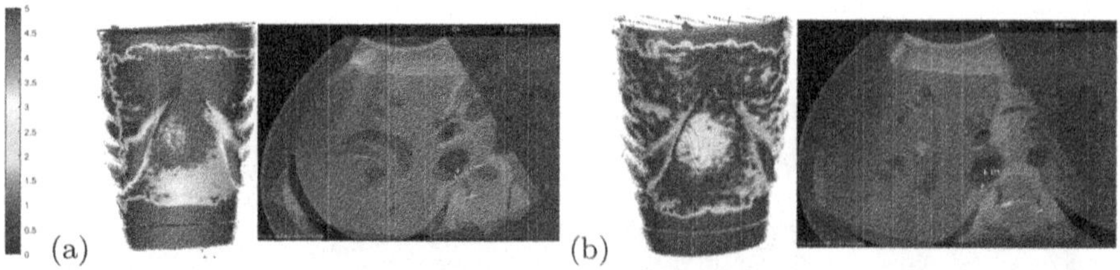

Fig. 6. Coregistration error distribution and fusion imaging result inside the volume at (a) perpendicular position and (b) 30° tilt angle on the morphological phantom. The fusion error in (a) 5 mm, (b) 9,6 mm. At 45°, the error is 17,6 mm. The colormap is in mm.

skin segmented from the MR/CT images. The output of the co-registration is a translation vector and a rotation matrix, which co-register the segmented surface to the 3D surface acquired by the camera (i.e., the Intel RealSense in the experimental setup), minimising the corresponding misalignment. The co-registration takes the segmented surface extracted from the anatomical images (MR/CT), the 3D surface acquired by the camera, and the reference virtual landmark as input. The segmented surface is oriented consistently (i.e., head-feet, right-left) to avoid errors related to body symmetries. The pipeline starts with a first orientation adjustment through *Principal Component Analysis* (PCA) computed on all the vertices of the surface extracted from the image and used to obtain the first rough alignment with the camera surface. Then, a translation is computed to superimpose the virtual landmark identified on the segmented surface with the one on the surface acquired by the camera. The final part consists of surface sub-region selection and tuning: a bounding box of fixed dimension (defined empirically) is identified on both surfaces and centred in the virtual landmark. All the vertices that belong to the bounding box (i.e. the Region of Interest) are considered for a finer alignment through *Iterative Closest Point algorithm* (ICP). Finally, a second ICP is run on the 80% of the first Region of Interest allowing for an accurate alignment of the segmented surface with the surface acquired by the camera. Then, the computed roto-translation is applied to the volumetric data to put the MR/CT images in the same reference system of the US probe (Fig. 2).

To optimise computation time, only the anterior surface of the patient is considered. In this way, the relevant part of the segmented surface undergoes registration. Visualising the registration error between the segmented and acquired skin gives valuable insight into the accuracy of the fusion obtained. The co-registration error is computed as the Hausdorff distance between the co-registered surfaces. Higher co-registration errors present a higher Hausdorff distance [11]. The distance distribution is mapped to RGB colours, and each vertex is assigned the corresponding colour according to its distance from the other surface. Vertices with a distance equal to or higher than 10 mm are coloured red. Vertices that have a null distance are shown in blue. The other distances are mapped to the shades between red and blue.

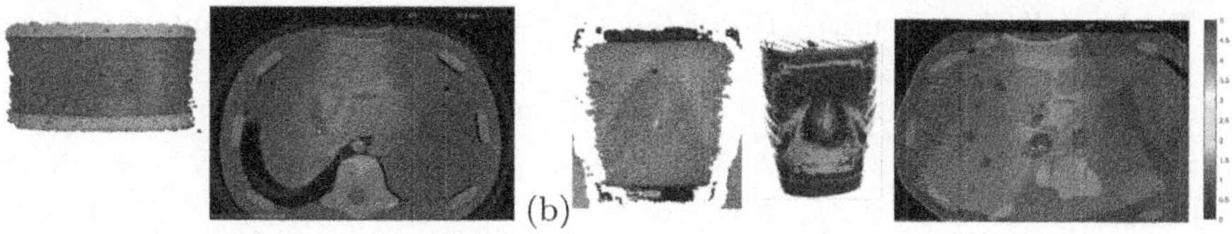

(a) (b)

Fig. 7. Coregistration error and fusion imaging result inside the volume. The blue dot represents the position of the landmark on the segmented surface, and the green dot is the misplaced landmark selected on the surface acquired by the camera. (a) CIRS Model 057 phantom (fusion error 13,4 mm) (b) Kyoto Kagaku dual modality human abdomen phantom (fusion error 9,7 mm). The colormap is in mm. (Color figure online)

3 Experimental Results and Validation

We describe the results of the robustness of the skin co-registration to noise, selected parameters, and the accuracy of the image fusion. The tests were performed through the integrated system prototype in a setup miming the clinical environment. We used phantoms to test US-CT fusion imaging and a small set of volunteers to evaluate US-MRI.

Phantom Tests. The co-registration experimental tests were performed on two phantoms representing the abdominal district: CIRS Model 057 and the Kyoto Kagaku dual modality human abdomen (Fig. 3(a)). The co-registration is accurate, and the error localises in those regions where the morphology of the surfaces differs due to the different resolutions of the MR and 3D cameras (Fig. 3(b)). The concatenation of multiple transformations challenges the fusion accuracy. These transformations include the patient surface to the RGB-D camera, the EM sensor on the RGB-D camera, the EM field generator to the EM sensor on the US probe, and finally, the US image. Other sources of error are related mainly to the acquired area's limited dimension and symmetric shapes. Smaller errors can be quickly corrected with manual tuning to produce a more accurate result (if needed). Despite these aspects, the fusion system remains in a range of errors that, in the worst-case scenario, reach 17 mm.

Distance from the 3D Camera. To verify the robustness of the co-registration, the skin surface was captured by placing the 3D camera at different distances from the skin. The co-registration remains stable against the acquisition distance. At very high distances, the noise acquired by the camera increases, confirming the algorithm's robustness to the noise and symmetries (Figs. 4, 5). Indeed, the co-registration includes the automatic identification of regions of interest both on the segmented surface and the 3D surface acquired by the camera. These Regions of Interest identify correspondent regions on the two surfaces and notably reduce the inclusion of noise from the 3D camera.

Camera Tilting. We tested how much tilting the 3D camera at various angles during the acquisition would affect the co-registration. We kept the camera's

Table 1. Fusion accuracy: the distance between camera and phantom is 35 cm.

Landmark	Orientation (°)	Mean error on surface (mm)	Error in volume (mm)
Sternum	0	2,67	6,20
Sternum	90	2,02	4
Sternum	180	2,35	5
Sternum	270	2,45	6,80
First rib	0	2,24	4,70
Second rib	0	2,07	4,57
Third rib	0	1,76	3,70
Umbilicus	0	3,23	6.46

distance from the surface equal to 35 cm. Then, we tilted the camera with varying angles, starting from the camera perpendicular to the phantom surface, then tilting it 30 and then 45° from the original position. Figure 6 shows the co-registration distance distribution and error inside the volume. The higher error at 45° could be linked to distortions in the morphology of the surface acquired by the camera due to the tilted position.

Virtuar Landmark Displacement. To verify the influence of the selection of corresponding virtual landmarks on the co-registration, we evaluated how much the perturbation of the position of the virtual reference influences the fusion quality. In this experiment, the camera was fixed at 35 cm from the phantom and perpendicular to it. We selected slightly displaced landmarks along the X, the Z axes, and diagonal (X and Y displacement) directions and measured the changes of the angles between the corresponding co-registration matrices. Significant differences in the rotation angles were not found. The robustness of the co-registration to a virtual landmarks misplacement confirms that the algorithm will not be user-dependent; i.e., users can apply different approaches to selecting the landmark. Indeed, the accuracy of matching the landmark in the segmented image with that in the camera mesh has minimal impact on the final fusion result. Therefore, even an inexperienced user, who may select landmarks with less precision, will still achieve a satisfactory fusion outcome (Fig. 7).

Camera Orientation. The 3D surface obtained by the camera was captured by moving the camera around the phantom to simulate a better possible clinical configuration, where the EM transmitter and camera are placed around the patient bed. The accuracy result is better in the 0° and 180° with respect to the lateral one (90–270). We evaluated the precision and accuracy of the system considering different camera orientations and other virtual reference positions on the morphological phantom (Table 1). We considered the mean distance between the two surfaces in the same Region of interest and the error inside the volume between the US image and the CT image once the fusion was performed (i.e., after the surface co-registration). These errors are in the order of the millimetre.

(a)
(b)
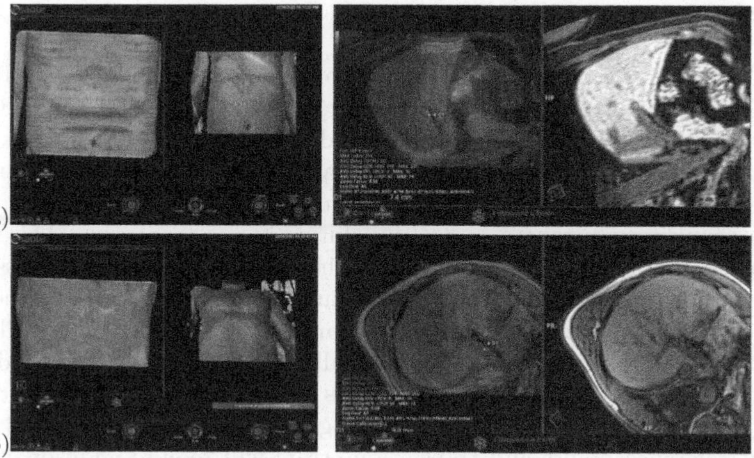

Fig. 8. (a) Volunteer 1 MR images and skin surface acquired at 0° (left), (right) co-registration with accuracy error of 7.4 mm, (b) Volunteer 2 MR image and skin surface 0-degree acquisition (left) co-registration result with accuracy error 9 mm (right).

Fig. 9. From left to right: surface error visualisation; volume error visualised on image slice; volume error visualised with the 3D model of the interested organ. The colormap is in mm.

The higher error inside the volume is coherent with the higher distance of the internal structures from the skin surface, which represented the reference for the coregistration.

Volunteers Tests. The error accuracy in the image fusion results was evaluated on three volunteers considering the distance between the portal vein bifurcation in the two imaging modalities (US and MRI). Volunteer 1 (MR Serie T2 AX MVXD), resulted in an accuracy error of 7.4 mm; Volunteer 2 (Siemens MR Serie T2 HASTE), of 9 mm; Volunteer 3 (Siemens MR Serie T2 TRUFI), 7.4 mm (Fig. 8). The results confirm that the fusion imaging system is robust to the minimal difference in posture and on the technician performing the exam. Moreover, the system can be used also with older images, which is important to avoid harmful radiation (e.g., CT images).

4 Conclusions and Future Work

The paper has introduced a method for combining volumetric CT/MR images with US images using a 3D depth sensor, without requiring extensive training on large datasets and without the need for external physical markers. The co-registration process demonstrates satisfactory accuracy (on the order of millimetres) and is robust against noise, virtual landmark misalignment, varying camera acquisition distances, and tilting during image acquisition. This seamless integration simplifies fusion imaging, making it accessible even for less experienced operators. As the first limitation regarding the co-registration, the system still requires the user to select a virtual landmark, making the overall process only partially automatic. The primary challenges in removing the need for virtual landmarks arise from body symmetries and regions lacking distinctive morphological features. Achieving accurate alignment of two well-oriented surfaces without reference points and translation errors will require additional analysis in future research. Secondly, the co-registration error localises in those regions where the morphology of the surfaces differs due to the different resolutions of the MR/CT and 3D cameras which is a problem linked to the heterogeneous nature of the input data. The segmentation presents a few challenges as well. Indeed, there is the need for the skin isovalue as input, which is easily obtainable but does not allow for fully automatic segmentation. The isovalue in input permits to focus specifically on the skin and, thus, to obtain a patient-specific segmentation. However, in the presence of dresses or drapes on the patients, whose isovalue in the image is higher than the skin's isovalue, the segmentation could follow the dress and not the skin introducing errors in the following pipeline.

Future work will concentrate on achieving a completely automatic coregistration and segmentation, reducing the computational time by sub-sampling the volumetric image, and utilising a 3D camera with lower resolution. Furthermore, efforts will be directed towards developing an augmented system to visualise errors and potential misalignment within the volume image, not just on the surface (refer to Fig. 9). Preliminary tests and results have been promising and offer valuable information for planning minimally invasive interventions.

Acknowledgments. Funded by the European Union - NextGenerationEU and by the Ministry of University and Research (MUR), National Recovery and Resilience Plan (NRRP), Mission 4, Component 2, Investment 1.5, project "RAISE -Robotics and AI for Socio-economic Empowerment" (ECS00000035). Martina Paccini, Giacomo Paschina, Stefano De Beni, and Giuseppe Patanè are part of the RAISE Innovation Ecosystem.

References

1. Azam, M.A., Khan, K.B., Salahuddin, S., Rehman, E., Khan, S.A., Khan, M.A., Kadry, S., Gandomi, A.H.: A review on multimodal medical image fusion: Compendious analysis of medical modalities, multimodal databases, fusion techniques and quality metrics. Computers in biology and medicine **144**, 105253 (2022)

2. Beare, R., Yang, J.Y.M., Maixner, W.J., Harvey, A.S., Kean, M.J., Anderson, V.A., Seal, M.L.: Automated alignment of perioperative mri scans: A technical note and application in pediatric epilepsy surgery. Tech. rep., Wiley Online Library (2016)
3. Bjurlin, M.A., Mendhiratta, N., Wysock, J.S., Taneja, S.S.: Multiparametric mri and targeted prostate biopsy: Improvements in cancer detection, localization, and risk assessment. Central European Journal of Urology **69**(1), 9 (2016)
4. Faragallah, O.S., El-Hoseny, H., El-Shafai, W., Abd El-Rahman, W., El-Sayed, H.S., El-Rabaie, E.S.M., Abd El-Samie, F.E., Geweid, G.G.: A comprehensive survey analysis for present solutions of medical image fusion and future directions. IEEE Access **9**, 11358–11371 (2020)
5. Gayet, M., van der Aa, A., Beerlage, H.P., Schrier, B.P., Mulders, P.F., Wijkstra, H.: The value of magnetic resonance imaging and ultrasonography (mri/us)-fusion biopsy platforms in prostate cancer detection: A systematic review. BJU international **117**(3), 392–400 (2016)
6. Lee, C.Y., Chang, T.F., Chang, N.Y., Chang, Y.C.: An automated skin segmentation of breasts in dynamic contrast-enhanced magnetic resonance imaging. Scientific Reports **8**(1), 6159 (2018)
7. Lorensen, W.E., Cline, H.E.: Marching cubes: A high resolution 3d surface construction algorithm. ACM Siggraph Computer Graphics **21**(4), 163–169 (1987)
8. Markova, V., Ronchetti, M., Wein, W., Zettinig, O., Prevost, R.: Global multimodal 2d/3d registration via local descriptors learning. In: International Conference on Medical Image Computing and Computer-Assisted Intervention. pp. 269–279. Springer (2022)
9. Ognard, J., Mesrar, J., Benhoumich, Y., Misery, L., Burdin, V., Ben Salem, D.: Edge detector-based automatic segmentation of the skin layers and application to moisturization in high-resolution 3 tesla magnetic resonance imaging. Skin Research and Technology **25**(3), 339–346 (2019)
10. Peng, C., Cai, Q., Chen, M., Jiang, X.: Recent advances in tracking devices for biomedical ultrasound imaging applications. Micromachines **13**(11), 1855 (2022)
11. Taha, A.A., Hanbury, A.: An efficient algorithm for calculating the exact hausdorff distance. IEEE transactions on pattern analysis and machine intelligence **37**(11), 2153–2163 (2015)
12. Tu, M., Jung, H., Moghadam, A., Raythatha, J., Hsu, J., Kim, J.: Exploring the performance of geometry-based markerless registration in a simulated surgical environment: A comparative study of registration algorithms in medical augmented reality. In: 2023 45th Annual International Conference of the IEEE Engineering in Medicine & Biology Society (EMBC). pp. 1–4. IEEE (2023)
13. Wang, Y., Fu, T., Wu, C., Xiao, J., Fan, J., Song, H., Liang, P., Yang, J.: Multimodal registration of ultrasound and mr images using weighted self-similarity structure vector. Computers in Biology and Medicine **155**, 106661 (2023)
14. Wang, Y., Qiu, Y., Thai, T., Moore, K., Liu, H., Zheng, B.: A two-step convolutional neural network based computer-aided detection scheme for automatically segmenting adipose tissue volume depicting on ct images. Computer Methods and Programs in Biomedicine **144**, 97–104 (2017)
15. Weston, A.D., Korfiatis, P., Kline, T.L., Philbrick, K.A., Kostandy, P., Sakinis, T., Sugimoto, M., Takahashi, N., Erickson, B.J.: Automated abdominal segmentation of ct scans for body composition analysis using deep learning. Radiology **290**(3), 669–679 (2019)
16. Yang, M., Ding, H., Kang, J., Cong, L., Zhu, L., Wang, G.: Local structure orientation descriptor based on intra-image similarity for multimodal registration of liver ultrasound and mr images. Computers in biology and medicine **76**, 69–79 (2016)

Deep Learning Multi-channel Structural and Diffusion Tensor Neonatal Image Registration

Irina Grigorescu[1(✉)], Alena Uus[1], Daan Christiaens[1,2], Lucilio Cordero-Grande[1,4], Jana Hutter[1], Dafnis Batalle[1,3], A. David Edwards[1], Joseph V. Hajnal[1], Marc Modat[1], and Maria Deprez[1]

[1] School of Biomedical Engineering and Imaging Sciences, King's College London, London, UK
irina.grigorescu@kcl.ac.uk
[2] Departments of Electrical Engineering, ESAT/PSI, KU Leuven, Leuven, Belgium
[3] Department of Forensic and Neurodevelopmental Science, Institute of Psychiatry, Psychology and Neuroscience, King's College London, London, UK
[4] Biomedical Image Technologies, ETSI Telecomunicación, Universidad Politécnica de Madrid and CIBER-BNN, ISCIII, Madrid, Spain

Abstract. Investigating the maturation process of the brain through clinical studies is crucial for advancing our understanding of the developing brain. Neuroimaging studies employing magnetic resonance images (MRI) often necessitate pre-alignment of images to a standardized space. Registration of diffusion tensor images (DTI) has the potential to better align white matter (WM) structures than using structural MRI only, as it enables alignment of fiber orientation at each voxel. However, microstructural data are not well suited to accurately register the cortical gray matter (cGM) ribbon. In spite of this, in many studies the alignment of subjects is primarily guided by a single modality. In this work, we propose a multi-channel attention-based deep learning registration approach that selects the most salient features from multiple image modalities to improve alignment of individual MR images to a common atlas space. We apply the technique to align multi-channel datasets composed of structural T_2-weighted (T_2w) MRI and DTI maps into atlas space. The quantitative and qualitative evaluation confirmed that when we use the two modalities we obtain good alignment of anatomical structures, while also improving the alignment of the underlying white matter tracts. Moreover, we show that learning a spatially-varying attention map for weighting the two modalities obtains superior performance than the baseline multi-channel method that does not incorporate attention.

Keywords: multi-channel registration · attention · deep learning registration · diffusion tensor imaging

M. Modat et al. (Eds.): WBIR 2024, LNCS 15249, pp. 192–204, 2024.
https://doi.org/10.1007/978-3-031-73480-9_15

1 Introduction

The neonatal brain undergoes dramatic changes during early life, such as cortical folding, development of white matter (WM) tracts and myelination [8]. Magnetic resonance imaging (MRI) provides snapshots of the evolving morphology and tissue properties in the developing brain across multiple subjects and time-points [23]. However, neuroimaging studies often require that the images are pre-aligned to a common space so that development or pathologies can be studied longitudinally. Accurate anatomical alignment remains one of the key challenges for image registration algorithms [8].

Structural and microstructural MRI modalities offer complementary information about morphology and tissue properties of the developing neonatal brain, but often inter-subject alignment is driven by a single modality, such as structural [4], or diffusion [31] MRI. Moreover, the alignment of both structural T_2-weighted (T_2w) scans and diffusion tensor imaging (DTI) maps to a common reference space holds significant importance in neuroimaging, as T_2w scans are valuable for morphometric purposes, such as brain segmentation, while microstructural data can provide information about the underlying tissue properties.

Multi-channel registration aims to find inter-subject anatomical correspondences using sets of images of multiple modalities for each subject. Most such approaches rely on minimizing the weighted sum of dissimilarity measures calculated for each channel individually [3,6,7,12,18]. One drawback of this approach is that each channel is weighted on a global scale, thus not accounting for local variations within the images. To address this, studies have attempted to weight the deformation fields with certainty maps calculated from normalised spatial gradients correlated to structural content [11,29]. Rohde *et al.* [21] proposed a framework for multi-channel registration using a multivariate correlation approach, and applied it to DTI data. Guimond *et al.* [12] and Park *et al.* [18] introduced the multi-channel Demons algorithm [26], while Peyrat *et al.* [20] further incorporated trajectory constraints for registering 4D time-series cardiac computed tomography (CT) images. Finally, multi-channel registration has also been helpful in multi-modal applications where the target and the moving images consisted of different modalities, and in these cases [5,6] proposed the use of image synthesis [13,15] to generate synthetic images of the missing channels.

In this work, we propose a multi-channel deep learning attention-based image registration framework which combines T_2w images with DTI maps (see Fig. 1). Our proposed solution is based on a diffeomorphic non-rigid registration with stationary velocity field (SVF) representation [14]. Our method aims to optimise the multi-channel alignment through learned spatial attention maps which provide an optimal spatial weighting for the predicted modality-specific velocity fields. More specifically, we propose a lightweight convolutional neural network (CNN) termed **attention velocity network** which produces attention maps which identify salient features in the individual channels in order to drive the multi-channel registration process. To evaluate the proposed framework, we compare networks trained on single modality to multi-channel models, either with or without attention. Our results show that by using attention in a multi-channel

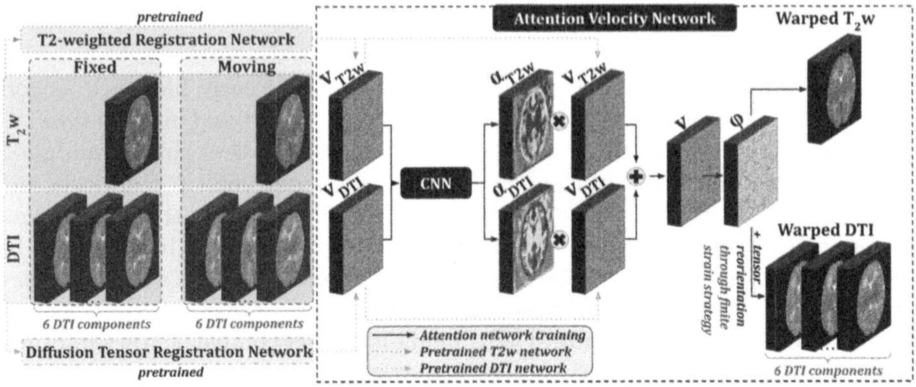

Fig. 1. Pipeline of our proposed multi-channel structural and microstructural **attention**-driven image registration framework which takes as input the modality-specific velocity fields v_{T2w} and v_{DTI}, and outputs the attention maps α_{T2w} and α_{DTI}. The black arrows represent the training and inference flow of data through the proposed attention velocity network, where the final deformation field φ is used to interpolate the moving images into their warped counterparts. Note that the data also undergoes tensor reorientation (*finite strain* strategy [2]). Network architecture details can be found in Fig. 2.

image registration setting we achieve superior overall alignment considering brain morphology as well as the WM microstructure.

2 Methodology

Our proposed **attention**-driven multi-channel deep learning image registration framework aims to combine information from T_2w neonatal scans with diffusion-weighted imaging (DWI)-derived microstructural maps. The overall pipeline is presented in Fig. 1, where the two channels are T_2w and DTI. The **attention velocity network** aims to learn the necessary spatial attention maps which select the most salient features from both structural and microstructural channels. More specifically, given a pair of fixed and moving T_2w and DTI volumes, the network produces two 3D modality-specific spatial attention maps α_{T2w} and α_{DTI} which are used to locally weigh the velocity fields v_{T2w} and v_{DTI}. The latter are obtained from pre-trained T_2w and DTI single-modality image registration networks which are based on a conditional variational autoencoder (CVAE) as proposed by [14]. Note that in Fig. 1 we show our proposed pipeline for T_2w and DTI data, however, these can be replaced with different modalities, such as tissue segmentation maps, or scalar-valued fractional anisotropy (FA) maps.

Throughout this work we use 3D structural (T_2w) and diffusion (DTI) MRI brain scans acquired as part of the developing Human Connectome Project (dHCP) [10] for the moving images, and a 36 weeks old neonatal atlas [29] of the same modalities as the fixed images. Details about the aquisition parameters are found in [10].

2.1 Data Preprocessing and Selection

Data Preprocessing. We affinely pre-registered all data to a common 36 weeks gestational age atlas space [29] using MIRTK [24], and then we resampled both structural and microstructural volumes to be 1 mm^3 isotropic resolution. To obtain the diffusion tensor (DT) maps, we used the DWI2TENSOR MRtrix3 [27] command, and we performed skull-stripping using the available dHCP brain masks [17]. All images were cropped to a 128 × 128 × 128 size.

Data Selection. To train our image registration models, we divided our dataset using an 80%–10%–10% split (see Table 1), using stratified sampling in order to keep the distribution of post-menstrual age (PMAs) at scan and the male-to-female ratio close to the original distribution. We report all of our results on the test set.

Table 1. Number of scans used for training, validation and testing the models.

Dataset	Subjects	GA (weeks)	PMA (weeks)
Train	350 (164♀+ 186♂)	38.0 (±3.8)	40.6 (±1.9)
Validate	34 (14♀ + 20♂)	39.7 (±1.4)	40.7 (±1.7)
Test	30 (12♀ + 18♂)	39.8 (±1.5)	40.6 (±1.9)

2.2 Model Training

Let F, M represent the fixed (target) and the moving (source) magnetic resonance (MR) volumes, respectively, and let φ be the deformation field. In this work, the focus is on scalar-valued structural data (F^{T2w} and M^{T2w}) and tensor-valued microstructural DTI volumes (F^{DTI} and M^{DTI}) which are 6-component data. The moving images (M^{T2w} and M^{DTI}, co-registered) are acquired from the same subject, while the fixed images (F^{T2w} and F^{DTI}, co-registered) are a 36 weeks old neonatal atlas [29].

Scalar-Valued. The **baseline** image registration network [14] is trained with T_2w data only, and aims to minimize the scalar-valued image registration loss $\mathcal{L}^{\mathbf{S}}_{reg} = \mathcal{L}_{\mathbf{KLD}} + \lambda_{\mathbf{BE}}\,\mathcal{L}_{\mathbf{BE}}(\varphi) + \lambda\,\mathcal{L}^{\mathbf{S}}_{dsim}$, where $\lambda_{\mathbf{BE}} = 0.01$ and $\lambda = 5000$, as proposed in [14]. The three terms are: the Kullback-Leibler (KL) divergence $\mathcal{L}_{\mathbf{KLD}}$ [9,14] (to reduce the gap between the prior and the encoded distributions), the bending energy (BE) regularisation penalty $\mathcal{L}_{\mathbf{BE}}$ [22] which regularizes the predicted deformation field φ, and the scalar-valued image dissimilarity loss $\mathcal{L}^{\mathbf{S}}_{dsim} = \lambda_{T2w}\,\mathcal{L}_{\mathbf{NCC}}\left(\mathbf{F}^{T2w}, \mathbf{M}^{T2w}\right)$, where $\mathcal{L}_{\mathbf{NCC}}$ is the symmetric normalised cross correlation (NCC) dissimilarity measure.

Tensor-Valued. For the models which use DTI data, we develop layers capable of dealing with the higher-order microstructural data and introduce them as part of the **baseline** network. Registration of DT images is not as straightforward to perform as scalar-valued data. When transforming the latter, the intensities in the moving image are interpolated at the new locations determined by the deformation field φ and assigned to the corresponding location in the target image space. However, after interpolating DT images, the diffusion tensors need to be reoriented to remain anatomically correct [2]. This is done by applying a rotation matrix R to each resampled diffusion tensor $\mathbf{D}$, such that: $\mathbf{D}' = R\mathbf{D}R^T$. When the transformation is non-linear the reorientation matrix can be computed at each point in the deformation field φ through a polar decomposition of the local Jacobian matrix J. This factorisation transforms the non-singular matrix J into a unitary matrix R (the pure rotation) and a positive-semidefinite Hermitian matrix P, such that $J = RP$ [25]. The rotation matrices R are then used to reorient the tensors without changing the local microstructure. This is known as the *finite strain* strategy [2]. We train this model either single (DTI-only) or multi-channel (T_2w+DTI). In this case, the optimizer aims to minimize the scalar and tensor-valued image registration loss $\mathcal{L}^{\mathbf{D}}_{reg} = \mathcal{L}_{\mathbf{KLD}} + \lambda_{\mathbf{BE}}\,\mathcal{L}_{\mathbf{BE}}(\varphi) + \lambda\,\mathcal{L}^{\mathbf{D}}_{dsim}$, where $\mathcal{L}^{\mathbf{D}}_{dsim} = \lambda_{T2w}\,\mathcal{L}_{\mathbf{NCC}}\left(\mathbf{F}^{T2w}, \mathbf{M}^{T2w}\right) + \lambda_{DTI}\,\mathcal{L}_{\mathbf{EDS}}\left(\mathbf{F}^{DTI}, \mathbf{M}^{DTI}\right)$. Here, $\mathcal{L}_{\mathbf{EDS}}$ is the Euclidean distance squared (EDS) [31].

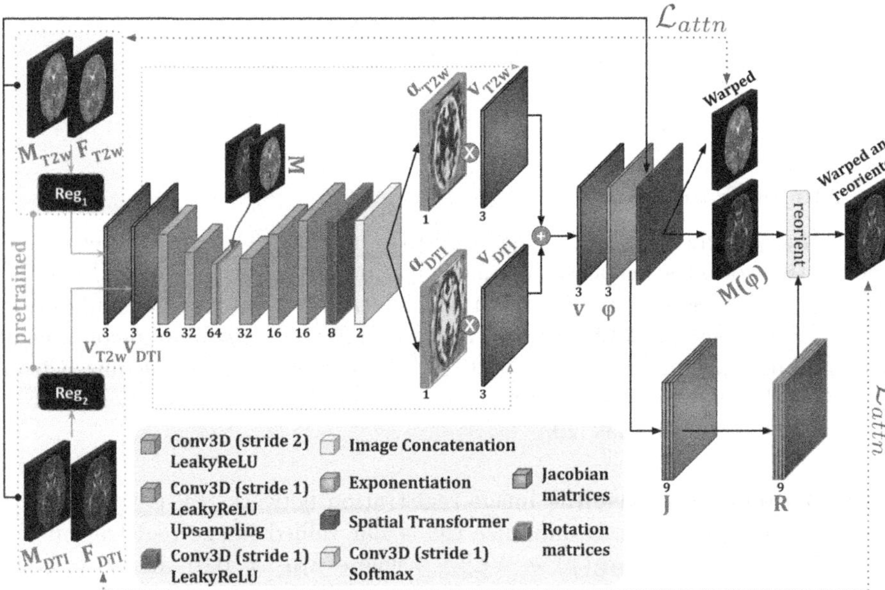

Fig. 2. Network architecture of our proposed **attention**-driven multi-channel model. Note that the same diffusion tensor reorientation layers are added to the **baseline** DTI-only image registration network.

Attention Velocity Network. For the **attention**-driven registration task, we construct a lightweight CNN which uses pairs of modality-specific velocity fields as input, and outputs a combined velocity field which aims to align both structural and microstructural data simultaneously. The proposed network learns the attention maps α_{T2w} and α_{DTI}, for which $\alpha_{T2w} + \alpha_{DTI} = \mathbf{1}$ at every voxel. The input velocity fields are weighted with the attention maps and combined to create a final velocity field $v = v_{T2w} \odot \alpha_{T2w} + v_{DTI} \odot \alpha_{DTI}$, where $\odot$ represents the element-wise multiplication, and $+$ represents the element-wise addition. The final dense deformation field φ is then obtained through *scaling and squaring* layers applied to the velocity field v. The architecture of our proposed attention-based network is presented in Fig. 2. The loss function for training this network is $\mathcal{L}_{attn} = \lambda_{T2w}\,\mathcal{L}_{\mathbf{NCC}}(\mathbf{F}^{T2w}, \mathbf{M}^{T2w}) + \lambda_{DTI}\,\mathcal{L}_{\mathbf{EDS}}(\mathbf{F}^{DTI}, \mathbf{M}^{DTI})$.

Training Details. We train 12 models: 2 single-modality (T_2w-only or DTI-only), and 10 multi-channel models (5 **baseline** and 5 **attention**-driven models) with the following set of hyperparameters $(\lambda_{T2w}, \lambda_{DTI}) = \{(1.0, 0.5), (1.0, 1.0), (1.0, 1.5), (1.0, 2.0), (1.0, 2.5)\}$. We train each model until convergence (using the Adam optimizer with its default parameters), with a maximum of 300 epochs, and use the models which performed best on the validation set. All networks were implemented in PyTorch (v1.10.2), with TorchIO (v0.18.73) [19] for data preprocessing (intensity normalisation) and loading, and training was performed on a 12 GB Titan XP. Average training times were: ∼5 h for the scalar-valued networks (T_2w-only), and ∼100 h for the tensor-valued networks due to their increased complexity (99 h for the DTI-only, 104 h for the **baseline** T_2w+DTI, and 99 h for the **attention** T_2w+DTI).

2.3 Evaluation of Image Registration Models

The quantitative evaluation is carried out on our test dataset of 30 subjects. Each subject and the atlas had the following tissue label segmentations obtained from T_2w images using the Draw-EM pipeline [16]: cortical gray matter (cGM), WM, deep gray matter (dGM) and ventricles. Additionaly, a WM structure called the internal capsule (IC) was manually segmented on the FA maps of all test subjects. These labels were propagated from each subject into the atlas space using the predicted deformation fields. A two-sample, two-sided paired t-test with a significance level of 5% was used to compare pairs of trained models.

Hyperparameter Tuning. To evaluate performance of the registration, we calculate Dice scores between the warped labels and the atlas labels, for all of our models. To choose the best set of hyperparameters for the multi-channel models, we look at the performance of our trained models on two structures: the cGM and the IC. We chose these two structures because the cGM ribbon is a difficult-to-align structure due to the inherent inter-individual variability in cortical folding across the cohort, while the IC is a white matter structure which is prominent in the microstructural data, and barely visible in T_2w structural images. We also produce average α_{DTI} attention maps obtained from 10 neonatal subjects scanned around 40 weeks PMA.

Alignment of Labels After Registration. After choosing the best overall set of hyperparameters based on the two selected structures, we report Dice scores for the remaining labels.

Evaluation of White Matter Alignment. We also carry out an evaluation of the alignment of WM microstructure, using two metrics proposed in [1]: 1) the cross correlation (CC) between the trace of tensors, which looks at the overall diffusivity, and 2) the overlap of eigenvalue–eigenvector pairs (OVL), which looks at the directional components of the DTIs.

Diffusion Tensor Maps. Finally, we present a qualitative result showcasing warped DTI maps and T_2w images obtained after registering a subject scanned around 40 weeks gestational age (GA) to the atlas.

3 Results

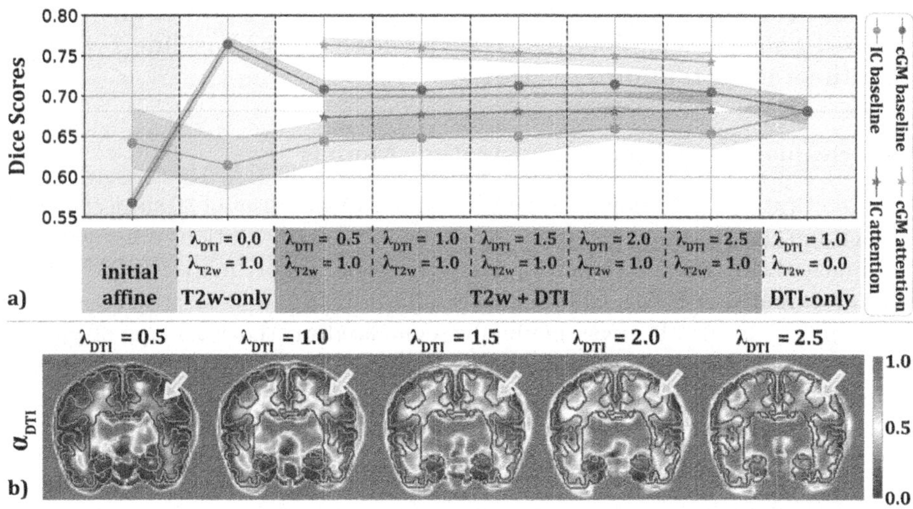

Fig. 3. Hyperparameter tuning results showing: a) mean Dice scores for cGM and IC structures where the shading around each line is the interquartile range (IQR), and b) average α_{DTI} attention maps.

Hyperparameter Tuning. Figure 3a summarizes our results in terms of Dice scores for the 7 **baseline** models (1 T_2w-only, 1 DTI-only and 5 T_2w+DTI models with different λ_{DTI} values), and the 5 proposed T_2w+DTI **attention**-based networks. The T_2w-only model (second column) obtains scores which are worse than the initial affine alignment (first column) for the IC, and best for the cGM label. On the other hand, the DTI-only model (last column) obtains the best overall IC alignment, and the worst for the cGM. The **attention** networks always outperform ($p - value < 0.05$) the **baseline** multi-channel models, for all values of λ_{DTI} (columns 3–7), for both the cGM and IC tissue types. Figure 3b shows the midbrain coronal slices of average α_{DTI} attention maps for increasing values of λ_{DTI}. It is evident from the figure that for $\lambda_{DTI} = 0.5$ the network uses the DTI channel mainly around the dGM region of the brain, while at the same time, for $\lambda_{DTI} > 1.0$, the α_{DTI} maps seem to *spill over* onto the cortical ribbon (as highlighted by the green arrows), *i.e.*, the models use some information from the DTI channel when aligning the cGM. We found the middle ground to be at $(\lambda_{T2w}, \lambda_{DTI}) = (1.0, 1.0)$, in order to not lose performance on cGM alignment, while also produced good results in terms of the IC label. All of our following results are based on this set of hyperparameters.

Table 2. Mean (±standard deviation) Dice scores of the single- and multi-channel ($\lambda_{DTI} = 1.0$) experiments, together with the initial affine alignment. Statistically significant differences ($p - value < 0.05$) are reported in terms of best/worst overall score (highlighted in bold/shaded in gray, respectively), and best amongst the multi-channel models (highlighted with a black box). ***A*** - *"attention"*.

Model	cGM	WM	Ventricles	dGM	IC
affine	.57±.02	.70±.03	.63±.05	.90±.02	.64±.07
T2w-only	**.76±.01**	**.84±.02**	.80±.02	**.93±.01**	.61±.04
DTI-only	.68±.02	.77±.02	.76±.03	.91±.01	**.68±.03**
T2w+DTI	.71±.01	.75±.03	.76±.03	.90±.01	.65±.03
A T2w+DTI	.76±.01	.83±.02	**.80±.02**	.93±.01	**.68±.03**

Alignment of Labels After Registration. Table 2 shows the Dice scores obtained by our trained models for all tissue types (cGM, WM, ventricles, dGM and IC). The proposed **attention**-based network systematically outperforms the **baseline** T_2w+DTI models for all structures, with the latter obtaining significantly worse scores ($p - value < 0.05$) than its **attention**-driven counterpart. These results also highlight the importance of adding the T_2w structural channel to DTI registration in order to improve structural alignment; specifically, the DTI-only model drops in performance when compared to the T_2w-only model in terms of aligning the cGM, WM, ventricles and dGM tissue types, but regains its performance when using both **attention** and the extra structural channel.

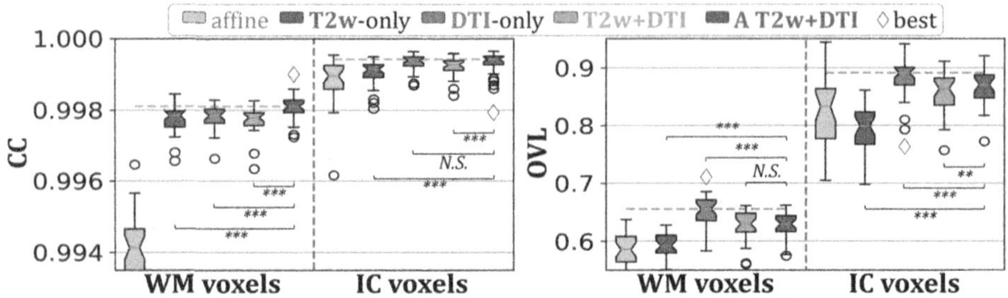

Fig. 4. Test dataset CC scores (left) and average OVL values (right) calculated between the warped DT images with reoriented tensors and the fixed DTI atlas. Best performing models ($p - value < 0.05$) are marked with a yellow diamond. $***$-*p-value* < 0.001, $**$-*p-value* < 0.01, $* - $ *p-value* < 0.05, *N.S.*-"*not significant*". (Color figure online)

Evaluation of White Matter Alignment. Figure 4 summarizes our results in terms of CC scores (left) and average OVL values (right), either computed across the WM voxels (using the atlas WM label map), or within the IC voxels (using the atlas IC label map). In terms of the CC scores, our proposed **attention** T_2w+DTI model outperforms all other models ($p - value < 0.05$), followed closely by the DTI-only model which performs similarly well for the IC voxels. For the average OVL values, the best overall scores are obtained by the DTI-only model, followed by our proposed **attention** model, which outperforms its non-attention counterpart ($p - value < 0.05$) for the IC voxels. Both **baseline** and **attention** T_2w+DTI models perform similarly well for the WM voxels. On the other hand, the T_2w-only model consistently underperforms in these two metrics, showing that the DTI channel is essential for correct alignment of WM microstructure.

Diffusion Tensor Maps. Figure 5 shows an example of warped DTI maps (rows 1–2) and T_2w images (row 3) obtained after registering a subject scanned around 40 weeks GA to the atlas. The IC is strikingly less coherent in the T_2w-only model (yellow arrows, first row) when compared to the atlas, while the overall shape of the corpus callosum (white arrows, second row) is less preserved. This is also the case for the **baseline** T_2w+DTI model where the IC seems to be thinner than obtained by the **attention** T_2w+DTI and DTI-only models (yellow arrows, first row), while the corpus callosum exhibits an uneven shape (white arrows, second row). The cortical ribbon (last row) is not well aligned for the DTI-only model, it is generally "*overspilling*" into the WM for the **baseline** T_2w+DTI, while being similarly well aligned with the fixed (blue outline) cortex in the T_2w-only and our proposed **attention** T_2w+DTI models.

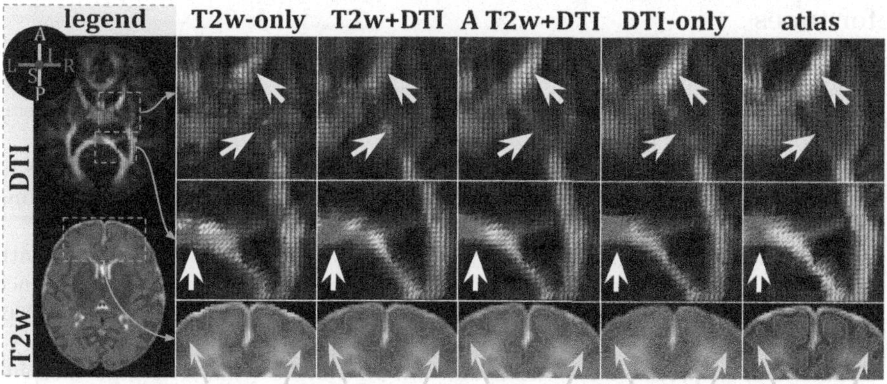

Fig. 5. Mid-brain axial slices of the warped DTI and T_2w maps of a 40 weeks GA neonate for 4 of our trained models: T_2w-only, **baseline** T_2w+DTI ($\lambda_{DTI} = 1.0$), **attention** T_2w+DTI ($\lambda_{DTI} = 1.0$) and DTI-only, together with the fixed DTI atlas. Yellow arrows (first row) point to the IC, while the white arrows (second row) point to the corpus callosum. The warped T_2w images (last row) are shown together with their corresponding highlighted cortical ribbon (in red), as well as an outline (in blue) of the fixed (target) cortex. Cyan arrows (last row) point to mis-registered regions of the cortical ribbon. (Color figure online)

4 Conclusion

In this work we presented a novel solution for multi-channel registration, which combines structural and microstructural tensor-valued MRI data based on learned spatially varying attention maps that optimise the multi-channel alignment. Our evaluation showed that the proposed **attention**-driven image registration network improves overall alignment when compared to models trained on multi-channel data without attention, while maintaining the performance of the single-channel registration for the structures salient on that channel. Key limitations of our work include the reliance on accurately co-registered structural and microstructural data, and the DTI model's inability to resolve crossing fibers [28,30]. This has the potential to be explored in future work.

Acknowledgements. This work was supported by the Academy of Medical Sciences Springboard Award [SBF004\1040], Medical Research Council (Grant no. [MR/K006355/1]), European Research Council under the European Union's Seventh Framework Programme [FP7/20072013]/ERC grant agreement no. 319456 dHCP project, the EPSRC Research Council as part of the EPSRC DTP (grant Ref: [EP/R513064/1]), the Wellcome/EPSRC Centre for Medical Engineering at King's College London [WT 203148/Z/16/Z], the NIHR Clinical Research Facility (CRF) at Guy's and St Thomas', and by the National Institute for Health Research Biomedical Research Centre based at Guy's and St Thomas' NHS Foundation Trust and King's College London.

References

1. Adluru, N., Zhang, H., Fox, A.S., Shelton, S.E., Ennis, C.M., Bartosic, A.M., Oler, J.A., Tromp, D.P., Zakszewski, E., Gee, J.C., et al.: A diffusion tensor brain template for rhesus macaques. Neuroimage **59**(1), 306–318 (2012)
2. Alexander, D.C., Pierpaoli, C., Basser, P.J., Gee, J.C.: Spatial transformations of diffusion tensor magnetic resonance images. IEEE transactions on medical imaging **20**(11), 1131–1139 (2001)
3. Avants, B., Duda, J.T., Zhang, H., Gee, J.C.: Multivariate normalization with symmetric diffeomorphisms for multivariate studies. In: International Conference on Medical Image Computing and Computer-Assisted Intervention. pp. 359–366. Springer (2007)
4. Avants, B.B., Epstein, C.L., Grossman, M., Gee, J.C.: Symmetric diffeomorphic image registration with cross-correlation: evaluating automated labeling of elderly and neurodegenerative brain. Medical image analysis **12**(1), 26–41 (2008)
5. Chen, M., Carass, A., Jog, A., Lee, J., Roy, S., Prince, J.L.: Cross contrast multichannel image registration using image synthesis for mr brain images. Medical image analysis **36**, 2–14 (2017)
6. Chen, M., Jog, A., Carass, A., Prince, J.L.: Using image synthesis for multi-channel registration of different image modalities. In: Medical Imaging 2015: Image Processing. vol. 9413, pp. 462–468. SPIE (2015)
7. Daga, P., Winston, G., Modat, M., Cardoso, M.J., Stretton, J., Symms, M., McEvoy, A.W., Hawkes, D., Duncan, J., Ourselin, S.: Integrating structural and diffusion mr information for optic radiation localisation in focal epilepsy patients. In: 2011 IEEE International Symposium on Biomedical Imaging: From Nano to Macro. pp. 353–356. IEEE (2011)
8. Dubois, J., Alison, M., Counsell, S.J., Hertz-Pannier, L., Hüppi, P.S., Benders, M.J.: Mri of the neonatal brain: a review of methodological challenges and neuroscientific advances. Journal of Magnetic Resonance Imaging **53**(5), 1318–1343 (2021)
9. Duchi, J.: Derivations for linear algebra and optimization. Berkeley, California **3**(1), 2325–5870 (2007)
10. Edwards, A.D., Rueckert, D., Smith, S.M., Seada, S.A., Alansary, A., Almalbis, J., Allsop, J., Andersson, J., Arichi, T., Arulkumaran, S., et al.: The developing human connectome project neonatal data release. Frontiers in Neuroscience **16** (2022)
11. Forsberg, D., Rathi, Y., Bouix, S., Wassermann, D., Knutsson, H., Westin, C.F.: Improving registration using multi-channel diffeomorphic demons combined with certainty maps. In: International Workshop on Multimodal Brain Image Analysis. pp. 19–26. Springer (2011)
12. Guimond, A., Guttmann, C.R., Warfield, S.K., Westin, C.F.: Deformable registration of dt-mri data based on transformation invariant tensor characteristics. In: Proceedings IEEE International Symposium on Biomedical Imaging. pp. 761–764. IEEE (2002)
13. Jog, A., Carass, A., Pham, D.L., Prince, J.L.: Random forest flair reconstruction from t 1, t 2, and p d-weighted mri. In: 2014 IEEE 11th International Symposium on Biomedical Imaging (ISBI). pp. 1079–1082. IEEE (2014)
14. Krebs, J., Mansi, T., Mailhé, B., Ayache, N., Delingette, H.: Unsupervised probabilistic deformation modeling for robust diffeomorphic registration (2018)

15. Liu, X., Jiang, D., Wang, M., Song, Z.: Image synthesis-based multi-modal image registration framework by using deep fully convolutional networks. Medical & Biological Engineering & Computing **57**, 1037–1048 (2019)
16. Makropoulos, A., Gousias, I.S., Ledig, C., Aljabar, P., Serag, A., Hajnal, J.V., Edwards, A.D., Counsell, S.J., Rueckert, D.: Automatic whole brain MRI segmentation of the developing neonatal brain. IEEE transactions on medical imaging (2014)
17. Makropoulos, A., Robinson, E.C., Schuh, A., Wright, R., Fitzgibbon, S., Bozek, J., Counsell, S.J., Steinweg, J., Vecchiato, K., Passerat-Palmbach, J., et al.: The developing human connectome project: A minimal processing pipeline for neonatal cortical surface reconstruction. Neuroimage (2018)
18. Park, H.J., Kubicki, M., Shenton, M.E., Guimond, A., McCarley, R.W., Maier, S.E., Kikinis, R., Jolesz, F.A., Westin, C.F.: Spatial normalization of diffusion tensor mri using multiple channels. Neuroimage **20**(4), 1995–2009 (2003)
19. Pérez-García, F., Sparks, R., Ourselin, S.: TorchIO: a Python library for efficient loading, preprocessing, augmentation and patch-based sampling of medical images in deep learning. arXiv:2003.04696 [cs, eess, stat] (Mar 2020), http://arxiv.org/abs/2003.04696, arXiv: 2003.04696
20. Peyrat, J.M., Delingette, H., Sermesant, M., Xu, C., Ayache, N.: Registration of 4d cardiac ct sequences under trajectory constraints with multichannel diffeomorphic demons. IEEE transactions on medical imaging **29**(7), 1351–1368 (2010)
21. Rohde, G.K., Pajevic, S., Pierpaoli, C., Basser, P.J.: A comprehensive approach for multi-channel image registration. In: International Workshop on Biomedical Image Registration. pp. 214–223. Springer (2003)
22. Rueckert, D., Sonoda, L.I., Hayes, C., Hill, D.L.G., Leach, M.O., Hawkes, D.J.: Nonrigid registration using free-form deformations: application to breast MR images. IEEE Transactions on Medical Imaging (1999)
23. Rutherford, M.: MRI of the neonatal brain. Magnetic resonance imaging of the brain in preterm infants: 24 weeks' gestation to term pp. 25–49 (2002)
24. Schuh, A., Makropoulos, A., Robinson, E.C., Cordero-Grande, L., Hughes, E., Hutter, J., Price, A.N., Murgasova, M., Teixeira, R.P.A., Tusor, N., et al.: Unbiased construction of a temporally consistent morphological atlas of neonatal brain development. BioRxiv p. 251512 (2018)
25. Shoemake, K., Duff, T.: Matrix animation and polar decomposition. In: Graphics Interface. vol. 92, pp. 258–264. Citeseer (1992)
26. Thirion, J.P.: Image matching as a diffusion process: an analogy with maxwell's demons. Medical image analysis **2**(3), 243–260 (1998)
27. Tournier, J.D., Smith, R., Raffelt, D., Tabbara, R., Dhollander, T., Pietsch, M., Christiaens, D., Jeurissen, B., Yeh, C.H., Connelly, A.: MRtrix3: A fast, flexible and open software framework for medical image processing and visualisation. NeuroImage **202**, 116137 (2019)
28. Tuch, D.S., Reese, T.G., Wiegell, M.R., Makris, N., Belliveau, J.W., Wedeen, V.J.: High angular resolution diffusion imaging reveals intravoxel white matter fiber heterogeneity. Magnetic Resonance in Medicine: An Official Journal of the International Society for Magnetic Resonance in Medicine **48**(4), 577–582 (2002)
29. Uus, A., Grigorescu, I., Pietsch, M., Batalle, D., Christiaens, D., Hughes, E., Hutter, J., Cordero Grande, L., Price, A.N., Tournier, J.D., Rutherford, M.A., Counsell, S.J., Hajnal, J.V., Edwards, A.D., Deprez, M.: Multi-channel 4D parametrized atlas of macro- and microstructural neonatal brain development. Frontiers in Neuroscience **15**, 721 (2021)

30. Wiegell, M.R., Larsson, H.B., Wedeen, V.J.: Fiber crossing in human brain depicted with diffusion tensor MR imaging. Radiology **217**(3), 897–903 (2000)
31. Zhang, H., Yushkevich, P.A., Alexander, D.C., Gee, J.C.: Deformable registration of diffusion tensor MR images with explicit orientation optimization. Medical image analysis **10**(5), 764–785 (2006)

Registration by Regression (RbR): A Framework for Interpretable and Flexible Atlas Registration

Karthik Gopinath[1(✉)], Xiaoling Hu[1], Malte Hoffmann[1], Oula Puonti[1,2], and Juan Eugenio Iglesias[1,3,4]

[1] Massachusetts General Hospital and Harvard Medical School, Boston, USA
kgopinath@mgh.harvard.edu
[2] Danish Research Centre for Magnetic Resonance, Copenhagen University Hospital, Copenhagen, Denmark
[3] Centre for Medical Image Computing, University College London, London, UK
[4] Computer Science and AI Laboratory, Massachusetts Institute of Technology, Cambridge, USA

Abstract. In human neuroimaging studies, atlas registration enables mapping MRI scans to a common coordinate frame, which is necessary to aggregate data from multiple subjects. Machine learning registration methods have achieved excellent speed and accuracy but lack interpretability and flexibility at test time (since their deformation model is fixed). More recently, keypoint-based methods have been proposed to tackle these issues, but their accuracy is still subpar, particularly when fitting nonlinear transforms. Here we propose Registration by Regression (RbR), a novel atlas registration framework that: is highly robust and flexible; can be trained with cheaply obtained data; and operates on a single channel, such that it can also be used as pretraining for other tasks. RbR predicts the (x, y, z) atlas coordinates for every voxel of the input scan (i.e., every voxel is a keypoint), and then uses closed-form expressions to quickly fit transforms using a wide array of possible deformation models, including affine and nonlinear (e.g., Bspline, Demons, invertible diffeomorphic models, etc.). Robustness is provided by the large number of voxels informing the registration and can be further increased by robust estimators like RANSAC. Experiments on independent public datasets show that RbR yields more accurate registration than competing keypoint approaches, over a wide range of deformation models.

1 Introduction

Image registration seeks to find a spatial mapping between two images, and is a crucial component of human neuroimaging pipelines, e.g., for measuring change between different timepoints of the same subject [16]; building population atlases

---

K. Gopinath and X. Hu—Contributed equally as first authors.
O. Puonti and J. E. Iglesias—Contributed equally as senior authors.

M. Modat et al. (Eds.): WBIR 2024, LNCS 15249, pp. 205–215, 2024.
https://doi.org/10.1007/978-3-031-73480-9_16

or subject-specific templates via co-registration of multiple scans [24]; measuring structural differences via tensor-based morphometry [19]; mapping pre-, intra-, and post-operative images [1]; or automated segmentation [21]. One important application of these methods is registering brain MRI scans to digital atlases, such as MNI. Atlas registration enables aggregation of spatial information of multiple subjects into a common coordinate frame (e.g., for voxel-based morphometry [5]), as well as mapping of population-wide information to the native space of a specific subject, e.g., label priors for image segmentation [11]. Atlas registration typically relies on the same methods as pairwise registration.

Most classical pairwise registration methods use numerical optimization [31] to minimize a cost function, often combining an image similarity term with a regularizer; the former seeks to align the images to each other, while the latter prevents excessively convoluted (unrealistic) deformation. Decades of research have provided medical image registration with features such as diffeomorphisms [6], symmetry [8], or inter-modality mapping [32].

In 2017, faster registration methods based on deep learning were initially published. The first attempts were supervised and sought to predict ground truth fields obtained, e.g., with accurate classical methods [39]. These algorithms then evolved into unsupervised methods, trained with the same losses used by classical algorithms – possibly augmented with segmentation losses [7,9]. By now, deep learning registration methods have incorporated many of the features of their classical counterparts, e.g., diffeomorphisms [25], symmetry [20], progressive warping [27], or inter-modality support [13]. However, deep learning registration also has drawbacks, including: lack of interpretability (as many other deep learning domains); lack of flexibility of the deformation model, which is fixed and cannot be modified at test time (beyond tweaking the regularizer weight with hypernetworks [17]); and lack of robustness of the affine component.

Very recently, keypoint methods [14,37] have been proposed, which predict blob-like feature maps such that their centers of gravity can be used to analytically compute a transform between the two images. While the feature prediction remains opaque, this approach partly paves the way for addressing the lack of interpretability by providing insight into the factors driving the alignment, i.e., the keypoints. Keypoints also have potential in dynamic selection of deformation model at test time to warp one point cloud to the other, e.g., affine or thin-plate spline (TPS) transforms in KeyMorph [37]. Finally, keypoints also effectively tackle the lack of robustness of the affine component. Earlier learning approaches for nonlinear registration assumed that the inputs were already affinely aligned; some tried to predict the 12-parameter affine matrix [9,29], but these predictions were generally sensitive to initialization and often failed to generalize well to new datasets.

Here we present Registration by Regression (RbR), a novel framework for flexible, robust registration of brain MRI scans to atlases. RbR is an evolution of keypoint methods where *every voxel* is a keypoint: it is a supervised convolutional neural network (CNN) that predicts (x, y, z) atlas coordinates for every input voxel. RbR has a number of advantages: *(i)* Since the number of keypoints is very high ($\sim 10^6$ vs $\sim 10^2$ in KeyMorph), nonlinear transforms can be fitted more accurately. *(ii)* After a forward pass of the CNN, a wide array of popular transforms and regularizers can be fit in closed-form, including affine and nonlinear models like Bsplines [34], Demons [35], or the diffeomorphic log-polyaffine model [2], among others. *(iii)* It supports robust estimators like RANSAC [12]. *(iv)* Every hyperparameter of the deformation model can be readily modified; this is in contrast with hypernetwork approaches [17], where only specific parameters can be tuned as the space of deformations is constant (e.g., control point spacing cannot be specified). *(v)* Being a prediction on a single image (channel), it can be directly and easily used as pretraining ("feature extractor") for other tasks. *(vi)* Compared with KeyMorph, it does not need pretraining to avoid clustering of keypoints in the center of the image. *(vii)* Compared with classical registration, it is much faster (particularly when considering that one fit different transforms "for free" after the forward pass) and can be used as building block in deep learning architectures (as a frozen differentiable block to align images, or as a feature extractor as mentioned above).

2 Methods

2.1 Data Preparation

RbR (Fig. 1) is a regression CNN that estimates, for every voxel of the input scan (with discrete coordinates x, y, z), its corresponding atlas coordinates x', y', z'. RbR uses supervised training with accurate nonlinear deformations obtained with a classical registration approach $\mathcal{R}$. If $I(x, y, z)$ is the input image and A is an atlas, the ground truth registration is:

$$\mathbf{F} : \mathbb{R}^3 \rightarrow \mathbb{R}^3, \text{where} \quad \mathbf{F}(x, y, z) = [x', y', z'] = \mathcal{R}(x, y, z; I, A).$$

Therefore, a large labeled dataset can be obtained "for free" by simply registering a set of unlabeled images to the atlas at hand using a classical registration algorithm – ideally a diffeomorphic methods that yields smooth and invertible fields [6]. We note that unlabeled images exist in abundance in public datasets – particularly 1mm isotropic MPRAGEs typically used in neuroimaging research.

We also note that, compared with using synthetic fields, the performance of our strategy is bounded by that of the classical algorithm $\mathcal{R}$. However, synthesis of realistic *(atlas, image, deformation)* triplets is not trivial, since atlases are much smoother than images. Moreover, classical algorithms still provide state-of-the-art accuracy in atlas registration – speed limitations aside.

2.2 Training

Given a 3D voxel-wise regression CNN $\boldsymbol{N}(\cdot)$ with parameters $\boldsymbol{\theta}$, an atlas A, and a set of J images $\{I_j\}_{j=1,\ldots,J}$, the goal is to minimize the loss:

$$\hat{\boldsymbol{\theta}} = \underset{\boldsymbol{\theta}}{\operatorname{argmin}} \sum_{j=1}^{J} \frac{1}{|\Omega_j|} \sum_{(x,y,z)\in\Omega_j} \ell_1[\mathcal{R}(x, y, z; I_k, A) - \boldsymbol{N}(x, y, z; I_k, \boldsymbol{\theta})], \quad (1)$$

where Ω_j is the foreground of I_j (i.e., voxels inside the brain), and we use ℓ_1 rather than ℓ_2 due to its higher robustness. We note that no spatial regularization is needed: the CNN just does its best in a voxel-by-voxel basis, and smoothness is provided by the deformation model and its parameters at test time. Nevertheless, augmentation is crucial to endow the trained models with the best possible generalization ability. We use aggressive augmentation during training, including random blurring, bias field, noise, and deformation (affine and elastic).

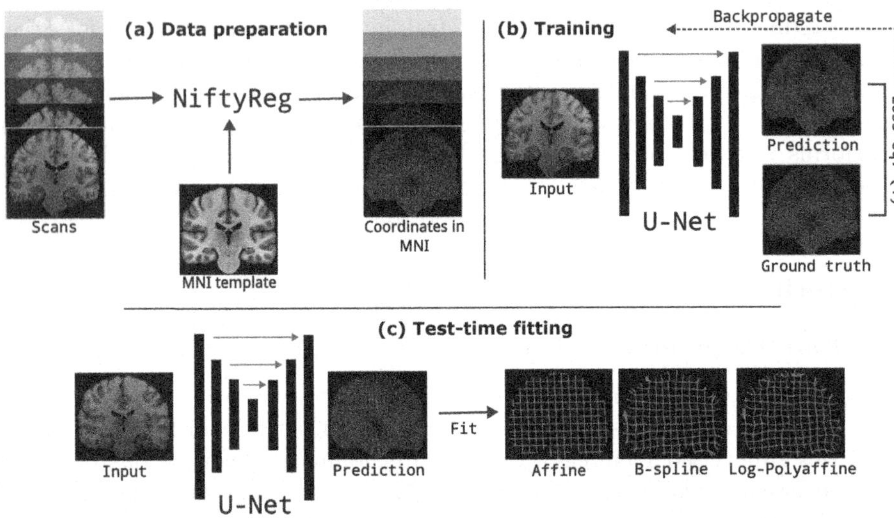

Fig. 1. (a) Training data are prepared by nonlinearly registering T1w scans from HCP and ADNI to the MNI template using NiftyReg. (b) A U-Net is trained to predict the MNI coordinates for every voxel of an input scan. (c) At test time, a wide array of transformations can be fitted between the input scan and predicted MNI coordinates.

2.3 Test-Time Fitting

Given a CNN prediction $[x', y', z'] = \boldsymbol{N}(x, y, z; I, \hat{\boldsymbol{\theta}})$ (which does *not* require the atlas), one can fit a deformation model to $[x', y', z']$ to obtain the final atlas registration. We consider three different types of models in this article: a family of models based on basis functions, a model based on the Demons algorithm [35], and a diffeomorphic model based on the log-polyaffine framework [2].

Basis Functions: a large class of transforms can be written as a sum of spatial basis functions and can be fitted to $[x', y', z']$ using regularized least squares, one spatial coordinate at the time. Let $\boldsymbol{d}$ be the flattened column vector of x', y', or z' coordinates. Let $\boldsymbol{\phi}$ be a set of B spatial functions evaluated on the discrete image domain Ω, such that each row corresponds to a spatial location (flattened), and each column corresponds to a basis function. Finally, let $\boldsymbol{c}$ be a column vector with B elements, corresponding to the coefficients of the basis functions. Then, the fitted transform is $\boldsymbol{\phi c}$ and its squared coordinate error is:

$$E_{coord} = [\boldsymbol{d} - \boldsymbol{\phi c}]^t[\boldsymbol{d} - \boldsymbol{\phi c}] = \boldsymbol{d}^t\boldsymbol{d} + \boldsymbol{c}^t\boldsymbol{\phi}^t\boldsymbol{\phi c} - 2\boldsymbol{d}^t\boldsymbol{\phi c}.$$

This fit can be regularized with quadratic penalty terms while remaining closed form. Here we choose the membrane energy, which penalizes the squared norm of the gradient of the deformation field. Let $\boldsymbol{G}_x, \boldsymbol{G}_y, \boldsymbol{G}_z$ be the spatial gradients of the basis functions; as for $\boldsymbol{\phi}$, every row corresponds to a spatial location, and every column to a basis function. The regularizer is then given by:

$$E_{reg} = [\boldsymbol{G}_x\boldsymbol{c}]^t[\boldsymbol{G}_x\boldsymbol{c}] + [\boldsymbol{G}_y\boldsymbol{c}]^t[\boldsymbol{G}_y\boldsymbol{c}] + [\boldsymbol{G}_z\boldsymbol{c}]^t[\boldsymbol{G}_z\boldsymbol{c}] = \boldsymbol{c}^t[\boldsymbol{G}_x^t\boldsymbol{G}_x + \boldsymbol{G}_y^t\boldsymbol{G}_y + \boldsymbol{G}_z^t\boldsymbol{G}_z]\boldsymbol{c}.$$

E_{coord} and E_{reg} are combined into an objective function E using a relative weight λ. Setting the gradient to zero yields the regularized least square estimate:

$$\nabla E = \nabla(E_{coord} + \lambda E_{reg}) = 2\boldsymbol{\phi}^t\boldsymbol{\phi c} - 2\boldsymbol{\phi}^t\boldsymbol{d} + 2\lambda[\boldsymbol{G}_x^t\boldsymbol{G}_x + \boldsymbol{G}_y^t\boldsymbol{G}_y + \boldsymbol{G}_z^t\boldsymbol{G}_z]\boldsymbol{c} = 0,$$
$$\Rightarrow \boldsymbol{c} = [\boldsymbol{\phi}^t\boldsymbol{\phi} + \lambda(\boldsymbol{G}_x^t\boldsymbol{G}_x + \boldsymbol{G}_y^t\boldsymbol{G}_y + \boldsymbol{G}_z^t\boldsymbol{G}_z)]^{-1}\boldsymbol{\phi}^t\boldsymbol{d}.$$

Within this family of transforms, we consider three in this article:

- **Affine**: a standard affine transform, fitted with all available voxels. We note that, in this case, the basis function matrix is simply $\boldsymbol{\phi} = [\boldsymbol{x}, \boldsymbol{y}, \boldsymbol{z}, \mathbb{1}]$ (where $\mathbb{1}$ is the all-ones vector), and the gradients of the basis functions are ignored.
- **Affine-RANSAC**: an affine transform fitted robustly with RANSAC.
- **Bsplines** [34]: with control point spacing specified by the user.

Other basis previously used in registration that would be straightforward to implement include lower-order polynomials [38] or the discrete cosine transform [4].

Demons-Like Model: The demons algorithm [35] computes a nonlinear registration by alternating between: *(i)* estimating force vectors via an optical-flow-like algorithm [18]; *(ii)* smoothing these force vectors with a Gaussian kernel; and *(iii)* applying the deformation before recomputing the force vectors. Here, we propose a similar algorithm with the difference that, since the output of the CNN is constant, we only take one step, i.e., we just filter the output of the CNN with a Gaussian kernel. The standard deviation is specified by the user.

Log-Polyaffine Model: Diffeomorphic models that can be analytically inverted are desirable in registration. Here we explore a log-polyaffine model [2] that is fitted to the output of the CNN as follows. First, we subdivide the image domain into cubic supervoxels of fixed, user-defined width W; this parameter controls the flexibility of the transform. Next, we compute an affine transform for every supervoxel s, assuming that it contains a minimum number of voxels in the brain mask. If we denote this affine transform by $\boldsymbol{T}_s$ in homogeneous coordinates, if can be shown that [2]:

$$\log(\boldsymbol{T}_s) = \begin{pmatrix} \boldsymbol{L}_s & \boldsymbol{v}_s \\ 0 & 0 \end{pmatrix},$$

and it can be shown that the affine transform can be represented by a stationary velocity field (SVF) given by: $\boldsymbol{\Psi}_s(x, y, z) = \boldsymbol{v}_s + \boldsymbol{L}_s[x, y, z]^t$. The log-polyaffine framework computes a global SVF $\boldsymbol{\Psi}$ as a weighted sum of SVFs:

$$\boldsymbol{\Psi}(x, y, z) = \sum_s w_s(x, y, z)(\boldsymbol{v}_s + \boldsymbol{L}_s[x, y, z]^t),$$

where $w_s(x, y, z)$ are normalized non-negative weights obtained with a Gaussian function of the distance between (x, y, z) and the center of supervoxel s; we set the standard deviation of this Gaussian to $W/2$. The final SVF $\boldsymbol{\Psi}$ can be integrated with the scale-and-square algorithm to obtain the final deformation field [3]; the inverse field can be obtained by integrating the negated SVF $(-\boldsymbol{\Psi})$.

2.4 Implementation Details

The CNN is a standard U-net [33], with a design inspired by nnU-net [22]. It has four resolution levels with two convolutional layers (comprising $3 \times 3 \times 3$ convolutions and a ReLu) followed by $2 \times 2 \times 2$ max pooling (in the encoder) or upconvolution (decoder). The final activation layer is linear, to regress the atlas coordinates in decimeters (which roughly normalizes them from -1 to 1). In addition to predicting atlas coordinates, the CNN also learns to predict a brain mask, which is used to define the domain Ω (i.e., to mask the loss during training, and to select the voxels that are used to fit the transforms at inference). For this, we add a segmentation loss to Eq. 1 with relative weight 0.5; the segmentation loss itself combines a Dice loss (weight 0.75) and cross-entropy loss (weight 0.25). The model is trained with stochastic gradient descent with learning rate 0.01, weight decay 3e$-$5, and momentum 0.99 for 100 epochs, setting aside 20% of the data for validation (selecting the best model). Input images are normalized with their median value inside the brain mask. At test time, nonlinear models are fitted on the residual of the Affine-RANSAC fit (clipped at 10mm for outlier rejection). The RANSAC algorithm uses 50,000 randomly selected voxels, a maximum of 100 iterations, tests 500 voxels in every attempt, and requires 20% of voxels to be inliers to consider any given solution. The minimum number of voxels per supervoxel in the log-polyaffine model is 100.

3 Experiments and Results

The training data consists of high-resolution, isotropic, T1-weighted scans of 897 subjects from the HCP dataset [36] and 1148 subjects from the ADNI [23]. The test data consists of high-resolution, isotropic, T1 of the first 100 subjects from both the ABIDE [10] and OASIS3 [26] datasets, for a total of 200 test subjects. All scans were segmented into 36 regions using FreeSurfer [11]. Details on the acquisitions can be found in the corresponding publications.

3.1 Data Preparation for RbR Training

The scans in the training dataset were masked and registered to the ICBM 2009b Nonlinear Symmetric MNI template using NiftyReg [28]. Specifically, we first ran the block matching algorithm for affine registration (`reg_aladin`) with the `-noSym` option, and subsequently used the fast free-form deformation algorithm (`f3d`) to compute the nonlinear registration. `f3d` was run in diffeomorphic mode (`-vel`) and local normalized correlation coefficient ($\sigma = 5$) as similarity metric. The processing time for the whole dataset was less than 24h on a 64-core desktop.

3.2 Experimental Setup

We quantify registration accuracy with average Dice scores between the segmentations of the scans and the segmentation of the atlas, deformed with the estimated registrations. We further evaluate the regularity of the deformations using the membrane energy, which is one of the smoothness metrics used in the registration literature. Using these metrics, we compared RbR with:

NiftyReg: We use the same parameters as for the training data. Since we used it as ground truth in training, NiftyReg provides a ceiling for the performance that RbR can achieve.

KeyMorph [37]**:** Our main competing method is KeyMorph. We test KeyMorph variants as trained by the original authors (2024-03-06 version). For affine registration, we use 128 keypoints with weights optimized for affine Dice overlap. To estimate non-linear transforms, we use the TPS model with 512 keypoints, Dice-specific weights, and $\lambda_{\text{TPS}} = 0$ (which yields the highest Dice in their article). We intensity normalize and skull-strip scans as in the original publication.

SynthMorph [13]**:** for completeness, we also consider a learning-based state-of-the-art algorithm trained to maximize segmentation overlap (Dice scores) with synthetic images. SynthMorph has an affine model based on keypoints [14] and a nonlinear model based on regression [15]. We finetuned SynthMorph with the MNI atlas permanently selected as moving image, and evaluated its affine and nonlinear modules separately. We used its default trade-off parameter $\lambda = 0.5$.

3.3 Results

Figure 2 shows qualitative registration results, while Fig. 3 shows box and significance plots for the average Dice and membrane energy. KeyMorph with TPS yields an average Dice of 0.60, which is 7 points lower than the value reported in the original publication. This could be a domain gap issue – including both the test datasets and the MNI atlas, which is blurrier than a regular scan. Errors are apparent in Fig. 2, where KeyMorph fails to correct the rotation about the A-P axis in linear mode and introduces wrong deformations in nonlinear mode.

Compared with KeyMorph, RbR does not require skull-stripping and yields better Dice, with medians of ~0.70 (affine) and ~0.73 (nonlinear) – higher than

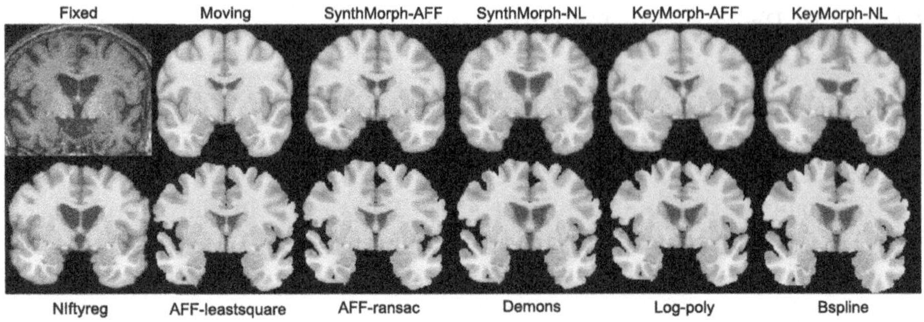

Fig. 2. Coronal slice of sample fixed image and corresponding registered MNI slice, using the different approaches.

Keymorph's ∼0.60 on our dataset or even the value reported in their publication (∼0.67). While the improvement from linear to nonlinear models in RbR may not be large in terms of median, it is very noticeable in terms of the first quartile and outliers. In Fig. 2, the difference between the nonlinear versions is subtle, but noticeable e.g., around the third ventricle (less aggressively "closed" by the Bspline model). The ability to more accurately follow contours is paid in terms of membrane energy, which is noticeably higher for the more flexible models.

Finally, we note that, while RbR-affine and Synthmorph-affine (both based on keypoints) achieve the same Dice as NiftyReg in the affine model, keypoint methods still trail classical and learning-based in nonlinear registration: NiftyReg and SynthMorph achieve Dice scores 10 points higher than RbR and ∼20 points higher than KeyMorph in this setting – and SynthMorph does so with similar levels of membrane energy.

Limitations. Initially, the philosophy behind regressing deformation fields directly was to leave out smoothness to fit a variety of other models. In the extended version of this article, we will incorporate smoothness constraints into the training phase and discuss the degree of smoothness as well as its impact on registration. Thorough comparisons with existing methods other than the current state of the art SynthMorph [13] and KeyMorph [37] methods will be considered, including Quicksilver [39] to provide a more comprehensive evaluation. Tuning the deformation model independently of the prediction is also critical. Comparing the proposed method against approaches like conditional LapIRN [30], which allows for post-hoc adjustments of regularization strength, will provide valuable insights into interpretable methods. This comparison will be included in the extended version to highlight the advantages and limitations of the current approach. Likewise, expanding the datasets and augmentation of training data with registration outcomes from multiple registration algorithms will ensure more reliable and robust deformation predictions.

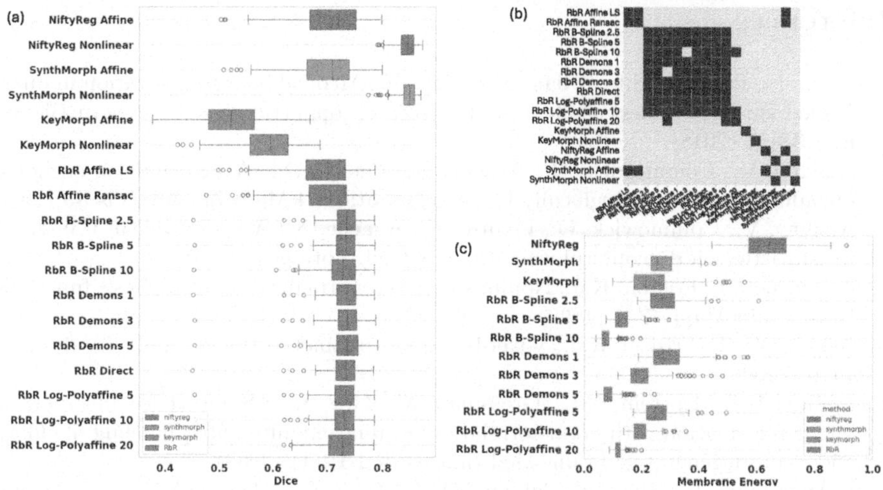

Fig. 3. (a) Box plot of average Dice between segmentations of scans and registered MNI. For RbR, the numbers indicate control point spacing (Bsplines), Gaussian σ (demons), or supervoxel width (log-polyaffine). (b) Significance matrix for two-tailed t-tests comparing the Dice scores; pink represents $p < 0.05$. (c) Membrane energy for nonlinear models. In the box plots, the center line is the median, the ends of the box are the 25th and 75th percentiles, and the whiskers extend to the furthest observations not considered outliers – which are marked with circles. Outliers are values more than 1.5 times the interquartile range away from the ends of the box. (Color figure online)

3.4 Discussion and Conclusion

RbR offers a new perspective on interpretable keypoint-based registration, posing it as a coordinate regression problem. Allowing every voxel to inform the model fit enables RbR to outperform standard keypoint approaches and fit less parsimonious nonlinear models. RbR also has disadvantages, particularly the need to retrain the model for each new atlas. Future work will include fitting other models, adding topological losses in training (e.g., penalizing negative Jacobians), investigating RbR's value as a feature extractor (i.e., as pretraining for other tasks), exploring other improvements that help close the gap with non-interpretable approaches, and integrating with domain randomization approaches for contrast- and resolution-agnostic registration (as in [14]). We believe that RbR will be particularly valuable in scenarios requiring robust and interpretable registration in presence of extreme rotations, such as fetal imaging.

Acknowledgments. Work primarily supported by NIH grants 1R01AG070988 and 1UM1MH130981. Support also provided by NIH 1RF1MH123195, 1RF1AG080371, and R00-HD101553. Oula Puonti was supported by a grant from Lundbeckfonden (grant number R360-2021-395).

References

1. Alam, F., Rahman, S.U., Ullah, S., Gulati, K.: Medical image registration in image guided surgery: Issues, challenges and research opportunities. Biocybern Biomed Eng **38**(1) (2018)
2. Arsigny, V., Commowick, O., Ayache, N., Pennec, X.: A fast and log-euclidean polyaffine framework for locally linear registration. J Math Imaging Vis **33** (2009)
3. Arsigny, V., Commowick, O., Pennec, X., Ayache, N.: A log-euclidean framework for statistics on diffeomorphisms. In: MICCAI. Springer (2006)
4. Ashburner, J., Friston, K.J.: Nonlinear spatial normalization using basis functions. Hum Brain Mapp **7**(4) (1999)
5. Ashburner, J., Friston, K.J.: Voxel-based morphometry–the methods. Neuroimage **11**(6) (2000)
6. Avants, B.B., Epstein, C.L., Grossman, M., Gee, J.C.: Symmetric diffeomorphic image registration with cross-correlation: evaluating automated labeling of elderly and neurodegenerative brain. Med Image Anal **12**(1) (2008)
7. Balakrishnan, G., Zhao, A., Sabuncu, M.R., Guttag, J., Dalca, A.V.: VoxelMorph: a learning framework for deformable medical image registration. IEEE T Med Imaging **38**(8) (2019)
8. Christensen, G.E., Johnson, H.J.: Consistent image registration. IEEE T Med Imaging **20**(7) (2001)
9. De Vos, B.D., Berendsen, F.F., Viergever, M.A., Sokooti, H., Staring, M., Išgum, I.: A deep learning framework for unsupervised affine and deformable image registration. Med Image Anal **52** (2019)
10. Di Martino, A., et al.: The autism brain imaging data exchange: towards a large-scale evaluation of the intrinsic brain architecture in autism. Mol Psychiatry **19**(6) (2014)
11. Fischl, B., Salat, D.H., et al.: Whole brain segmentation: automated labeling of neuroanatomical structures in the human brain. Neuron **33**(3) (2002)
12. Fischler, M.A., Bolles, R.C.: Random sample consensus: a paradigm for model fitting with applications to image analysis and automated cartography. Comm of the ACM **24**(6) (1981)
13. Hoffmann, M., Billot, B., Greve, D.N., Iglesias, J.E., Fischl, B., Dalca, A.V.: SynthMorph: learning contrast-invariant registration without acquired images. IEEE T Med Imaging **41**(3) (2021)
14. Hoffmann, M., Hoopes, A., Fischl, B., Dalca, A.V.: Anatomy-specific acquisition-agnostic affine registration learned from fictitious images. In: Medical Imaging 2023: Image Processing. vol. 12464. SPIE (2023)
15. Hoffmann, M., Hoopes, A., Greve, D.N., Fischl, B., Dalca, A.V.: Anatomy-aware and acquisition-agnostic joint registration with synthmorph. arXiv preprint arXiv:2301.11329 (2023)
16. Holland, D., Dale, A.M., Initiative, A.D.N., et al.: Nonlinear registration of longitudinal images and measurement of change in regions of interest. Med Image Anal **15**(4) (2011)
17. Hoopes, A., Hoffmann, M., et al.: Learning the effect of registration hyperparameters with hypermorph. Machine Learning for Biomedical Imaging **1** (2022)
18. Horn, B.K., Schunck, B.G.: Determining optical flow. Artif Intell **17**(1-3) (1981)
19. Hua, X., Lee, S., Yanovsky, I., et al.: Optimizing power to track brain degeneration in alzheimer's disease and mild cognitive impairment with tensor-based morphometry: an adni study of 515 subjects. Neuroimage **48**(4) (2009)

20. Iglesias, J.E.: A ready-to-use machine learning tool for symmetric multi-modality registration of brain MRI. Scientific Reports **13**(1) (2023)
21. Iglesias, J.E., Sabuncu, M.R.: Multi-atlas segmentation of biomedical images: a survey. Med Image Anal **24**(1) (2015)
22. Isensee, F., Jaeger, P.F., Kohl, S.A., Petersen, J., Maier-Hein, K.H.: nnU-Net: a self-configuring method for deep learning-based biomedical image segmentation. Nat Methods **18**(2) (2021)
23. Jack Jr, C.R., Bernstein, M.A., Fox, N.C., Thompson, P., Alexander, G., Harvey, D., Borowski, B., Britson, P.J., L. Whitwell, J., Ward, C., et al.: The Alzheimer's disease neuroimaging initiative (ADNI): MRI methods. J. Magn Reson Imaging **27**(4) (2008)
24. Joshi, S., Davis, B., Jomier, M., Gerig, G.: Unbiased diffeomorphic atlas construction for computational anatomy. NeuroImage **23** (2004)
25. Krebs, J., Delingette, H., Mailhé, B., Ayache, N., Mansi, T.: Learning a probabilistic model for diffeomorphic registration. IEEE T Med Imaging **38**(9), 2165–2176 (2019)
26. LaMontagne, P.J., et al.: Oasis-3: longitudinal neuroimaging, clinical, and cognitive dataset for normal aging and alzheimer disease. MedRxiv (2019)
27. Lv, J., Wang, Z., et al.: Joint progressive and coarse-to-fine registration of brain MRI via deformation field integration and non-rigid feature fusion. IEEE T Med Imaging **41**(10) (2022)
28. Modat, M., Ridgway, G.R., Taylor, Z.A., Lehmann, M., Barnes, J., Hawkes, D.J., Fox, N.C., Ourselin, S.: Fast free-form deformation using graphics processing units. Comput Methods Programs Biomed **98**(3) (2010)
29. Mok, T.C., Chung, A.: Affine medical image registration with coarse-to-fine vision transformer. In: Proceedings of the IEEE/CVF on CVPR (2022)
30. Mok, T.C., Chung, A.C.: Conditional deformable image registration with convolutional neural network. In: Medical Image Computing and Computer Assisted Intervention–MICCAI 2021: 24th International Conference, Strasbourg, France, September 27–October 1, 2021, Proceedings, Part IV 24. pp. 35–45. Springer (2021)
31. Nocedal, J., Wright, S.J.: Numerical optimization. Springer (1999)
32. Pluim, J.P., Maintz, J.A., Viergever, M.A.: Mutual-information-based registration of medical images: a survey. IEEE T Med Imaging **22**(8) (2003)
33. Ronneberger, O., Fischer, P., Brox, T.: U-net: Convolutional networks for biomedical image segmentation. In: MICCAI. Springer (2015)
34. Rueckert, D., Sonoda, L.I., Hayes, C., Hill, D.L., Leach, M.O., Hawkes, D.J.: Non-rigid registration using free-form deformations: application to breast MR images. IEEE T Med Imaging **18**(8) (1999)
35. Thirion, J.P.: Image matching as a diffusion process: an analogy with Maxwell's demons. Med Image Anal **2**(3) (1998)
36. Van Essen, D.C., Smith, S.M., Barch, D.M., Behrens, T.E., Yacoub, E., Ugurbil, K., Consortium, W.M.H., et al.: The WU-minn human connectome project: an overview. Neuroimage **80** (2013)
37. Wang, A.Q., Evan, M.Y., Dalca, A.V., Sabuncu, M.R.: A robust and interpretable deep learning framework for multi-modal registration via keypoints. Med Image Anal **90** (2023)
38. Woods, R.P., Grafton, S.T., Watson, J.D., Sicotte, N.L., Mazziotta, J.C.: Automated image registration: II. intersubject validation of linear and nonlinear models. J Comput Assist Tomogr **22**(1) (1998)
39. Yang, X., Kwitt, R., Styner, M., Niethammer, M.: Quicksilver: Fast predictive image registration–a deep learning approach. NeuroImage **158**, 378–396 (2017)

Diffusion Model-Based Hierarchical Registration of Whole-Body PET/CT Images

Jia Mi[1], Caiwen Jiang[1], and Dinggang Shen[1,2,3](✉)

[1] School of Biomedical Engineering and State Key Laboratory of Advanced Medical Materials and Devices, ShanghaiTech University, Shanghai, China
dgshen@shanghaitech.edu.cn

[2] Shanghai United Imaging Intelligence Co., Ltd., Shanghai, China

[3] Shanghai Clinical Research and Trial Center, Shanghai, China

Abstract. Whole-body PET/CT imaging technology can capture entire human-body images in a single scan to provide comprehensive metabolic and anatomical information. By analyzing these whole-body images, physiological and pathological changes in the whole body can be better understood. Since images can be acquired from different time points of the same subject (or different subjects), it is important to register them together for accurate measurement of signal changes in the same location. However, current registration methods are often designed for specific organs or regions, and thus cannot be effectively applied to whole-body image registration. To address this issue, we design a hierarchical registration framework based on diffusion model for whole-body images. The key components of our proposed framework include 1) designing a multi-part partitioning and fusion strategy to divide the whole-body image into multiple parts for reducing difficulty and computational burden of registration, 2) employing a diffusion model based deformable registration network to achieve precise registration for each corresponding part, and 3) further incorporating various topology constraints to preserve topological details for achieving more reliable registration. Extensive evaluations on 64 samples of whole-body images demonstrate that our proposed hierarchical framework can obtain robust registration across all human body parts, leading to accurate whole-body registration.

Keywords: Whole-body PET/CT · Hierarchical Registration · Diffusion Model · Topology Constraint

1 Introduction

Whole-body PET/CT is an advanced technology which enables simultaneous imaging of the entire body and offers comprehensive metabolic and anatomical information. This is particularly important for applications that require systematic analysis of the entire body, i.e., pharmacological analysis and systematic

M. Modat et al. (Eds.): WBIR 2024, LNCS 15249, pp. 216–225, 2024.
https://doi.org/10.1007/978-3-031-73480-9_17

diseases research [1–4]. To ensure data consistency and comparability in whole-body image analysis, registration is often necessary to establish spatial correspondences between images taken at different time points of the same subject (or different subjects). Therefore, developing a robust registration algorithm is crucial for accurate and reliable whole-body image analysis.

In the past decades, many medical image registration methods have been proposed, which can be categorized into two classes, i.e., traditional methods and deep learning-based methods. Traditional methods adopt a series of iterative processes to find the optimal transformation for each pair of images. These methods can achieve robust performance, but they are often computationally expensive and time-consuming. In contrast, deep learning-based methods have recently excelled in studies due to their ability to perform fast and fully automatic registration [5–7]. For instance, VoxelMorph [8] is a representative unsupervised registration network that can predict deformation fields in a single step, greatly accelerating the registration process while maintaining high accuracy. Many advanced methods [9–13] employ coarse-to-fine registration strategy to hierarchically refine the registration performance, either from the entire image to local small patches [9–11], or by repeatedly running a single network [12,13]. However, existing learning-based registration algorithms are often designed for specific organs or body regions, and cannot be effectively applied to whole-body image registration. Therefore, a robust algorithm is needed for fast and accurate whole-body registration. Achieving this goal presents several challenges. First, whole-body image contains a large number of anatomical structures, leading to significant anatomical variability in terms of shape and contrast. Second, positions of different body parts can change largely among different subjects. These challenges make it difficult to use a single model for whole-body registration.

Recently, powerful diffusion models [14,15] have demonstrated significant potential and success in registration tasks. For example, DiffuseMorph [16,17] employs diffusion models to extract latent features, and apply them to the deformation network to predict more optimal deformation fields. FSDiffReg [18] further introduces feature-wise and score-wise diffusion guidance to exploit the diffusion model's ability to guide the registration in 4D cardiac images. Since the diffusion process corrupts the anatomical structure of the original target image, the estimation of deformation fields is not well constrained. Moreover, these methods are just validated on 4D cardiac images and cannot directly used for whole-body images.

In this work, we develop a diffusion-based hierarchical registration framework for whole-body images. Instead of directly registering images at the whole-body level, our framework first divides the whole-body image into three body parts. Each part undergoes separate deformable registration, and then the registration results are integrated to obtain the final result. The hierarchical framework uses three structurally identical registration sub-networks to register different body parts, accommodating variations between them while reducing registration complexity. The registration sub-networks, utilizing diffusion models, can effectively extract rich semantic features for precise registration. Additionally, by incorporating topology constraints (i.e., segmentation loss and diffeomorphic registration), we can better preserve topological properties. Evaluations conducted on 64 samples of PET/CT data validate that our proposed hierarchical registration framework can achieve robust registration for whole-body.

2 Method

Figure 1(A) shows the schematic illustration of our registration framework. It consists of three parts including one multi-part partitioning module, three diffusion model-based registration sub-networks, and one multi-part fusion module. Given a pair of whole-body images, they are first divided into three body parts (namely, head-neck, thorax, and abdomen). Each part is then independently registered using a diffusion model-based registration network. Finally, the multi-part fusion module is used to fuse the three deformation fields for forming a whole-body transformation.

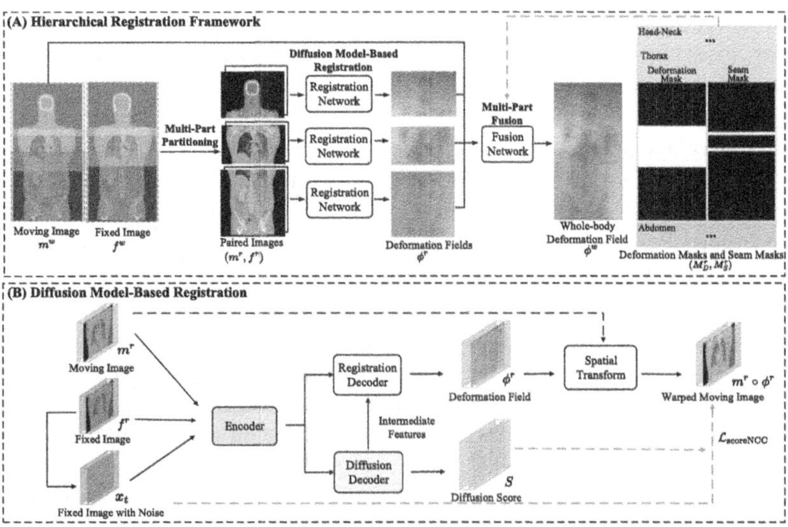

Fig. 1. Overview of our proposed method, where (A) provides the pipeline of the hierarchical registration framework including multi-part partitioning, diffusion model-based registration, and multi-part fusion, and (B) shows the details of diffusion model-based registration network.

2.1 Hierarchical Registration Framework

We adopt a hierarchical registration strategy to reduce the complexity of whole-body registration. This strategy involves multi-part partitioning and multi-part fusion processes. First, during the multi-part partitioning process, we divide the whole-body images along the z-axis into three body parts (denoted as $r = 1, 2, 3$) according to the masks of anatomical structures, i.e., 1) head-neck extends from head to upper boundary of ribs, 2) thorax spans from upper boundary of ribs to lower boundary of lungs, and 3) abdomen spans from upper boundary of liver to just above lower extremities. Each part is slightly extended beyond its core anatomical boundaries to keep features complete, leading to overlapped regions between parts. The use of region overlapping mitigates discontinuities at boundaries across body parts in the estimation of the whole-body deformation field.

Second, the three parts are then fed into three registration sub-networks for deformable registration. Details of registration network are given in Sect. 2.2.

Finally, the multi-part fusion process is adopted to combine the deformation fields ϕ^r ($r = 1, 2, 3$) output by the three registration sub-networks to obtain the final whole-body deformation field ϕ^w. This process is carried out by a fusion network. We aim to ensure smooth transitions at the boundary regions of the three parts while keeping the rest of the areas unchanged. To locate these regions in ϕ^w, we design the binary seam masks M_S^r and deformation masks M_D^r for each part. The seam masks are used to identify the boundary regions of each part, while the deformation masks locate the corresponding areas between ϕ^w and each ϕ^r. Figure 1(A) shows M_D^r and M_S^r of thorax, the voxels in thorax part (white region) are assigned a value of 1 while others (in black region) are assigned 0. To obtain M_S^r for a specific part, the deformation mask of the adjacent part is first convolved with a Sobel edge detection operator. Next, it is convolved three times with an all-one filter of kernel size 3. Finally, all elements are binarized to values of 0 and 1 to generate M_S^r. Thus we define loss function as follows:

$$\mathcal{L}_{\text{dfm}} = \sum^{r} \mathcal{L}_1(\phi^w \odot M_D^r, \phi^r \odot M_D^r) + \sum^{r} \mathcal{L}_1(\phi^w \odot M_S^r, \phi^r \odot M_S^r), \quad (1)$$

where $\odot$ is the element-wise multiplication and $\mathcal{L}_1$ denotes the L_1 loss. The total loss function for multi-part fusion is defined as:

$$\mathcal{L}_{\text{fusion}} = \lambda_d \mathcal{L}_{\text{dfm}} + \mathcal{L}_{\text{NCC}}(m^w \circ \phi^w, f^w) + \lambda \sum \|\nabla \phi^w\|^2, \quad (2)$$

where λ_d and λ are hyperparameters (which are set to 0.1 and 0.01, respectively) , L_{NCC} is the Normalized Cross Correlation loss, and $\sum \|\nabla \phi^w\|^2$ is the deformation field regularization term [8].

2.2 Diffusion Model-Based Registration Network

Due to the success of diffusion model used in different applications, we propose to employ diffusion model for image registration. The features of diffusion model can be used to enhance the deformation field estimation [16,17].

Inspired by [18], the registration and diffusion models are incorporated with a shared encoder. To take advantage of the informative semantics of the diffusion model, both the diffusion features and scores are used to guide image registration. Specifically, given a pair of fixed image f^r and moving image m^r, a perturbed fixed image x_t is generated by adding noise to f^r. The shared encoder takes the concatenated m^r, f^r, and x_t as input. The registration decoder predicts the deformation field ϕ^r, and the diffusion decoder predicts the diffusion score S. The intermediate features of the UNet-shaped diffusion decoder and the corresponding registration decoder are fused using cross attention in every layer to provide feature-level guidance. The diffusion score S is also used to reweight the similarity loss to provide score-level guidance. The final similarity loss is formulated as follows:

$$\mathcal{L}_{\text{scoreNCC}}(\phi^r \circ m^r, f^r, S) = \left(\frac{1}{1+e^{-S}}\right) \odot ((m^r \circ \phi^r) \odot f^r). \tag{3}$$

2.3 Auxiliary Topological Constraints on Registration

To further enhance the registration accuracy, additional topological constraints are taken into account for deformation field estimation. Since the diffusion process corrupts the anatomical structure of the original target image, which complicates the registration task, we integrate two auxiliary topological constraints to regularize the prediction of the deformation fields. To be specific, first, segmentation masks of several anatomical structures including bones and organs are used as auxiliary loss to impose structural constraints on the deformation field. Here, we integrate the DSC loss [19] below to ensure similarity between the segmentation masks $s_m \circ \phi$ and s_f of the deformed moving image and fixed image, respectively.

$$\text{DSC}(s_f, s_m, \phi) = 1 - \frac{1}{K}\sum_k \frac{2\sum_{p\in\Omega} s_f^k(p)[s_m^k \circ \phi](p)}{\sum_{p\in\Omega}(s_f^k(p))^2 + \sum_{p\in\Omega}([s_m^k \circ \phi](p))^2}, \tag{4}$$

where p denotes the coordinates of the voxel, and Ω represents the image domain. In addition, to ensure a diffeomorphic deformation which is continuous and topology-preserving, the deformation field is modeled by time-stational velocity fields (SVFs) which are defined through the following ordinary differential equation (ODE) [20]:

$$\frac{\partial\phi(\tau)}{\partial\tau} = v(\phi(\tau)), \tag{5}$$

where $\phi(0) = Id$ is the identity transformation and τ is time. By integral of the stationary velocity field v over $\tau = [0, 1]$, the final registration field ϕ can be obtained. Using the SVFs with an efficient scaling-and-squaring approach, ϕ can be obtained under end-to-end training.

Therefore, the total loss for registration sub-network training can be written as:

$$\mathcal{L}_{\text{reg}} = \mathcal{L}_{\text{diffusion}}(m^r, f^r, x_t, t) + \lambda_{\text{s}}\mathcal{L}_{\text{scoreNCC}} + \lambda_\phi \sum \|\nabla\phi^r\|^2 + \lambda_{\text{seg}}\text{DSC}(s_f, s_m, \phi^r), \tag{6}$$

where λ_s and λ_ϕ are hyperparameters and both are empirically set as 20, andλ_{seg} is set as 0.1. $L_{\text{diffusion}}$ is the diffusion loss for training the diffusion model [14], and t is the noise level of x_t.

3 Experiments and Results

3.1 Dataset

To evaluate our proposed method, we collect 64 whole-body PET images along with their corresponding CT images from the uEXPLORER PET/CT scanner. To normalize the dataset, we calculate the group's average in terms of height, volume, and center of mass using body masks as the reference image. All the other images are affinely registered to the reference, and their divided body parts are also aligned to the reference. Since there is no standardized positioning protocol within the dataset, limbs are removed from the images. The images are cropped to the size of $128 \times 176 \times 320$ by excluding the background, and resampled to a voxel size of $2.734 \times 2.734 \times 3.0\,\text{mm}^3$. The intensity of all the images is normalized to $[-1, 1]$. We split the data into training and testing with 50 images and 14 images, respectively.

3.2 Implementation Details

In this study, we use a well-developed segmentation algorithm to segment a total of 13 organ and bone masks from whole-body CT images [21], which include brain, mandible, rib, left lung, right lung, heart, spinal cord, spinal canal, vertebrae, spleen, liver, left kidney, and right kidney. The proposed framework is implemented using the PyTorch [22] platform and trained on an NVIDIA A100 GPU with a batch size of $\mathcal{B} = 1$ and a learning rate of 2×10^{-4}.

Table 1. Quantitative evaluation of the whole-body image registration in terms of HD95, DSC, and percentage of voxels with a non-positive Jacobian determinant. Standard deviations are provided in parentheses.

Method	Whole-body		
	HD95	DSC	$\lvert J\rvert \leq 0(\%)$
Affine	9.73(6.68)	0.49(0.24)	0
SyN	7.45(7.34)	0.66(0.20)	1.36(0.84)
VM-segs-diff	7.26(6.76)	0.67(0.21)	0.00(0.00)
FSD-segs-diff	7.48(7.66)	0.70(0.18)	0.00(0.00)
Ours	7.08(7.97)	0.73(0.17)	0.00(0.00)

3.3 Evaluation Metrics

We use three metrics to evaluate the registration performance including: 1) Dice Similarity Coefficient (DSC), 2) the 95th percentile of the Hausdorff Distance (HD95), and 3) $\lvert J\rvert \leq 0(\%)$, representing the percentage of voxels with a non-positive Jacobian determinant. DSC and HD95 are calculated on the paired

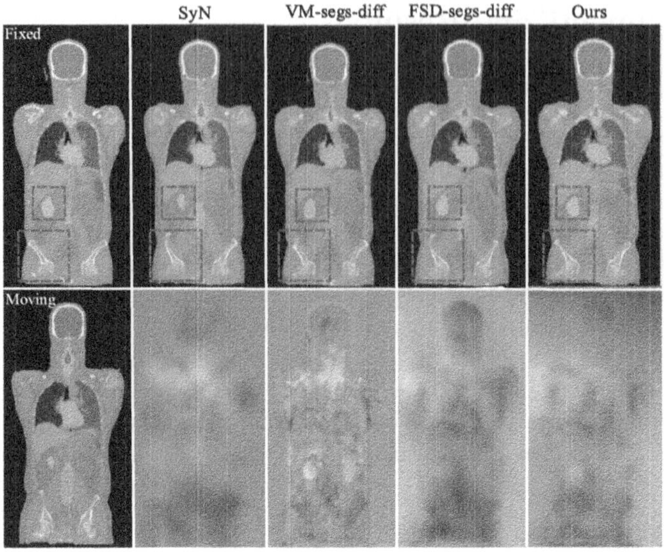

Fig. 2. Visual comparison for whole-body registration by different methods. The registered moving images (top) and the estimated deformation fields (bottom) are provided, where segmentation maps of several anatomical structures are overlaid.

masks of the fixed image and warped moving image. Evaluations are conducted for whole-body and three specific body parts. For whole-body registration, all 13 masks are utilized, while for each body part only masks of the anatomical structures included in the particular body part are considered.

3.4 Comparison with the State-of-the-Art Methods

Our hierarchical framework is compared with three other methods that directly employ deformable registration on whole-body images, including the traditional method SyN [23], the general ConvNets-based method VoxelMorph [8], and the diffusion model-based method FSDiffReg [18]. Note that, for fair comparison, both VoxelMorph and FSDiffReg are implemented with segmentation DSC loss and diffeomorphic registration, thus denoted as VM-segs-diff and FSD-segs-diff. Table 1 provides the quantitative summary of various methods for whole-body registration. 13 testing samples are registered to the reference image. We can see that VoxelMorph achieves improved registration accuracy compared to SyN. Our method outperforms the others by a large margin, achieving 0.73 in average DSC and 7.08 in HD95 for whole body.

In addition, we also demonstrate visual comparison in Fig. 2. Our method turns out to have the most accurate registration compared to the other methods, especially at regions with large deformation highlighted by red squares. Besides accuracy, the deformation field generated by our method seems more natural and plausible.

Table 2. Quantitative evaluation of registration of three body parts in terms of HD95, DSC, and percentage of voxels with a non-positive Jacobian determinant. Standard deviations are provided in parentheses.

Configurations			Head-Neck			Thorax			Abdomen		
Diffusion Model	Segmentation Loss	Diffeomorphic Registration	HD95	DSC	$\|J\| \leq 0(\%)$	HD95	DSC	$\|J\| \leq 0(\%)$	HD95	DSC	$\|J\| \leq 0(\%)$
	✓	✓	2.09(0.88)	0.80(0.13)	0.00(0.01)	6.72(5.70)	0.63(0.21)	0.00(0.00)	6.37(4.77)	0.66(0.20)	0.00(0.00)
✓			1.98(0.99)	0.81(0.13)	0.07(0.15)	7.58(6.69)	0.65(0.21)	0.70(0.24)	8.41(4.64)	0.67(0.16)	0.99(0.31)
✓	✓		2.33(1.21)	0.80(0.14)	0.05(0.07)	7.84(6.61)	0.66(0.19)	0.65(0.20)	8.26(4.63)	0.68(0.16)	0.80(0.33)
✓	✓	✓	2.40(1.09)	0.79(0.13)	0.00(0.00)	6.83(6.94)	0.72(0.16)	0.00(0.00)	6.63(4.92)	0.72(0.13)	0.00(0.00)

3.5 Ablation Study

To verify the effectiveness of our registration framework in terms of diffusion model-based registration network and two auxiliary topological constraints, we compare our model with another three variants, including 1) ours without diffusion model, which is implemented as a UNet-based registration network with two topological constraints, 2) ours without two topological constraints, and 3) ours without diffeomorphic registration. All methods are tested under the same experimental settings, with quantitative results presented in Table 2.

From the ablation study in Table 2, we can draw the following conclusions: 1) the use of diffusion model achieves improved registration performance by feature-level and score-level guidance from diffusion model; 2) the use of segmentation loss provides structural information, improving alignment according to DSC values, and 3) the use of diffeomorphic registration generates better deformation fields with less folding, as indicated by the Jacobian matrix. Diffeomorphic registration regularizes the estimated deformation field to balance registration accuracy and topology preservation. These three comparisons conjointly verify the effective design of our proposed framework, where both the diffusion model-based registration and the two auxiliary topology constraints contribute to the improvement of our registration task.

4 Conclusion

In this paper, we introduce a diffusion model-based hierarchical registration framework for whole-body image registration. To mitigate the complexity of whole-body registration, a multi-part partitioning and fusion strategy is used to break down the whole-body registration problem into multiple sub-problems of body-part registrations. Meanwhile, we propose a diffusion model-based registration network to perform each body-part registration. Experimental evaluation on 64 PET/CT whole-body samples show that, by incorporating semantic features extracted by diffusion model and topological constraints based on segmentation labels, our model achieves reliable whole-body registration with preserved topology.

Acknowledgments. This work was supported in part by National Natural Science Foundation of China (grant numbers U23A20295, 62131015, 62250710165), the STI 2030-Major Projects (No. 2022ZD0209000), Shanghai Municipal Central Guided Local Science and Technology Development Fund (grant number YDZX20233100001001), the China Ministry of Science and Technology (STI2030-Major Projects-2022ZD0213100), and The Key R&D Program of Guangdong Province, China (grant numbers 2023B0303040001, 2021B0101420006).

Declaration of Competing Interest. The authors declare that they have no known competing financial interests or personal relationships that could have appeared to influence the work reported in this paper.

References

1. Katal, S., Eibschutz, L.S., Saboury, B., Gholamrezanezhad, A., Alavi, A.: Advantages and applications of total-body PET scanning. Diagnostics 12(2), 426 (2022)
2. Ng, Q.K.-T., Triumbari, E.K.A., Omidvari, N., Cherry, S.R., Badawi, R.D., Nardo, L.: Total-body PET/CT-first clinical experiences and future perspectives. In: Seminars in Nuclear Medicine, 52(3), 330-339 (2022)
3. Slart, R.H., Bengel, F.M., Akincioglu, C., Bourque, J.M., Chen, W., Dweck, M.R., Hacker, M., Malhotra, S., Miller, E.J., Pelletier-Galarneau, M., Packard, R.R.: Total-Body PET/CT Applications in Cardiovascular Diseases: A Perspective Document of the SNMMI Cardiovascular Council. Journal of Nuclear Medicine, 65(4), 607-616 (2024)
4. Vandenberghe, S., Moskal, P., Karp, J.S.: State of the art in total body PET. EJNMMI Phys. 7(1), 1-33 (2020). https://doi.org/10.1186/s40658-020-00290-2
5. Avants, B.B., Epstein, C.L., Grossman, M., Gee, J.C.: Symmetric diffeomorphic image registration with cross-correlation: evaluating automated labeling of elderly and neurodegenerative brain. Med. Image Anal. 12(1), 26-41 (2008)
6. Christensen, G.E., Johnson, H.J.: Consistent image registration. IEEE Trans. Med. Imaging 20(7), 568-582 (2001)
7. Leow, A., Huang, S.C., Geng, A., Becker, J., Davis, S., Toga, A., Thompson, P.: Inverse Consistent Mapping in 3D Deformable Image Registration: Its Construction and Statistical Properties. In: Christensen, G.E., Sonka, M. (eds) IPMI 2005. LNCS, vol. 3565, pp. 493-503. Springer, Berlin, Heidelberg (2005). https://doi.org/10.1007/1150573041
8. Balakrishnan, G., Zhao, A., Sabuncu, M.R., Guttag, J., Dalca, A.V.: VoxelMorph: a learning framework for deformable medical image registration. IEEE Trans. Med. Imaging 38(8), 1788-1800 (2019)
9. Zhao, Y., Chen, X., McDonald, B., Yu, C., Mohamed, A.S., Fuller, C.D., Court, L.E., Pan, T., Wang, H., Wang, X., Phan, J.: A transformer-based hierarchical registration framework for multimodality deformable image registration. Comput. Med. Imaging Graph. 108, 102286 (2023)
10. Huang, Y., Ahmad, S., Fan, J., Shen, D., Yap, P.-T.: Difficulty-aware hierarchical convolutional neural networks for deformable registration of brain MR images. Med. Image Anal. 67, 101817 (2021). https://doi.org/10.1016/j.media.2020.101817
11. Mok, T.C.W., Chung, A.C.S.: Large deformation diffeomorphic image registration with laplacian pyramid networks. In: Martel, A.L., et al. (eds.) MICCAI 2020. LNCS, vol 12263, pp. 211-221 Springer, Cham (2020) https://doi.org/10.1007/978-3-030-59716-0_21

12. De Vos, B.D., Berendsen, F.F., Viergever, M.A., Sokooti, H., Staring, M., Išgum, I.: A deep learning framework for unsupervised affine and deformable image registration. Med. Image Anal. 52, 128-143 (2019)
13. Hu, B., Zhou, S., Xiong, Z., Wu, F.: Recursive decomposition network for deformable image registration. IEEE J. Biomed. Health Inform. 26(10), 5130-5141 (2022)
14. Ho, J., Jain, A., Abbeel, P.: Denoising diffusion probabilistic models. In: Advances in Neural Information Processing Systems, vol. 33, pp. 6840-6851 (2020)
15. Sohl-Dickstein, J., Weiss, E., Maheswaranathan, N., Ganguli, S.: Deep unsupervised learning using nonequilibrium thermodynamics. In: International Conference on Machine Learning (ICML), pp. 2256-2265. PMLR (2015)
16. Kim, B., Han, I., Ye, J.C.: DiffuseMorph: unsupervised deformable image registration using diffusion model. In: European Conference on Computer Vision, pp. 347-364. Springer (2022)
17. Kim, B., Ye, J.C.. Diffusion Deformable Model for 4D Temporal Medical Image Generation. In: Wang, L., Dou, Q., Fletcher, P.T., Speidel, S., Li, S. (eds) MICCAI 2022. LNCS, vol. 13431, pp. 539-548. Springer, Cham (2022). https://doi.org/10.1007/978-3-031-16431-6_51
18. Qin, Y., Li, X.: FSDiffReg: Feature-Wise and Score-Wise Diffusion-Guided Unsupervised Deformable Image Registration for Cardiac Images. In: Greenspan, H., et al.(eds) MICCAI 2023. LNCS, vol. 14229, pp.655-665. Springer, Cham (2023). https://doi.org/10.1007/978-3-031-43999-5_62
19. Milletari, F., Navab, N., Ahmadi, S.-A.: V-Net: Fully Convolutional Neural Networks for Volumetric Medical Image Segmentation. In: 2016 Fourth International Conference on 3D Vision (3DV), pp. 565-571. Stanford, CA, USA (2016). https://doi.org/10.1109/3DV.2016.79
20. Dalca, A.V., Balakrishnan, G., Guttag, J., Sabuncu, M.R.: Unsupervised Learning for Fast Probabilistic Diffeomorphic Registration. In: Frangi, A., Schnabel, J., Davatzikos, C., Alberola-López, C., Fichtinger, G. (eds) MICCAI 2018. LNCS, vol. 11070, pp. 729-738. Springer, Cham (2018). https://doi.org/10.1007/978-3-030-00928-1_82
21. Wasserthal, J., Breit, H.C., Meyer, M.T., Pradella, M., Hinck, D., Sauter, A.W., Heye, T., Boll, D.T., Cyriac, J., Yang, S. and Bach, M.: TotalSegmentator: robust segmentation of 104 anatomic structures in CT images. Radiology: Artificial Intelligence, 5(5) (2023)
22. Paszke A, Gross S, Massa F, Lerer A, Bradbury J, Chanan G, Killeen T, Lin Z, Gimelshein N, Antiga L, Desmaison A.: Pytorch: An imperative style, high-performance deep learning library. In: Advances in Neural Information Processing Systems, vol. 32, pp. 8026-8037 (2019)
23. Avants, B.B., Epstein, C.L., Grossman, M., Gee, J.C.: Symmetric diffeomorphic image registration with cross-correlation: evaluating automated labeling of elderly and neurodegenerative brain. Med. Image Anal. 12(1), 26-41 (2008)

Feature/Similarity Learning

Unsupervised Similarity Learning for Image Registration with Energy-Based Models

Daniel Grzech[1], Loïc Le Folgoc[1,5], Mohammad Farid Azampour[1,2,3], Athanasios Vlontzos[1,4], Ben Glocker[1], Nassir Navab[3,7], Julia Schnabel[3,6], and Bernhard Kainz[1,8](✉)

[1] Imperial College London, London, UK
{d.grzech17,bkainz}@imperial.ac.uk
[2] Munich Center for Machine Learning, Munich, Germany
[3] Technische Universität München, Munich, Germany
[4] Spotify Technology SA, London, UK
[5] Télécom Paris, Palaiseau, France
[6] King's College London, London, UK
[7] Johns Hopkins University, Baltimore, MD, USA
[8] Friedrich-Alexander-Universität Erlangen-Nürnberg, Erlangen, Germany

Abstract. We present a new model for deformable image registration, which learns in an unsupervised way a data-specific similarity metric. The proposed method consists of two neural networks, one that maps pairs of input images to transformations which align them, and one that provides the similarity metric whose maximisation guides the image alignment. We parametrise the similarity metric as an energy-based model, which is simple to train and allows us to improve the accuracy of image registration compared to other models with learnt similarity metrics by taking advantage of a more general mathematical formulation, as well as larger datasets. We also achieve substantial improvement in the accuracy of inter-patient image registration on MRI scans from the OASIS dataset compared to models that rely on traditional functions.

Keywords: energy-based models · image registration

1 Introduction

Image registration seeks to determine a transformation between a source and a target image, ensuring that corresponding regions in the transformed source and target images align and contain identical anatomical or physiological data. The main limitation of deformable image registration in medical imaging pipelines is accuracy, which remains unsatisfactory in case of many challenging problems, *e.g.* inter-patient registration of large organs or multi-modal registration. State-of-the-art image registration models, *e.g.* DLIR [35] and VoxelMorph [2,3,8,9],

M. Modat et al. (Eds.): WBIR 2024, LNCS 15249, pp. 229–240, 2024.
https://doi.org/10.1007/978-3-031-73480-9_18

use deep learning to learn a function that maps a pair of input images to a transformation which aligns them. Neural networks offer a speed up of many orders of magnitude over traditional iterative methods for image registration and can benefit from training on labelled data, *e.g.* organ segmentations [23]. However, most deep learning-based image registration models rely on traditional similarity metrics, *i.e.* sum of squared differences (SSD), local cross-correlation (LCC), and mutual information (MI), which means that without labelled training data, they show only moderate improvement in terms of accuracy over traditional methods.

We propose a novel methodology to improve the accuracy of image registration by learning a similarity metric in an unsupervised way. In previous literature, methods for unsupervised similarity learning in medical image registration usually rely on ad-hoc adversarial training of generative adversarial networks (GANs) [11,12,27]. Methods which focus on mono-modal image registration do not take advantage of their full potential to increase accuracy due to constraints on the model size and training data imposed by the mathematical formulation [15]. In this work, we adopt a fast, general, and mathematically rigorous approach. We formulate the similarity metric as an energy-based model (EBM) and make use of recent advancements in maximum likelihood (ML) training in order to learn a data-driven function from raw images.

1.1 Contributions

1. We propose a new formulation for unsupervised similarity learning in deformable medical image registration based on an energy-based model (EBM);
2. The similarity metric is trained using stochastic gradient Langevin dynamics (SGLD) and leads to improved image registration accuracy over neural networks trained with traditional metrics, with few additional parameters compared to other image registration models based on deep learning;
3. Unlike previous approaches to unsupervised similarity learning, we use full-sized images rather than image patches and enable similarity learning both in case of atlas-based and pairwise image registration.

1.2 Related Work

EBMs use concepts developed in order to study the statistical properties of gases in thermal equilibrium. They are an unsupervised learning model that makes it possible to learn the underlying probability distribution of the data and draw new samples based on the trained model [22,29].

In machine learning, EBMs were originally used for the efficient learning of representations with sparse auto-encoders [22,29]. The first work to propose training an EBM parametrised by a deep neural network via Langevin sampling was used for image generation [37]. EBMs trained in this way were also used *e.g.* for point cloud [38] and video generation [39,40]. Some works reinterpreted other types of machine learning models as EBMs, *e.g.* classifiers [14] and GANs [6]. However, the success of generative modelling in computer vision with variational autoencoders (VAEs) [20], GANs [13], and most recently diffusion models [30], limited the adoption of EBMs. These types of generative models were also applied in the context of medical image registration, *e.g.* VoxelMorph which used the ideas behind VAEs [9], the unsupervised similarity learning methods based on GANs [11,12,27], and DiffuseMorph which was based on diffusion models [19].

Non-standard similarity metrics in image registration tend to be either tailored to the problem through the insight of human experts [16] or learnt in a supervised way [4,5,7,33]. There is limited existing literature on unsupervised similarity learning for deformable image registration. Most such works are based on GANs [11,12,27] and rely on image patches generated from the dataset in an ad-hoc way. The pairs of input image patches are labelled as "well-registered" (positive examples x^+) or "unregistered" (negative examples x^-). x^+ do not involve the model output—instead they are defined as patches of the target image and of the weighted voxel-wise sum of the source and target images—while x^- are defined as patches of the target and the warped source images.

The unsupervised similarity learning method most similar to ours uses a variational Bayesian formulation and contrastive divergence in order to train a neural network which parameterises the similarity metric [15]. However, it suffers from slow speed due to the iterative nature of the registration algorithm. Moreover, the capacity of the neural network is constrained by intricate initialisation and memory, which makes it complicated to take advantage of large datasets and limits the improvement in accuracy.

Despite their relative lack of popularity, EBMs trained using SGLD continue to attract interest because they are simpler than other generative models [37,40] and generate samples without the use of an explicit generator [10,26]. To the best of our knowledge, this is the first time that EBMs are successfully used in the context of medical image registration. The code and the model weights will be made available in a public repository after the review period.

2 Method

2.1 Background

Let $\mathcal{I}{:=}\{I_k \mid k \in \{1, \ldots, N\}\}$ be a dataset of images, where $I_k \colon \Omega_k \to [0, 1]$. We denote the dataset of image pairs by $\mathcal{D}{:=}\mathcal{I} \times \mathcal{I} = \{(I_t, I_s) \mid t, s \in \{1, \ldots, N\}\}$, where I_t and I_s are respectively the target and the source image. In case of mono-modal image registration, we wish to align the image domains Ω_t and Ω_s using a transformation $\varphi\left(w_{ts}\right) : \Omega_s \to \Omega_t$ parameterised by w_{ts}, *i.e.*, to find parameters such that $I_t \simeq I_s\left(w_{ts}\right) {:=} I_s \circ \varphi^{-1}\left(w_{ts}\right)$.

Diffeomorphic transformations preserve topology and are invertible, which are important properties in medical imaging applications. In order to guarantee diffeomorphic properties and to parameterise the transformations in a memory-efficient way, we use stationary velocity free-form deformations based on cubic B-splines [25,28,32]. The use of spline functions implicitly ensures smoothness of the output. The numerical integration of spatially smooth stationary velocity fields via scaling-and-squaring produces diffeomorphisms [1].

Unsupervised image registration models based on deep learning compute a function $f_\theta\colon I_t \times I_s \mapsto \varphi^{-1}(w_{ts})$, which is implemented as a neural network with parameters θ. The loss function is similar to that used in traditional iterative image registration methods:

$$\hat{\theta} = \arg\min_{\theta} \mathcal{L}_{\text{sim}}\Big(I_t, I_s\left(w_{ts}\right)\Big) + \lambda_{\text{reg}}\mathcal{L}_{\text{reg}}\left(w_{ts}\right) \tag{1}$$

where $\mathcal{L}_{\text{sim}}$ is a similarity metric, λ_{reg} is a scalar regularisation weight, and $\mathcal{L}_{\text{reg}}$ ensures that the output transformations are spatially smooth. The usual choices of similarity metrics include SSD, LCC, and MI, *i.e.*, aside from improved generalisation and a regularisation effect, unsupervised deep learning image registration models are comparable to traditional algorithms in terms of accuracy.

2.2 Energy-Based Model

To maximise accuracy, we propose not only to fit a function which aligns the images in the dataset, but also to learn a data-specific similarity metric $\mathcal{L}_{\text{data}}$ to use as part of the loss function in Eq. (1).

In most deep learning-based image registration models, the function $f_\theta{:=}f^{(1)}_{\theta_D} \circ f^{(2)}_{\theta_E}$ is a convolutional neural network (CNN) which consists of an encoder $f^{(2)}_{\theta_E}$ with parameters θ_E and a decoder $f^{(1)}_{\theta_D}$ with parameters θ_D. The encoder $f^{(2)}_{\theta_E}$ maps input images to a latent space where similar images are close with respect to some distance [21]. Similarly, a suitable similarity metric maps similar image pairs to low energy values and disparate image pairs to high energy values.

We propose to use a learnable similarity metric implemented as an EBM. Let $x \in \mathcal{D}$ be a pair of input images and $f_{\theta_S}(x) \in \mathbb{R}_+$ an energy function, implemented as a neural network with weights θ_S. The energy function defines a Boltzmann probability distribution:

$$p_{\theta_S}(x) = \frac{1}{Z(\theta_S)} \cdot \exp\big(-f_{\theta_S}(x)\big) \tag{2}$$

$$Z(\theta_S) = \int_{\mathcal{D}} \exp\big(-f_{\theta_S}(x)\big)\mathrm{d}x \tag{3}$$

where $Z(\theta_S)$ is a normalisation constant. This energy function is the equivalent of the similarity term inside the loss function used by traditional image registration via iterative optimisation of transformation parameters individually for every pair of images in the dataset.

It is difficult to draw samples from this distribution with Markov chain Monte Carlo (MCM methods, *e.g.* random walk and Gibbs sampling, which suffer from long mixing times, especially for high-dimensional data like images [10]. To sample from Eq. (3), we adopt SGLD, which makes use of the energy function gradient to decrease the mixing time [36]. Equation (4) draws samples from the distribution defined by the energy function as $t \to \infty$ and $\tau \to 0$ [36]:

$$\tilde{x}^{(t)} \leftarrow \tilde{x}^{(t-1)} - \frac{\tau}{2}\nabla_x f_{\theta_S}\big(\tilde{x}^{(t-1)}\big) + \omega^{(t)}, \omega^{(t)} \sim \mathcal{N}\left(0, \tau\right) \tag{4}$$

We wish to find a similarity metric which maximises the likelihood of images in the dataset when registering them with the image registration network f_θ trained using Eq. (1). This is done by minimising the negative log-likelihood of the image pairs aligned with the image registration model. The loss function and its gradient are given by:

$$\begin{aligned}\mathcal{L}_{\text{ML}}(\theta_S) &= \mathbb{E}_{x\sim p_{\mathcal{D}}}\left[-\log p_{\theta_S}(x)\right]\\ &= \mathbb{E}_{x\sim p_{\mathcal{D}}}\left[f_{\theta_S}(x) - \log Z(\theta_S)\right] \qquad (5)\\ \nabla_{\theta_S}\mathcal{L}_{\text{ML}} &\approx \mathbb{E}_{x^+\sim p_{\mathcal{D}}}\left[\nabla_{\theta_S} f_{\theta_S}(x^+)\right] - \mathbb{E}_{x^-\sim q_{\theta_S}}\left[\nabla_{\theta_S} f_{\theta_S}(x^-)\right] \qquad (6)\end{aligned}$$

where q_{θ_S} is defined by the sampling in Eq. (4) [34]. This gradient decreases the energy of the registered image pairs x^+ and increases the energy of unregistered image pairs x^- [10]. Because the training data does not include ground truth transformations, the registered image pairs x^+ are sampled using Eq. (4), with ${x^+}^{(0)}$ set to the fixed and warped moving image, and the unregistered image pairs x^- are the fixed and the unwarped moving image, with the two images that comprise both image pairs concatenated along the channel dimension.

Algorithm 1: Model training

Data: Images $\mathcal{D} = \mathcal{I} \times \mathcal{I}$
Initialisation similarity metric f_{init}, *e.g.* SSD, LCC, or MI
Result: Image registration network $f_\theta = f^{(1)}_{\theta_D} \circ f^{(2)}_{\theta_E}$
Data-specific similarity metric f_{θ_S}

```
L_sim ← f_init                                                    /* Pre-training */
for Epoch do
    for (I_t, I_s) ∈ D do
        Update θ using Eq. (1) and the Adam optimiser, learning rate η_θ
        Update θ_S using Eq. (7) and the Adam optimiser, learning rate η^pre_θS
    end
end

L_sim ← f_θS                                                       /* Training */
for Epoch do
    for (I_t, I_s) ∈ D do
        Update θ using Eq. (1) and the Adam optimiser, learning rate η_θ
        Sample images using Eq. (4)
    end
    Update θ_S using Eq. (6) and the Adam optimiser, learning rate η_θS
end
```

2.3 Model Training

To successfully train the model and avoid issues with training stability, we need a suitable similarity function from the start. However, unlike [15] we avoid the need to manually initialise the neural network weights, which is not feasible *e.g.* in case of MI, and train the model in two phases. In the 1st phase, we pre-train the similarity metric f_{θ_S} to match the value of a similarity metric of choice f_{init} suitable for the task, *e.g.* SSD, LCC, or MI, by minimising the squared difference between the two values:

$$\hat{\theta}_S = \arg\min_{\theta_S} \Big(f_{\theta_S}(I_t, I_s) - f_{\text{init}}(I_t, I_s) \Big)^2 \tag{7}$$

The initialisation similarity metric f_{init} is also used to pre-train the image registration network. In the 2nd phase, we begin to use the data-specific similarity metric within the loss function in Eq. (1) by setting $\mathcal{L}_{\text{sim}} := f_{\theta_S}$, and train the EBM using Eq. (6). The training loops are summarised in Algorithm 1.

2.4 Implementation

Figure 1 shows the neural network architecture used in the experiments. The image registration network is based on the official implementation of VoxelMorph [2,3,8,9,18], which in turn is based on the 3D UNet [31], with the transformation control point spacing of two voxels along each dimension [28]. The similarity metric f_{θ_S} is pre-trained to match SSD or MI. When initialised to SSD, it takes as input the voxel-wise squared differences between the intensities of the fixed and warped moving images. When initialised to MI, it takes as input the raw fixed and warped moving images.

We calibrate the model hyperparameters to get zero voxels with a non-positive Jacobian determinant of the transformation on the validation image pairs. The regularisation weight is set to $\lambda_{reg} = 1.5 \times 10^3$ and $\lambda_{reg} = 3.0 \times 10^5$ in case of initialisation to SSD and MI respectively. To ensure fair comparison, we train the baseline models with the regularisation weights set to the same values. We use the Adam optimiser with different learning rates for each part of the model. The similarity metric f_{θ_S} is pre-trained with a learning rate of $\eta_{\theta_S}^{\text{pre}} = 1 \times 10^{-4}$ and trained with learning rates of $\eta_{\theta_S} = 1 \times 10^{-5}$ and $\eta_{\theta_S} = 1 \times 10^{-6}$ in case of initialisation to SSD and MI respectively. The registration network is trained with a learning rate of $\eta_\theta = 2 \times 10^{-4}$. We use a step size of $\tau = 2 \times 10^{-3}$ to sample images with SGLD and stop after 2,500 iterations. We train the model with mini-batches of a single image pair, with 100 updates to the image registration network parameters θ per update to the similarity metric network parameters θ_S. The similarity metric f_{θ_S} is pre-trained to match the initialisation function f_{init} for 500 epochs and then trained for 1,000 epochs.

Training requires at least 16 GB GPU memory and on a system with a single GeForce RTX 3090 GPU takes approximately two and a half, and five days in case of the models initialised to SSD and MI respectively.

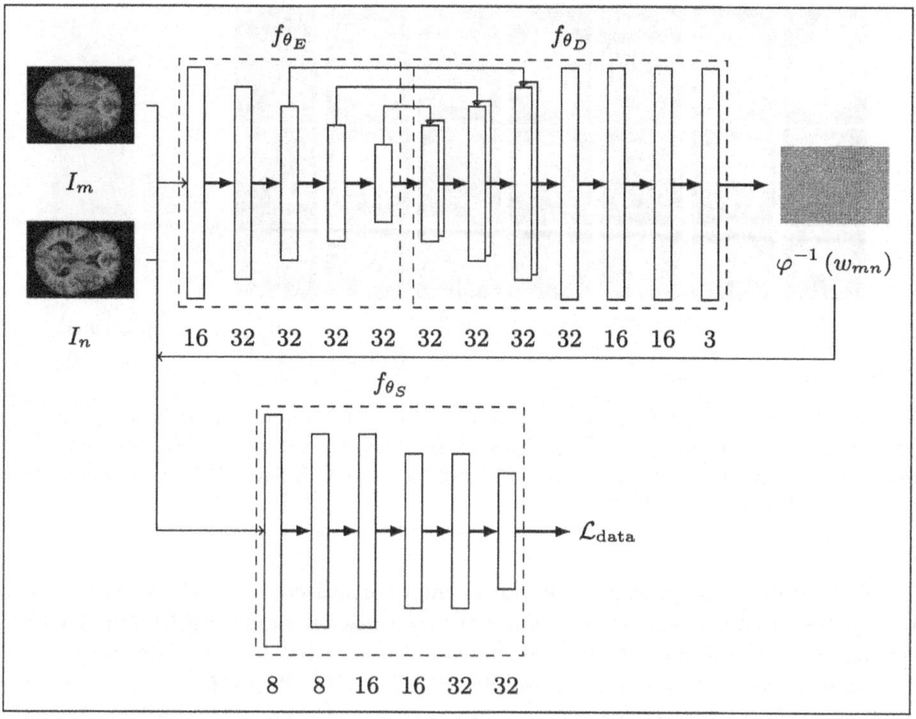

Fig. 1. Image registration network $f_{\theta_D} \circ f_{\theta_E}$ and similarity metric f_{θ_S} with f_{init} set to SSD. The arrows indicate 3D convolutional layers, with the rectangle height representing the output dimensions and the number of output channels displayed underneath. Each layer is followed by a leaky ReLU activation. When f_{init} is set to MI, the similarity metric sub-network f_{θ_S} has an encoder-decoder architecture similar to the image registration network, where the encoder has one additional layer with 64 channels and is followed by five decoder layers.

3 Results

In order to facilitate comparison to other image registration models, we train the model on data from the publicly available OASIS dataset [24], and in particular the images used in the mono-modal task at the MICCAI 2021 learn2reg challenge [17]. The dataset consists of 447 T1-weighted magnetic resonance imaging (MRI) brain scans and includes the anatomical labels of 35 subcortical structures. 408 scans are used for training and validation, with 10 validation and 9 test image pairs. We train and test the framework on full-resolution volumes of dimensions $160 \times 192 \times 224$ voxels and 1 mm voxel length along each dimension.

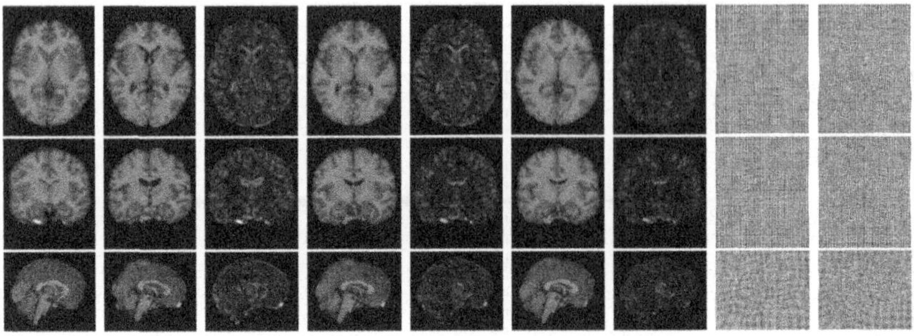

Fig. 2. The model output on a test image pair when the similarity metric is initialised to MI. Image registration models trained with the learnt similarity metrics output transformations that better align the source and the target image without decreasing the transformation smoothness.

To evaluate the proposed method when initialised to SSD, we compare it with VoxelMorph based on the same transformation model and trained with a traditional SSD similarity metric [28], as well as with a learnt similarity metric trained using SGLD but manually initialised to SSD [15]. To evaluate the method when initialised to MI, we compare it with VoxelMorph based on the same transformation model and trained with a traditional MI similarity metric. The model selected for evaluation has the best average accuracy among those that output zero voxels with a non-positive Jacobian determinant of the transformation on the validation image pairs. When initialised to SSD, the proposed method achieves 0.7537 average Dice score (DSC) on the test data, compared to 0.6292 in case of the image registration network trained with MI and 0.7361 in case of the learnt similarity metric manually initialised to SSD. When initialised to MI, the proposed method achieves 0.7278 average DSC on the test data, compared to 0.6831 in case of the image registration network trained with MI. These results are summarised in Fig. 2 and Fig. 3. Importantly, use of the learnt similarity metrics does not have a negative impact on the output transformations smoothness. The transformations output on the test data by each of the models have zero voxels where the transformation Jacobian determinant is non-positive.

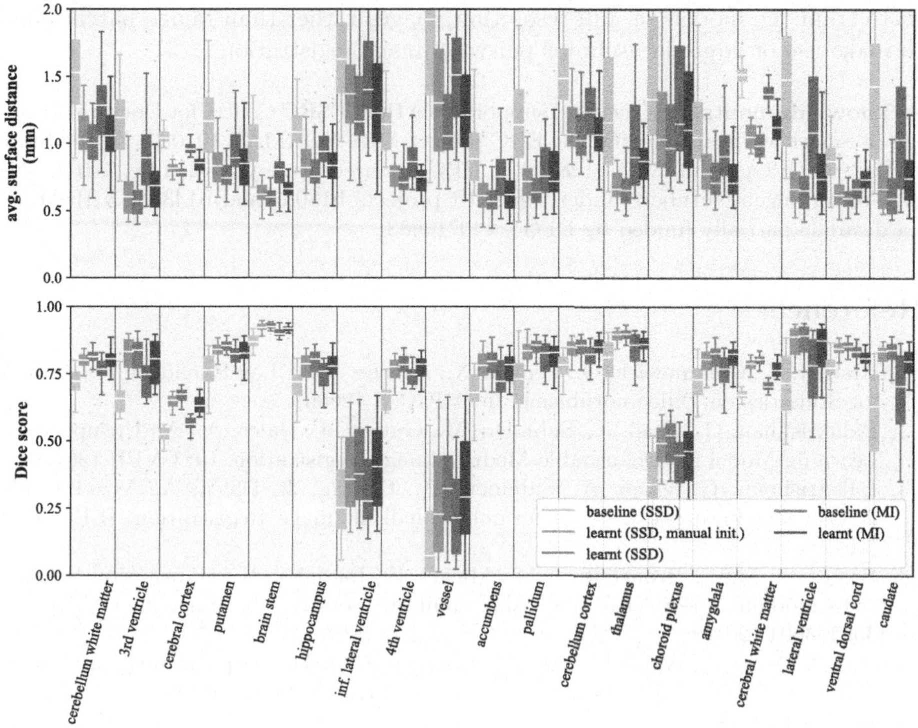

Fig. 3. Average surface distances (ASDs) and DSCs on subcortical structure segmentations when aligning test image pairs using the baseline model trained with SSD, the proposed method with the EBM manually initialised to SSD, and the proposed method, as well as the baseline model trained with MI and the proposed method. To test statistical significance we use a one-tailed Welch's t-tests at the 0.05 significance level. We found that improvement over the model trained with SSD is statistically significant for all 19/19 subcortical structure segmentations for ASD and DSC. In case of the model with manual initialisation to SSD, the improvement is statistically significant for 7/19 structures in terms of ASD and 6/19 structures in terms of DSC. For MI, the improvement is statistically significant for 16/19 structures in terms of ASD and 15/19 in terms of DSC.

4 Conclusion

We present a new framework for unsupervised similarity learning for deformable medical image registration. We show on the examples of SSD and MI that it can significantly improve the accuracy of image registration without a negative impact on the transformation smoothness. Because the models do not require manual initialisation of neural network weights, they can be used for a wider range of image registration tasks than previous unsupervised similarity learning models. The simplicity of our similarity metric formulation as an EBM allows us to train the models on full-resolution images rather than image patches and to make use of large datasets for pairwise image registration.

Acknowledgments. We received support from the EPSRC Centre for Doctoral Training in Smart Medical Imaging, the ERC project MIA-NORMAL 101083647, the State of Bavaria (HTA), and DFG 512819079. HPC resources were provided by NHR@FAU of FAU Erlangen-Nürnberg under the NHR projects b180dc and b143dc. NHR@FAU hardware is partially funded by DFG - 440719683.

References

1. Arsigny, V., Commowick, O., Pennec, X., Ayache, N.: A Log-Euclidean Framework for Statistics on Diffeomorphisms. In: MICCAI (2006)
2. Balakrishnan, G., Zhao, A., Sabuncu, M., Guttag, J., Dalca, A.: An Unsupervised Learning Model for Deformable Medical Image Registration. In: CVPR (2018)
3. Balakrishnan, G., Zhao, A., Sabuncu, M., Guttag, J., Dalca, A.: VoxelMorph: A Learning Framework for Deformable Medical Image Registration. IEEE TMI (2019)
4. Bronstein, M.M., Bronstein, A.M., Michel, F., Paragios, N.: Data fusion through cross-modality metric learning using similarity-sensitive hashing. In: CVPR. pp. 3594–3601 (2010)
5. Cao, X., Yang, J., Wang, L., Xue, Z., Wang, Q., Shen, D.: Deep learning based inter-modality image registration supervised by intra-modality similarity. In: MLMI MICCAI (2018)
6. Che, T., Zhang, R., Sohl-Dickstein, J., Larochelle, H., Paull, L., Cao, Y., Bengio, Y.: Your gan is secretly an energy-based model and you should use discriminator driven latent sampling. In: NeurIPS (2020)
7. Cheng, X., Zhang, L., Zheng, Y.: Deep similarity learning for multimodal medical images. In: Computer Methods in Biomechanics and Biomedical Engineering: Imaging & Visualization. vol. 6, pp. 248–252 (2018)
8. Dalca, A., Balakrishnan, G., Guttag, J., Sabuncu, M.: Unsupervised Learning for Fast Probabilistic Diffeomorphic Registration. In: MICCAI (2018)
9. Dalca, A., Balakrishnan, G., Guttag, J., Sabuncu, M.: Unsupervised Learning of Probabilistic Diffeomorphic Registration for Images and Surfaces. MedIA **57** (2019)
10. Du, Y., Mordatch, I.: Implicit Generation and Generalization in Energy-Based Models. In: NeurIPS (2019)
11. Fan, J., Cao, X., Wang, Q., Yap, P.T., Shen, D.: Adversarial learning for mono- or multi-modal registration. MedIA **58** (2019)
12. Fan, J., Cao, X., Xue, Z., Yap, P.T., Shen, D.: Adversarial Similarity Network for Evaluating Image Alignment in Deep Learning based Registration. In: MICCAI (2018)
13. Goodfellow, I.J., Pouget-Abadie, J., Mirza, M., Xu, B., Warde-Farley, D.: Generative Adversarial Nets. In: NIPS (2014)
14. Grathwohl, W., Wang, K.C., Jacobsen, J.H., Duvenaud, D., Norouzi, M., Swersky, K.: Your classifier is secretly an energy based model and you should treat it like one. In: ICLR (2020)
15. Grzech, D., Azampour, M.F., Glocker, B., Schnabel, J., Navab, N., Kainz, B., Folgoc, L.L.: A variational Bayesian method for similarity learning in non-rigid image registration. In: CVPR (2022)

16. Heinrich, M.P., Jenkinson, M., Bhushan, M., Matin, T., Gleeson, F.V., Brady, S.M., Schnabel, J.A.: MIND: Modality independent neighbourhood descriptor for multi-modal deformable registration. MedIA **16**, 1423–1435 (2012)
17. Hering, A., et al.: Learn2Reg: comprehensive multi-task medical image registration challenge, dataset and evaluation in the era of deep learning. IEEE TMI (2022)
18. Hoffmann, M., Billot, B., Greve, D.N., Iglesias, J.E., Fischl, B., Dalca, A.V.: SynthMorph: Learning contrast-invariant registration without acquired images. IEEE TMI **41**, 543–558 (2022)
19. Kim, B., Han, I., Ye, J.C.: Diffusemorph: Unsupervised deformable image registration using diffusion model. In: ECCV (2022)
20. Kingma, D.P., Welling, M.: Auto-Encoding Variational Bayes. In: ICLR (2014)
21. Krebs, J., Delingette, H., Mailhé, B., Ayache, N., Mansi, T.: Learning a Probabilistic Model for Diffeomorphic Registration. IEEE TMI **38**(9) (2019)
22. LeCun, Y., Chopra, S., Hadsell, R., Ranzato, M.A., Huang, F.J.: A Tutorial on Energy-Based Learning. Predicting Structured Data pp. 1–59 (2006)
23. Lee, M.C.H., Oktay, O., Schuh, A., Schaap, M., Glocker, B.: Image-and-Spatial Transformer Networks for Structure-Guided Image Registration. In: MICCAI (2019)
24. Marcus, D.S., Fotenos, A.F., Csernansky, J.G., Morris, J.C., Buckner, R.L.: Open access series of imaging studies: Longitudinal MRI data in nondemented and demented older adults. Journal of Cognitive Neuroscience **22** (2010)
25. Modat, M., Daga, P., Cardoso, M.J., Ourselin, S., Ridgway, G.R., Ashburner, J.: Parametric non-rigid registration using a stationary velocity field. In: MMBIA. IEEE (2012)
26. Nijkamp, E., Hill, M., Zhu, S.C., Wu, Y.N.: Learning Non-Convergent Non-Persistent Short-Run MCMC Toward Energy-Based Model. In: NeurIPS (2019)
27. Qin, C., Shi, B., Liao, R., Mansi, T., Rueckert, D., Kamen, A.: Unsupervised Deformable Registration for Multi-Modal Images via Disentangled Representations. In: IPMI (2019)
28. Qiu, H., Qin, C., Schuh, A., Hammernik, K., Rueckert, D.: Learning Diffeomorphic and Modality-invariant Registration using B-splines. In: MIDL (2021)
29. Ranzato, M., Poultney, C., Chopra, S., LeCun, Y.: Efficient Learning of Sparse Representations with an Energy-Based Model. In: NIPS (2006)
30. Rombach, R., Blattmann, A., Lorenz, D., Esser, P., Ommer, B.: High-resolution image synthesis with latent diffusion models. In: CVPR (2022)
31. Ronneberger, O., Fischer, P., Brox, T.: U-net: Convolutional networks for biomedical image segmentation. In: MICCAI (2015)
32. Rueckert, D., Sonoda, L.I., Hayes, C., Hill, D.L.G., Leach, M.O., Hawkes, D.J.: Nonrigid registration using free-form deformations: Application to breast MR images. IEEE TMI **18**(8), 712–721 (1999)
33. Simonovsky, M., Gutiérrez-Becker, B., Mateus, D., Navab, N., Komodakis, N.: A deep metric for multimodal registration. In: MICCAI. pp. 10–18 (2016)
34. Turner, R.: CD notes (2005)
35. de Vos, B.D., Berendsen, F.F., Viergever, M., Sokooti, H., Staring, M., Išgum, I.: A deep learning framework for unsupervised affine and deformable image registration. MedIA **52** (2019)
36. Welling, M., Teh, Y.W.: Bayesian Learning via Stochastic Gradient Langevin Dynamics. In: ICML. pp. 681–688 (2011)
37. Xie, J., Lu, Y., Zhu, S.C., Wu, Y.N.: A Theory of Generative ConvNet. In: ICML (2016)

38. Xie, J., Xu, Y., Zheng, Z., Zhu, S.C., Wu, Y.N.: Generative pointnet: Deep energy-based learning on unordered point sets for 3d generation, reconstruction and classification. In: CVPR (2021)
39. Xie, J., Zhu, S.C., Wu, Y.N.: Synthesizing dynamic patterns by spatial-temporal generative convnet. In: CVPR (2017)
40. Xie, J., Zhu, S.C., Wu, Y.N.: Learning Energy-Based Spatial-Temporal Generative ConvNets for Dynamic Patterns. IEEE Transactions on Pattern Analysis and Machine Intelligence **43**(2), 516–531 (2021)

Segmentation by Registration-Enabled SAM Prompt Engineering Using Five Reference Images

Yaxi Chen[1], Aleksandra Ivanova[2], Shaheer U. Saeed[3], Rikin Hargunani[4], Jie Huang[1], Chaozong Liu[2,4], and Yipeng Hu[3](✉)

[1] Mechanical Engineering Department, University College London, London, UK
[2] Institute of Orthopaedic and Musculoskeletal Science, University College London, London, UK
[3] Centre for Medical Image Computing, Wellcome/EPSRC Centre for Interventional and Surgical Sciences, Department of Medical Physics and Biomedical Engineering, University College London, London, UK
yipeng.hu@ucl.ac.uk
[4] Royal National Orthopaedic Hospital, Stanmore, UK

Abstract. The recently proposed Segment Anything Model (SAM) is a general tool for image segmentation, but it requires additional adaptation and careful fine-tuning for medical image segmentation, especially for small, irregularly-shaped, and boundary-ambiguous anatomical structures such as the knee cartilage that is of interest in this work. Repaired cartilage, after certain surgical procedures, exhibits imaging patterns unseen to pre-training, posing further challenges for using models like SAM with or without general-purpose fine-tuning. To address this, we propose a novel registration-based prompt engineering framework for medical image segmentation using SAM. This approach utilises established image registration algorithms to align the new image (to-be-segmented) and a small number of reference images, without requiring segmentation labels. The spatial transformations generated by registration align either the new image or pre-defined point-based prompts, before using them as input to SAM. This strategy, requiring as few as five reference images with defined point prompts, effectively prompts SAM for inference on new images, without needing any segmentation labels. Evaluation of MR images from patients who received cartilage stem cell therapy yielded Dice scores of 0.89, 0.87, 0.53, and 0.52 for segmenting femur, tibia, femoral- and tibial cartilages, respectively. This outperforms atlas-based label fusion and is comparable to supervised nnUNet, an upper-bound fair baseline in this application, both of which require full segmentation labels for reference samples. The codes are available at: https://github.com/chrissyinreallife/KneeSegmentWithSAM.git.

Keywords: Cartilage · SAM · MRI · Prompt Engineering

M. Modat et al. (Eds.): WBIR 2024, LNCS 15249, pp. 241–252, 2024.
https://doi.org/10.1007/978-3-031-73480-9_19

1 Introduction

Image segmentation forms the basis of various clinical tasks, including surgical planning [1,15], diagnostic assistance [2,32], and surgical navigation [20]. Tissue inhomogeneity [31], presence of various artefacts [25,26] and operator interpretation variability [4,7], are challenges that often negatively impact segmentation performance of automated or manual segmentation methods. When coupled with limited dataset sizes in specialised clinical applications, due to labelling cost or low disease prevalence, it may become prohibitively infeasible to capitalise advances in modern machine learning [33].

Atlas-based segmentation first constructs one or more manually segmented images as a reference atlas library, before segmenting new patients by image registration between them. While being effective for brain imaging [3], the performance of atlas-based methods significantly diminishes for small and low-contrast soft tissue structures [6]. The difficulty in hyaline cartilage segmentation, of interest in this work, arises from its thinness, relatively small size, curved morphology and challenging localisation of its interfaces (such as the interface between femoral and tibial cartilage) [28].

Learning-based methods, such as U-Net [23], V-Net [18] and their variants, have achieved state-of-the-art performance for several medical image segmentation tasks [29]. However, their effectiveness is highly dependent on the labelled training data availability, often sparse and costly in the medical domain [9], especially for pathological cases with their added complexity, variation and rarity [17]. Few-shot segmentation methods can be adapted to new tasks with a few labelled samples, proposed for abdominal MR images [30], skin images [8], and pelvic MRs [14]. However, due to limited labelled data and significant differences in data distributions compared to scans with greater availability for training, application to many specialised areas, such as knee cartilage segmentation, remains a challenge [16,27]. Additionally, conventional few-shot learning approaches also necessitate large datasets, for cross-task learning [34]. To summarise, unique data distribution and extreme availability render existing learning-based approaches, arguably, inadequate for our segmentation problem [28].

Vision foundation Models such as SAM have emerged as general-purpose tools for image segmentation [13]. Building on these advances, the MedSAM foundation model, developed on a large-scale medical image dataset, allows for medical image segmentation for a variety of unseen clinical tasks [16]. While SAM and other similar models demonstrate superior generalisation compared to traditional few-shot segmentation approaches [13], thanks to their greater training data diversity, they struggle to effectively segment rare out-of-training-data-distribution structures [16]. In addition, foundation models also enable segmentation through sparse annotation (such as selected points of interest) learnt from various training tasks, adaptable for out-of-distribution samples, with denser annotations leading to improved performance. The sparse annotation at inference alleviates, to a degree, hurdles of data scarcity in medical imaging, however, it limits clinical applicability due to time and expertise constraints during inference. To address this, prompt engineering approaches have been proposed [11], but may still require large amounts of weakly-labelled or unlabeled images to learn effective prompting strategies. While these methods reduce annotation burden by replacing full manual segmentation with weak annotations, such as points and relatively more laborious bounding boxes, it remains an open question to develop effective prompting strategies for individual applications.

In this work, our proposed SAM prompting framework is based on spatially aligning the input image (or the point prompt) to a limited set of weakly-labelled reference samples, as opposed to training a prompt engineering network. This alignment mechanism enables obtaining segmentation for new samples without any manually annotated prompts for the sample itself. The weak annotations are only required for a small reference set of 5 images. As shown in the presented experimental results, the image/prompt alignment to a reference set enables accurate segmentation to be obtained from SAM, for out-of-distribution samples. This method substantially reduces the data annotation cost associated with learning prompt engineering while enabling segmentation from foundation models using a limited set of weakly-labelled references, a practical scenario found in many potential clinical applications.

The contributions of this work are summarised: 1) we introduce a SAM prompt engineering framework, based on aligning samples to-be-segmented with a small reference set of weakly-annotated samples, without the need for any large weakly-annotated dataset for learning prompting; 2) we propose two effective image registration strategies to align either images or prompts to the reference set of samples; 3) we demonstrate improved performance compared to alternative approaches which require full segmentation of the reference set and to using SAM without the proposed registration strategy, for a clinically challenging task of pathological bone tissue image segmentation.

2 Methods

Our proposed prompt engineering method for segmentation foundation models relies on an alignment of the input image or prompt to a limited reference set of weakly-labelled samples. This alignment allows us to construct effective prompts for segmentation foundation models such that we can obtain accurate segmentation, even for out-of-distribution samples, using as few as five point-prompt-labelled reference images.

2.1 Segment Anything Model

In this work, since we use a pre-trained, fixed-weight foundation model for segmentation, without any fine-tuning, the segmentation model is considered a non-parametric function:

$$f(x,p) = y : \mathcal{X} \times \mathcal{P} \rightarrow \mathcal{Y}$$

where $x \in \mathcal{X}$ is an image sampled from the image domain (considered out-of-distribution of the foundation model training image domain), $p \in \mathcal{P}$ is the prompt sampled from the domain of prompts, and $y \in \mathcal{Y}$ is a binary pixel-level segmentation label predicted by the non-parametric foundation model f. With a prompt p_i, a sample $x_i \in \mathcal{X}$ is segmented $y_i = f(x_i, p_i)$. In this work, the prompt is a set of P points $p_i = \{p_{i,k}\}_{k=1}^{P}$ that specifies structures of interest in the segmentation (see examples in Fig. 1).

2.2 Reference and Segmentation Samples

A set of N reference images $\{x_i\}_{i=1}^{N} \in \mathcal{X}^N$ are used to form a reference set, with which to align the image that is to be segmented. For this reference set, we also have manually annotated prompts available $\{p_i\}_{i=1}^{N} \in \mathcal{P}^N$. For notational clarity, we define the image to be segmented, without any available prompts or labels, as $x_{new} \in \mathcal{X}$. The proposed approaches to obtain x_{new} are outlined in the following sections.

2.3 Image Registration

Image registration estimates a transform that warps a moving image to be aligned to a fixed image. The transform-predicting registration function is denoted as a non-parametric function:

$$r(x_{\text{ref}}, x_{\text{new}}) = d : \mathcal{X} \times \mathcal{X} \to \mathcal{D}$$

where $x_{\text{ref}} \in \mathcal{X}$ is the reference image (used as the fixed image in this case), x_{new} is the moving image (to be registered with the reference image), and $d \in \mathcal{D}$ is in spatial transformation domain $\mathcal{D}$ and warps the new image to be aligned with the reference image x_{ref}. the warped image x'_{new} thus is:

$$x'_{\text{new}} = x_{\text{new}} \circ \text{d}$$

2.4 Image Alignment to the Reference Set

Aligning the new image x_{new} to the reference x_i is denoted as $x_{new} \circ d = x'_{new}$, where the transform is $d = r(x_i, x_{new})$. In this work, this transform is assumed invertable and the inverse of the transformation is thus denoted as $x'_{new} \circ (d)^{-1} = x_{new}$. For nonlinear transformation that is non-invertable, numerical approximation may be sought and further validated.

Obtaining Pixel-Level Segmentation: For the set of reference images $\{x_i\}_{i=1}^{N}$, we can compute the transforms to warp the image to be segmented x_{new} to align with each of the reference images, denoted as $\{d_i\}_{i=1}^{N} = \{r(x_i, x_{new})\}_{i=1}^{N}$. We then obtain N warped images $\{x'_{new,i}\}_{i=1}^{N} = \{x_{new} \circ d_i\}_{i=1}^{N}$.

The warped images $\{x'_{new,i}\}_{i=1}^{N}$ together with the original point prompts $\{p_i\}_{i=1}^{N}$, for the reference images, are used as input to the foundation model, in

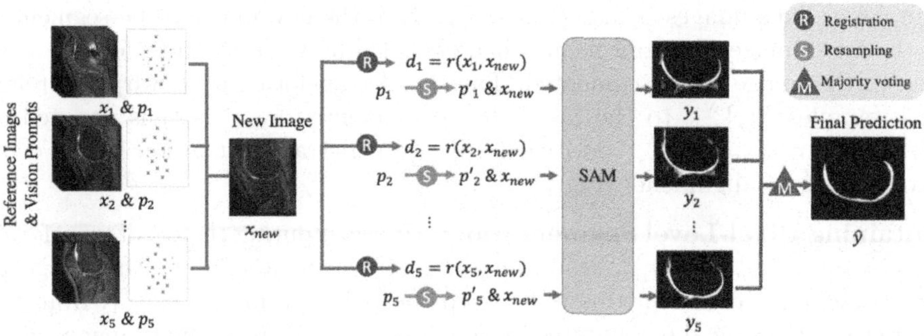

Fig. 1. The proposed Image Alignment strategy, the first type of the two registration-enabled segmentation pipelines based on SAM prompt engineering, illustrated with samples of the reference images, point prompts and new images.

order to obtain a set of pixel-level segmentations $\{y_i\}_{i=1}^N = \{f(x'_{new,i}, p_i)\}_{i=1}^N$. We then apply the inverse transform to reverse the warp on these masks, denoted as $\{y'_i\}_{i=1}^N = \{y_i \circ (d_i)^{-1}\}_{i=1}^N$. A pixel-level majority voting [21] between these N segmentations (after inverse transform applied), $\{y'_i\}_{i=1}^N$, then allows us to obtain the final prediction $\hat{y}$ for the sample x_{new}.

Algorithm 1. Segmentation Pipelines

(a) Image Alignment	**(b) Vision Prompt Alignment**
Input: New image x_{new},	**Input:** New image x_{new},
Reference images $\{x_i\}_{i=1}^N$,	Reference images $\{x_i\}_{i=1}^N$,
and Prompts $\{p_i\}_{i=1}^N$	and Prompts $\{p_i\}_{i=1}^N$
1. Registration:	1. Registration:
$\{d_i\}_{i=1}^N = \{r(x_{new}, x_i)\}_{i=1}^N$	$\{d_i\}_{i=1}^N = \{r(x_i, x_{new})\}_{i=1}^N$
2. Resampling:	2. Resampling:
$\{x'_{new,i}\}_{i=1}^N = \{x_{new} \circ d_i\}_{i=1}^N$	$\{p'_i\}_{i=1}^N = \{p_i \circ d_i\}_{i=1}^N$
3. SAM: $\{y_i\}_{i=1}^N = \{f(x'_{new}, p_i)_{i=1}^N$	3. SAM: $\{y_i\}_{i=1}^N = \{f(x_{new}, p'_i)_{i=1}^N$
4. Resampling $\{y'_i\}_{i=1}^N = \{y_i \circ (d_i)^{-1}\}_{i=1}^N$	4. Majority Voting
5. Majority Voting	**Output:** Final prediction $\hat{y}$
Output: Final prediction $\hat{y}$	

2.5 Prompt Alignment to the New Sample

In addition to the above-described mechanism of aligning the new image x_{new} to a reference set $\{x_i\}_{i=1}^N$, for prompt construction, this section describes a second proposed mechanism of prompt construction, based on the same reference set, by aligning the prompts of the reference set $\{p_i\}_{i=1}^N$ to the new image x_{new}.

To achieve prompt-alignment, image registration is again employed, similar to the image-alignment approach. However, instead of considering the reference samples as fixed images as in Sect. 2.4, we regard the new image to-be-segmented as the fixed image. Transforms are then obtained by warping the reference samples to the image to-be-segmented. The set of transforms that warp the reference samples $\{x_i\}_{i=1}^N$ to the image to be segmented x_{new} is thus denoted as $\{d_i\}_{i=1}^N = \{r(x_{new}, x_i)\}_{i=1}^N$ (switched order of registration function inputs compared to image alignment) (Fig. 2).

Obtaining Pixel-Level Segmentation: After computing the set of transforms $\{d_i\}_{i=1}^N = \{r(x_{new}, x_i)\}_{i=1}^N$ that warp the reference samples to the image to be segmented, we can, in this case, warp the point prompts $\{p_i\}_{i=1}^N$, from the reference set, using the computed transforms (instead of warping images as in the image-alignment strategy), which is denoted as $\{p_i'\}_{i=1}^N = \{p_i \circ d_i\}_{i=1}^N$.

The image to be segmented x_{new} along with warped point prompts $\{p_i'\}_{i=1}^N$ are then used as input to the foundation model, in order to obtain a set of pixel-level segmentations $\{y_i\}_{i=1}^N = \{f(x_{new}, p_i')\}_{i=1}^N$. Similar to image-alignment, a pixel-level majority voting between these N segmentations $\{y_i\}_{i=1}^N$ then allows us to obtain the final prediction $\hat{y}$ for the sample x_{new}.

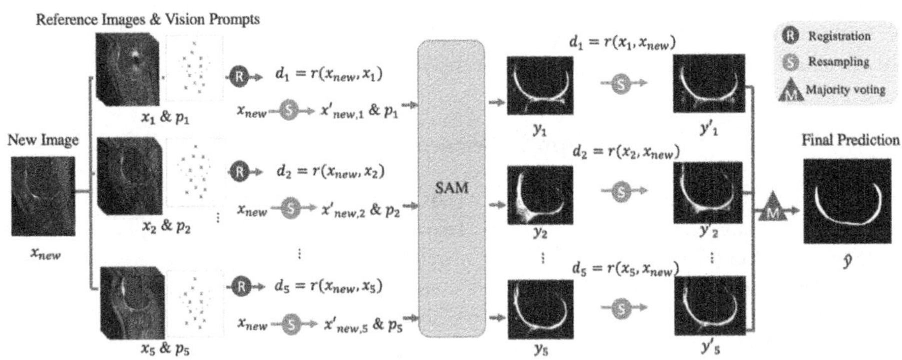

Fig. 2. The proposed Prompt Alignment strategy, the second type of the two registration-enabled segmentation pipelines based on SAM prompt engineering, illustrated with samples of the new image, reference image and point prompts.

3 Experiments and Results

3.1 Dataset and Implementation Details

Pathology Knee Joint MR Data and Pre-processing: MR images were acquired from 35 patients who had undergone cartilage surgery with stem cell transplantation techniques at Royal National Orthopaedic Hospital [5]. Proton density (PD) weighted MR scans in the sagittal view were manually segmented to acquire anatomical and pathological labels, verified by an experienced radiology team. All images were normalized, resampled, and center-cropped to a consistent size of 512×512×40 and an image intensity range of [0, 3000].

Vision Prompts: Out of the 35 samples, 5 were randomly chosen as reference images. Specific points prompts were manually selected for each reference image, guided by radiologists. These prompts encompass positive points (areas within the structure of interest) and negative points (areas outside the structure of interest). The same points prompt are used for comparisons between algorithms. Further details on these point prompts can be found in the open-source implementation of this project.

Registration and Resampling: In our study, NiftyReg [19] was utilized for both the registration and resampling processes. For registration, an initial affine registration, which aligns the images based on rotation, scale, and translation, was applied before the non-rigid algorithm, based on free-form deformation [24]. The bending energy weight was set to 0.65, which empirically represent the stiffness characteristics of bone tissue in our experiments, with the sum-of-square difference (SSD) as the intra-modality similarity measure function. The remainder parameters were kept default values. These configurations, including the prior affine alignment and the choice of the well-established NiftyReg package, used in the available repository were based on preliminary experiments with qualitatively satisfactory alignment between the PD-weighted MR images.

Post-registration Point Prompt Filtering: A simple, automatic filtering was applied for point prompts, to ensure removal of mislabelled or misaligned prompt points, based on the intensity value of point prompt samples. For the femur and tibia, we used clipping intensity values [0, 1200] for positive points. For femoral and tibial cartilage, we used an intensity range of [800, 3000] for positive points. Any points lying outside these intensity ranges were removed. Pairs of images and corresponding prompts were then input into the default ViT-H SAM model to obtain the pixel-level segmentation, as described in Sects. 2.4 and 2.5.

3.2 Comparison and Ablation Experiments

We compare our method with an atlas-based segmentation method [22] and learning-based methods, 2D and 3D nnUNet [12]. It is noteworthy that both the non-learning and learning-based approaches require full segmentation of the reference set of images and the same availability in labelled images reflects the real-world clinical scenarios as discussed in Sect. 1. In this study, therefore, these are considered fair comparison between the proposed methods and the alternatives that have access to more detailed reference labels.

Atlas-Based Segmentation: In our implementation of the atlas-based method, all 5 reference images needed to be fully segmented. The 5 masks $\{y_i\}_{i=1}^{N}$ for the 5 reference images were warped from the reference space to the new image space (similar to the prompt-alignment strategy), denoted as

Table 1. Femur and Tibia segmentation results (top); Femoral Cartilage and Tibial Cartilage segmentation results (bottom). AR: usage of affine registration; NR: usage of non-rigid registration; FL: usage of full segmentation labels (opposed to point prompts, in our method). I-Align is our strategy based on image alignment and P-align is our strategy based on prompt alignment. Comparison needs to take into account the label requirement, as further discussed in the text.

				Femur		Tibia	
Method	AR	NR	FL	RMSD (mm)	Dice	RMSD (mm)	Dice
Ours:I-align	✓	✓	✗	2.93 ± 2.08	0.89 ± 0.04	5.80 ± 2.71	0.87 ± 0.03
Ours:P-align	✓	✓	✗	4.67 ± 4.33	0.86 ± 0.09	5.80 ± 2.71	0.87 ± 0.03
I-align	✓	✗	✗	5.64 ± 3.82	0.87 ± 0.04	5.37 ± 2.72	0.87 ± 0.03
P-align	✓	✗	✗	5.02 ± 4.34	0.84 ± 0.11	4.11 ± 4.90	0.84 ± 0.16
NoReg	✗	✗	✗	6.37 ± 4.98	0.84 ± 0.12	3.19 ± 1.39	0.86 ± 0.07
Atlas	✓	✗	✗	7.71 ± 8.59	0.79 ± 0.14	6.18 ± 6.70	0.83 ± 0.12
Atlas	✓	✓	✗	7.80 ± 8.69	0.79 ± 0.14	6.21 ± 6.77	0.83 ± 0.12
nnUNet-2D	-	-	✓	7.68 ± 14.28	0.82 ± 0.29	5.52 ± 7.36	0.83 ± 0.20
nnUNet-3D	-	-	✓	15.50 ± 22.45	0.87 ± 0.22	29.56 ± 33.22	0.80 ± 0.26
				Femoral Cartilage		Tibial Cartilage	
Method	AR	NR	FL	VOE	Dice	VOE	Dice
Ours:I-align	✓	✓	✗	0.63 ± 0.14	0.53 ± 0.15	0.64 ± 0.10	0.52 ± 0.11
Ours:P-align	✓	✓	✗	0.65 ± 0.16	0.50 ± 0.18	0.66 ± 0.12	0.50 ± 0.14
I-align	✓	✗	✗	0.66 ± 0.14	0.51 ± 0.14	0.67 ± 0.07	0.50 ± 0.08
P-align	✓	✗	✗	0.65 ± 0.16	0.50 ± 0.17	0.68 ± 0.11	0.47 ± 0.14
NoReg	✗	✗	✗	0.60 ± 0.03	0.35 ± 0.19	0.92 ± 0.08	0.15 ± 0.12
Atlas	✓	✗	✗	0.86 ± 0.11	0.22 ± 0.17	0.84 ± 0.08	0.26 ± 0.12
Atlas	✓	✓	✗	0.76 ± 0.54	0.24 ± 0.17	0.84 ± 0.07	0.27 ± 0.10
nnUNet-2D	-	-	✓	0.54 ± 0.24	0.59 ± 0.27	0.61 ± 0.20	0.52 ± 0.24
nnUNet-3D	-	-	✓	0.46 ± 0.21	0.67 ± 0.20	0.49 ± 0.13	0.66 ± 0.13

$\{y'_i\}_{i=1}^{N=5} = \{y_i \circ d_i\}_{i=1}^{N}$. The final prediction $\hat{y}$ for x_{new} was obtained by pixel-level majority voting [21].

Learning-Based 2D&3D nnUNet Segmentation: For the purpose of providing a benchmark in this study, the dataset was divided into a training set comprising 200 images from 5 patients, for the comparison purpose based on the same label availability (but also requiring fully segmented labels), and the same testing set with 1200 images from 30 patients. The nnUNetv2 was employed [12], leveraging both its 2D and 3D low resolution U-Net architectures.

Ablation Studies: To assess the impact of various components of our proposed methods, we conducted ablation experiments, including configurations with or without affine registration (AR) and/or non-rigid registration (NR). The benefits of increased registration accuracy appear to diminish, in terms of achieving improved segmentation performance. Varying the number of reference images was also tested, with fewer reference images leading to a poorer segmentation performance as expected. Average dice scores of 0.11 for cartilage and 0.24 for the femur were obtained using only 3 reference images. More reference images lessen the label-to-performance efficiency, as can be verified using our open-sourced implementation and provided example images. However, future study may be interesting to explore the relationships between the required or optimal reference set size and the clinically required accuracy.

3.3 Results Based on Clinically-Reported Evaluation Metrics

A root-mean-square symmetric surface distance (RMSD) [10] is reported for bone segmentation. An additional volume overlap error (VOE), commonly reported for cartilage segmentation, is also included. All metrics were computed over slices containing bone tissue. The Dice coefficient is also reported for all experiments. These metrics were adopted for our post-procedure evaluation in current clinical and research practice for cartilage repair, where criteria such as MOCART (Magnetic Resonance Observation of Cartilage Repair Tissue) are based on the measurements on bone tissue surfaces, volumes and other morphological features.

Our method, with only five weakly-labelled reference images, achieves Dice scores of 0.89 and 0.87, for the femur and tibia segmentation, respectively. Comparatively lower Dice scores of 0.82 and 0.83, respectively, were observed for the nnUNet-2D. For femoral and tibial cartilage, the Dice scores of our proposed methods are 0.53 and 0.52, closely matching or slightly underperforming against nnUNet-2D's 0.59 and 0.52, which requires full segmentation labels for the reference images. In cases, such as tibia segmentation, our method yielded a superior Dice coefficient but relatively average RMSD, as shown in Table 1. This discrepancy may be due to the higher anatomical variability of the tibia between patients, whereas the proposed approach may prioritise overlap over edge detection, for achieving a globally less biased segmentation, which may worth quantifying for further understanding. The ANOVA test followed by Tukey HSD among all the methods confirm our method significantly surpasses atlas-based methods in bone and cartilage segmentation with p-values of 0.034 and 0.023 respectively, as well as ablated variants for cartilage segmentation. Further results are summarised in Table 1.

4 Discussion and Conclusion

We show statistically significant performance improvements for our method, compared to other tested methods and to ablated variants, demonstrating the utility of our framework for segmenting new images using only five weakly-labelled references. In contrast, other methods with comparable performance utilise full segmentation labels - a non-trivial expertise-demanding task even for a small reference set due to the relatively irregularly-shaped structures. It is worth noting that lower performance was observed in cartilage segmentation in our method compared to the nnUNet approach which uses full segmentation labels, which may require further test data set to show statistical significance and, if significant, can potentially be bridged by further development of the proposed approaches. Cartilage areas are typically curved, elongated and small, which makes them inherently difficult to segment accurately using only weak labels. This complexity suggests that additional information provided by full segmentation annotations could be beneficial. Future research could explore incorporating such supplementary data to address these challenges. Additionally, a targeted registration concentrating on bone and cartilage interfaces instead of the whole image, could potentially enhance performance.

Our results, for a real clinical patient dataset for knee MR segmentation, demonstrate higher performance and, perhaps more importantly, a more practical and feasible segmentation approach, compared to existing methods. The improved performance demonstrates the effectiveness of utilising weakly-labelled point prompts in combination of well-established image registration algorithms within our proposed framework, with potentials especially in many real-world clinical applications in which datasets can be extremely constrained by the availability of pathological images and pixel-level annotated samples.

Acknowledgement. This work was supported in part by EPSRC [EP/T029404/1], Wellcome/EPSRC Centre for Interventional and Surgical Sciences [203145Z/16/Z] and the International Alliance for Cancer Early Detection, a partnership between Cancer Research UK [C73666/A31378], Canary Center at Stanford University, the University of Cambridge, OHSU Knight Cancer Institute, University College London and the University of Manchester.

References

1. Alirr, O.I., Rahni, A.A.A., Golkar, E.: An automated liver tumour segmentation from abdominal ct scans for hepatic surgical planning. IJCARS **13**, 1169–1176 (2018)
2. Amyar, A., Modzelewski, R., Li, H., Ruan, S.: Multi-task deep learning based ct imaging analysis for covid-19 pneumonia: Classification and segmentation. COMPUT BIOL MED **126**, 104037 (2020)
3. Capobianco, E., Dominietto, M.: Assessment of brain cancer atlas maps with multimodal imaging features. J. Transl. Med. **21**(1), 1–11 (2023)
4. Chalcroft, L.F., Qu, J., Martin, S.A., Gayo, I.J., Minore, G.V., Singh, I.R., et al.: Development and evaluation of intraoperative ultrasound segmentation with negative image frames and multiple observer labels. In: MICCAI Workshop 2021. pp. 25–34. Springer (2021)
5. Chimutengwende-Gordon, M., Ahmad, M.A., Bentley, G., Brammah, J., Carrington, R., Miles, et al.: Stem cell transplantation for the treatment of osteochondral defects of the knee: Operative technique for a single-stage transplantation procedure using bone marrow-derived mesenchymal stem cells. The Knee **28**, 400–409 (2021)

6. Costea, M., Zlate, A., Durand, M., Baudier, T., Grégoire, V., Sarrut, D., et al.: Comparison of atlas-based and deep learning methods for organs at risk delineation on head-and-neck ct images using an automated treatment planning system. Radiotherapy and Oncology **177**, 61–70 (2022)
7. Czolbe, S., Arnavaz, K., Krause, O., Feragen, A.: Is segmentation uncertainty useful? In: IPMI 2021. pp. 715–726. Springer (2021)
8. Feyjie, A.R., Azad, R., Pedersoli, M., Kauffman, C., Ayed, I.B., Dolz, J.: Semi-supervised few-shot learning for medical image segmentation. arXiv preprint arXiv:2003.08462 (2020)
9. Haque, I.R.I., Neubert, J.: Deep learning approaches to biomedical image segmentation. IMU **18**, 100297 (2020)
10. Heimann, T., Morrison, B.J., Styner, M.A., Niethammer, M., Warfield, S.: Segmentation of knee images: a grand challenge. In: Proc. MICCAI Workshop on Medical Image Analysis for the Clinic. vol. 1. Beijing, China (2010)
11. Huang, J., Jiang, K., Zhang, J., Qiu, H., Lu, L., Lu, S., et al.: Learning to prompt segment anything models. arXiv preprint arXiv:2401.04651 (2024)
12. Isensee, F., Jaeger, P.F., Kohl, S.A., Petersen, J., Maier-Hein, K.H.: nnu-net: a self-configuring method for deep learning-based biomedical image segmentation. Nature methods **18**(2), 203–211 (2021)
13. Kirillov, A., Mintun, E., Ravi, N., Mao, H., Rolland, C., Gustafson, L., et al.: Segment anything. arXiv preprint arXiv:2304.02643 (2023)
14. Li, Y., Fu, Y., Gayo, I.J., Yang, Q., Min, Z., Saeed, S.U., et al.: Prototypical few-shot segmentation for cross-institution male pelvic structures with spatial registration. Med. Image Anal. **90**, 102935 (2023)
15. Litjens, G., Toth, R., Van De Ven, W., Hoeks, C., Kerkstra, S., Van Ginneken, B., et al.: Evaluation of prostate segmentation algorithms for mri: the promise12 challenge. Med. Image Anal. **18**(2), 359–373 (2014)
16. Ma, J., He, Y., Li, F., Han, L., You, C., Wang, B.: Segment anything in medical images. Nature Communications **15**(1), 654 (2024)
17. Malhotra, P., Gupta, S., Koundal, D., Zaguia, A., Enbeyle, W., et al.: Deep neural networks for medical image segmentation. J. Healthc. Eng. **2022** (2022)
18. Milletari, F., Navab, N., Ahmadi, S.A.: V-net: Fully convolutional neural networks for volumetric medical image segmentation. In: 3D vision 2016. pp. 565–571. Ieee (2016)
19. Modat, M., Ridgway, G.R., Taylor, Z.A., Lehmann, M., Barnes, J., Hawkes, D.J., et al.: Fast free-form deformation using graphics processing units. Comput Methods Programs Biomed **98**(3), 278–284 (2010)
20. Montaña-Brown, N., Ramalhinho, J., Allam, M., Davidson, B., Hu, Y., Clarkson, M.J.: Vessel segmentation for automatic registration of untracked laparoscopic ultrasound to ct of the liver. IJCARS **16**(7), 1151–1160 (2021)
21. Nguyen, N.T., Le, P.B.: Topological voting method for image segmentation. Journal of Imaging **8**(2), 16 (2022)
22. Rohlfing, T., Brandt, R., Menzel, R., Russakoff, D.B., Maurer Jr, C.R.: Quo vadis, atlas-based segmentation? In: Handbook of Biomedical Image Analysis: Volume III: Registration Models, pp. 435–486. Springer (2005)
23. Ronneberger, O., Fischer, P., Brox, T.: U-net: Convolutional networks for biomedical image segmentation. In: MICCAI 2015. pp. 234–241. Springer (2015)
24. Rueckert, D., Sonoda, L.I., Hayes, C., Hill, D.L., Leach, M.O., Hawkes, D.J.: Nonrigid registration using free-form deformations: application to breast mr images. IEEE transactions on medical imaging **18**(8), 712–721 (1999)

25. Saeed, S.U., Fu, Y., Stavrinides, V., Baum, Z.M., Yang, Q., Rusu, M., , et al.: Image quality assessment for machine learning tasks using meta-reinforcement learning. Med. Image Anal. **78**, 102427 (2022)
26. Saeed, S.U., Yan, W., Fu, Y., Giganti, F., Yang, Q., Baum, Z., et al.: Image quality assessment by overlapping task-specific and task-agnostic measures: application to prostate multiparametric mr images for cancer segmentation. arXiv preprint arXiv:2202.09798 (2022)
27. Schreiner, M.M., Raudner, M., Marlovits, S., Bohndorf, K., Weber, M., Zalaudek, M., et al.: The mocart (magnetic resonance observation of cartilage repair tissue) 2.0 knee score and atlas. Cartilage **13**(1_suppl), 571S–587S (2021)
28. Shan, L., Zach, C., Charles, C., Niethammer, M.: Automatic atlas-based three-label cartilage segmentation from mr knee images. Med. Image Anal. **18**(7), 1233–1246 (2014)
29. Siddique, N., Paheding, S., Elkin, C.P., Devabhaktuni, V.: U-net and its variants for medical image segmentation: A review of theory and applications. Ieee Access **9**, 82031–82057 (2021)
30. Tang, H., Liu, X., Sun, S., Yan, X., Xie, X.: Recurrent mask refinement for few-shot medical image segmentation. In: Proceedings of the IEEE/CVF international conference on computer vision. pp. 3918–3928 (2021)
31. Thippeswamy, P.B., Nedunchelian, M., Rajasekaran, R.B., Riley, D., Khatkar, H., Rajasekaran, et al.: Updates in postoperative imaging modalities following musculoskeletal surgery. JCOT **22**, 101616 (2021)
32. Vorontsov, E., Tang, A., Pal, C., Kadoury, S.: Liver lesion segmentation informed by joint liver segmentation. In: ISBI 2018. pp. 1332–1335. IEEE (2018)
33. Wang, S., Yang, D.M., Rong, R., Zhan, X., Xiao, G.: Pathology image analysis using segmentation deep learning algorithms. AJP **189**(9), 1686–1698 (2019)
34. Wang, Y., Yao, Q., Kwok, J.T., Ni, L.M.: Generalizing from a few examples: A survey on few-shot learning. ACM computing surveys (csur) **53**(3), 1–34 (2020)

Electron Microscopy Image Registration with Twin Axial Transformer and Progressive Training

Can Zhou[1], Haiqun Jin[2], Chunying Yin[2], Jun Guo[2], Futian Wang[1], Yueyi Zhang[2], and Ruobing Zhang[2,3,4](✉)

[1] AHU-IAI AI Joint Laboratory,Anhui University, Hefei 230601, China
[2] Institute of Artificial intelligence, Hefei Comprehensive National Science Center, Hefei 230088, China
[3] Jiangsu Key Laboratory of Medical Optics, Suzhou Institute of Biomedical Engineering and Technology, Chinese Academy of Sciences,Suzhou 215163, China
zhangrb@sibet.ac.cn
[4] School of Biomedical Engineering (Suzhou), Division of Life Sciences and Medicine, University of Science and Technology of China, Suzhou 215163, China

Abstract. Convolutional Neural Network (CNN) based approaches for electron microscopy (EM) image registration are limited in their effectiveness due to their locality characteristic. Recently, the Vision Transformer (ViT) has significantly advanced computer vision by effectively processing long-distance dependencies in images with its self-attention mechanism. Nevertheless, while ViT is proficient in handling a wide range of images, it encounters difficulties in processing images with large deformations and complex structures in image registration tasks. To address these challenges, in this paper, we present an EM image registration network that leverages the Twin Axial Transformer enhanced with a multi-scale feature extraction module. The integration of multi-level feature extraction with the ViT framework allows for a detailed analysis of visual information at various scales, significantly enhancing the network's ability to process and align EM images. Furthermore, we introduce a progressive training approach designed to overcome the difficulties associated with training on datasets with large deformations, making the training process more efficient and stable. Lastly, we automatically generate deformed images and their corresponding high-quality optical flow labels to effectively improve the registration accuracy. Experimental results on a private dataset and a publicly available dataset demonstrate that the network proposed in this study significantly improves registration accuracy and efficiency in handling large-scale deformations in EM image registration compared to existing technologies.

Keywords: Image Registration · Electron Microscopy · Vision Transformer · Progressive Training · Large Deformation

M. Modat et al. (Eds.): WBIR 2024, LNCS 15249, pp. 253–264, 2024.
https://doi.org/10.1007/978-3-031-73480-9_20

1 Introduction

Electron microscopes (EMs), utilizing a technique that employs a beam of electrons for high-resolution imaging, play a crucial role in medical and biology research. During the electron microscopy imaging process, samples need to be cut into very thin and continuous sections, a procedure that often introduces physical deformations. Each layer may become slightly twisted or deformed, making it difficult to directly align images from consecutive layers. Image registration technology uses algorithms to precisely adjust and align these images, compensating for these deformations and ensuring the correct position and orientation of each image layer. This enables researchers to accurately reconstruct three-dimensional structures ranging from individual cells to entire tissues. However, EM image registration, confronts significant challenges due to the inevitable deformations during sample preparation and imaging processes, and notably struggles with images exhibiting large variations or complex structures.

In the early stages of image registration development, traditional methods primarily relied on hand-crafted features such as SIFT [1] and ORB [2], achieving registration by identifying and matching key points within images. The RANSAC algorithm [3], as a model fitting method, provided an effective framework for handling outliers and ensuring registration accuracy, significantly enhancing the robustness of feature point matching. In the medical and biomedical image registration fields, the Elastix toolkit [4] and the elastic registration method [5] for tissue slice images, as well as the elastic volume reconstruction technique [6], have been widely used. While these methods show good precision and stability for small-scale deformations, they require complex parameter adjustments and manual intervention when faced with large-scale deformations and complex structures, limiting their flexibility and efficiency in processing large datasets.

With the rise of deep learning technology, the image registration field has undergone significant changes. Deep learning has shown great potential in feature extraction, pattern recognition, and image understanding. The DeepFlow method [7], applying deep matching techniques to optical flow estimation. The FlowNet [8] is the first fully CNN-based optical flow estimation network, and the SPyNet [9], which uses a pyramid structure to refine optical flow estimation layer by layer, not only improved the accuracy of optical flow estimation but also significantly enhanced computational efficiency by automatically learning deep features of image content for more precise and efficient image registration.

For specific types of image registration problems, such as EM image registration, deep learning methods have also demonstrated their potential and challenges. In the quest for efficient and accurate medical image registration methods, models proposed by Balakrishnan et al. [10] [11], adopting unsupervised learning strategies, directly learn deformation fields from unlabeled image pairs, effectively overcoming the dependence on registration ground truth labels inherent in traditional registration methods. SDHNet [12] introduced an innovative approach by generating Hierarchical Deformation Fields (HDFs) and utilizing a self-deformation distillation scheme, significantly enhancing registration accuracy and efficiency. The SsEMnet network [13], by combining spatial transformers and convolutional autoencoders, is specifically optimized for the dense and varying structural features in ssEM images. Zhou et al. [14] learned a global optimization function during the training phase, using pretrained autoencoders to extract high-level features from each image as a measure of similarity, avoiding pixel-level correlations and ensuring the independence of image content, and introduced 3D convolution to reduce cumulative errors during registration. EMReg [15] proposed the introduction of spatial attention modules into feature extraction and used Cost Volume to assist in optical flow estimation. However, these methods generally perform well only on small deformations and struggle with large deformations due to the intrinsic limitations of CNNs.

To address the challenges of complex structures and large-scale deformations faced in EM image registration, in this paper, we propose a deep learning framework that combines multi-scale feature extraction with the ViT to precisely capture the complex structures and deformations of images. Our framework employs a dual feature encoder TwinAxialFormer to extract a feature pyramid layer by layer and computes the residual optical flow, thereby achieving recursive optical flow field prediction from coarse to fine resolution. To overcome the training challenges posed by large deformations, we adopt a progressive training method, which is similar to the approach used by Han et al. [16] and is particularly suited for models with hierarchical structures, such as our pyramid network architecture. This training approach starts from the top layer, initially using datasets with smaller degrees of deformation for training to focus on the precise registration of fine features, and then progressively introduces datasets with larger deformations, expanding the model's range of deformation adaptation layer by layer. This phased training not only enhances the model's robustness but also optimizes the inter-layer relationships through a final global fine-tuning phase, enabling the model to achieve precise registration performance for that specific range of deformation.

2 Method

As shown in Fig. 1, given two consecutive EM images, namely the fixed image and the moving image, we use a twin feature encoder to extract a feature pyramid for both images layer by layer and compute the residual optical flow for each layer, which is combined with the initial optical flow of each layer to obtain the optical flow estimation for the current layer. Recursive calculations from coarse to fine resolutions of the optical flow fields are performed to ultimately obtain a full-resolution optical flow prediction.

2.1 TwinAxialFormer

As shown in Fig. 1, the TwinAxialFormer initially employs convolutional blocks to downsample the input to create image patches. These patches are then processed by a multi-head attention structure, a pivotal element of the network that enables simultaneous handling of multiple information representations. This

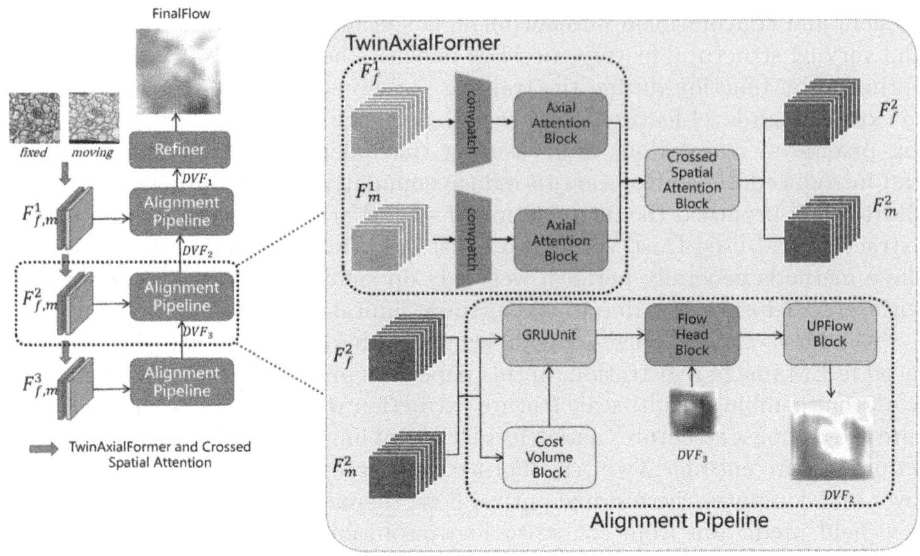

Fig. 1. Overview of the proposed image registration framework, including the TwinAxialFormer and Pyramid Alignment Pipeline for iterative refinement of optical flow fields.

functionality significantly boosts the network's capacity to effectively concentrate on various spatial dimensions. Building on this foundation, and inspired by the axial attention mechanism cited in Ho et al. [17], we have developed an axial attention module specifically for 2D image registration. Nested within the multi-head attention framework, this module is designed to process the information through two primary steps, optimizing it for enhanced registration accuracy.

1. **Row-wise Attention**:
   - Attention weights are calculated on a per-row basis, with each patch computing attention scores in relation to other patches within the same row. Here, q_i (queries), k_i (keys), and v_i (values) are specifically to the i-th row of the image patch. d_k is the dimensionality of the keys when calculated on a per-row basis.
   - The row-wise attention formula provided here calculates the attention scores specifically for the i-th row of the image patch:

   $$\text{Attn}_i = \text{softmax}\left(\frac{q_i k_i^T}{\sqrt{d_k}}\right) v_i$$

   This operation focuses the model's attention on interactions within each row, enhancing feature integration horizontally.
2. **Column-wise Attention**: Similarly, attention weights are then computed and applied across the columns. This step, following the row-wise attention, shifts the focus to vertical interactions and, in conjunction with the previously

applied row-wise attention, allows the model to capture global dependencies within the image, thereby integrating information across both dimensions comprehensively.

This hierarchical application of axial attention, focusing first on rows and then columns, enables direct interaction between patches within the same row or column, and indirect interaction between patches in different rows or columns. This effectively establishes global connections between patches while significantly reducing computational resource requirements. To finalize the feature integration process, the network utilizes a Multilayer Perceptron (MLP) module to further enhance and refine the transformed features. Similar to the classic Transformer architectur, the network also incorporates layer normalization, residual connections, and dropout strategies, ensuring the model's generalization ability and refine the feature learning trajectory.

After using the TwinAxialFormer for feature extraction, we employ the Alignment Pipeline to estimate optical flow. This process starts with the two sets of feature maps, denoted as $F^i_{f,m}$, which is obtained by TwinAxialFormer. We first input the feature maps $F^i_{f,m}$ into a Cost Volume Block to calculate their cost volume. It constructs a three-dimensional data structure that encapsulates the similarity between pairs of pixels or patches from two different images. For each pixel in the first image, the cost volume stores a series of costs that correspond to how well that pixel matches with various other pixels in the second image across a range of possible displacements. Next, we feed both the cost volume and feature maps into a GRU unit for further fusion and integration. Then, the initial optical flow of this layer DVF_{i+1} and the fused features are processed together in a Flow Head Block to obtain the current layer's flow estimate. The Flow Head Block consists of two convolutional layers with a ReLU activation in between. Finally, the flow estimate of this layer is passed through an UPFlow Block to generate the initial optical flow estimate for the next layer DVF_i. In this block, a mask is dynamically learned using two convolutional layers with a ReLU activation function in between, which adjusts the upsampling weights. This learned mask is then used to perform a convex combination, effectively upsampling the flow field to a higher resolution while maintaining smoothness and accuracy. By repeating the optical flow estimation process three times in a pyramid format, we obtain an optical flow estimate at the original resolution. Subsequently, we employ a Refiner module to further enhance the quality of the optical flow estimate. The Refiner consists of a series of seven convolutional layers with intermediate ReLU activations. This design progressively extracts and refines features, increasing and then decreasing the number of channels to capture detailed information while reducing noise. The final output is a high-resolution optical flow field with improved accuracy.

2.2 The Progressive Training Method

The progressive training method is specifically designed to address the challenge of achieving convergence to a satisfactory local optimum when training with large deformation data. This method mitigates the performance degradation typically associated with the complexity of the large-scale deformations. This approach adopts a phased strategy throughout the training process, particularly suitable for models with hierarchical structures, such as pyramid network architectures.

The training initially selecting datasets with lesser degrees of distortion for training the highest detail-capturing layers. This allows the model to concentrate on learning precise registration of fine features. As the model advances down to subsequent layers of the pyramid, progressively datasets with greater degrees of distortion are introduced. This stepwise training enables the model to accommodate a wider range of distortions, thereby enhancing its robustness with each phase. Upon completing the training of a layer and freezing its parameters, the model effectively locks in the features learned at that level of distortion. As the training extends to layers associated with increasingly significant distortions, the model gradually broadens its registration capabilities. Following the individual training and parameter freezing of each layer, the constraints on all layers are lifted, ushering the model into a global fine-tuning phase. This fine-tuning phase fine-tuned with data specific to a singular degree of distortion, preparing it to adeptly manage scenarios with that particular level of distortion.

2.3 Loss Function

Our loss function consists of three parts. The specific formula is as follows:

$$\begin{aligned} \mathrm{L} = & \frac{1}{N}\sum_{i=1}^{N}(\mathrm{HuberLoss}(I_m^i \circ G^i, I_m^i \circ FF^i)) \\ & + \alpha * \frac{1}{N}\sum_{i=1}^{N}(\mathrm{HuberLoss}(G^i, FF^i)) + \lambda * \frac{1}{N}\sum_{i=1}^{N}\left\|\nabla FF^i\right\|_2^2 \end{aligned} \quad (1)$$

where N represents the batch size, I_m^i represents the vector representation of the i-th moving image, G^i represents the ground-truth of optical flow field for the i-th pair, FF^i represents the final predicted optical flow field for the i-th pair. The operations $I_m^i \circ G^i$ and $I_m^i \circ FF^i$ denote the transformation of the moving image I_m^i by the optical flow fields G^i and FF^i, respectively. The first part, quantifies the similarity between the transformed moving image $I_m^i \circ G^i$ and the transformed moving image $I_m^i \circ FF^i$. The HuberLoss is chosen for its robustness in handling outliers compared to the traditional mean squared error (MSE), balancing accuracy and stability during the registration process. The second part, focuses on the consistency between the predicted optical flow field FF^i and the ground-truth flow field G^i. The parameter α adjusts the contribution of this loss component within the overall loss function, allowing the model to flexibly balance between the image similarity loss and the optical flow field similarity loss. The third part, introduces a smoothness loss by constraining the spatial gradient of the optical flow field to ensure the smoothness of the predicted flow. The weight λ controls the influence of the smoothness loss within the total loss function. This component helps suppress potential unnatural jumps in the optical flow prediction, maintaining the continuity and naturalness of the predicted optical flow field.

3 Experiments

3.1 Datasets

In this study, we employ two datasets to evaluate the proposed image registration method: the publicly available CREMI dataset [18] and our self-constructed MEC dataset. The CREMI-A dataset is part of a collection designed for detailed neural circuitry mapping in the adult fly brain. It offers two $(5\,\mu m)^3$ volumes of serial section electron microscopy, each with dimensions of $1250\,px \times 1250\,px \times 125\,px$ for both training and testing. The MEC dataset, derived from the mouse medial entorhinal cortex, features a volume of 0.6mm^3 with high-resolution electron microscopy data totaling 2PB. The dataset includes around 17,000 slices, each with a voxel resolution of 4nm by 4nm by 35nm, and is processed for alignment and registration.

To generate the unregistered image pairs required for training and testing, we randomly sampled 512×512 sized image blocks from both datasets. Specifically, we sampled 2480 pairs of image blocks from the CREMI dataset and 2500 pairs from the MEC dataset. Each pair of images includes one fixed image and one corresponding reference image. We then applied thin-plate spline deformation [19] to the reference image, and obtained high-quality optical flow fields as ground truth by registering the moving image with the reference image, using this flexible warping method to generate moving images that simulate misalignments caused by slicing or imaging errors typically encountered in practical scenarios.

During the registration process, the moving images will be transformed to match the fixed images, which is key to assessing the performance of the registration algorithm. To thoroughly train and validate our model, we divided the image pairs from the CREMI and MEC datasets into training, validation, and test sets, with ratio of 8:1:1.

3.2 Implementation Details

The experiments were conducted on a workstation equipped with four Nvidia Titan XP GPUs, ensuring the efficiency and reliability of the training process. We chose PyTorch [20] as the deep learning framework for implementing the network architecture, considering its flexibility and wide community support.

Each training session lasted for 200 epochs, a setting aimed at ensuring the network had sufficient time to learn the complex deformation relationships between images. The initial learning rate was set to 0.00125, combined with the Adam optimizer [21], because Adam can provide stable and rapid convergence when dealing with large datasets. To further optimize the adjustment of the learning rate, we employed the CosineAnnealingLR scheduler with a cycle set to 16 epochs and a minimum learning rate (η_{min}) set to 0.0001. This learning rate scheduling method helps the model refine its learning in the later stages of training, avoiding overfitting, and aids the optimization algorithm in escaping local minima, increasing the possibility of finding a better local optimum.

Moreover, we generated three datasets with different degrees of distortion on both the CREMI and MEC datasets. This was done to assess the model's performance on registration tasks of varying difficulty levels. Through this approach, we were able to gain a deeper understanding of the model's adaptability to various types of deformations and its accuracy in registration.

3.3 Ablation

We conducted a detailed ablation analysis to separately evaluate the effectiveness of the TwinAxialFormer and the proposed progressive training method. Initially, we explored the impact of utilizing the TwinAxialFormer by comparing it against a baseline where the TwinAxialFormer was replaced with a classic convolution-based feature extraction module. This comparison aimed to isolate the contribution of the TwinAxialFormer to the overall performance. Following this, we shifted our focus to the progressive training method. Here, we contrasted the outcomes of applying a straightforward, one-time direct training strategy against the progressive training approach.

As shown in Table 1, which encompassed both aspects of our ablation study, it became evident that each component, the TwinAxialFormer and progressive training, played a pivotal role in enhancing the model's performance. The TwinAxialFormer significantly boosted the model's ability to accurately capture and align features, as highlighted by the improved metrics when compared to the conventional convolution-based approach. Similarly, the adoption of progressive training further amplified the effectiveness of the training regimen, leading to a more robust and finely-tuned model.

Table 1. Ablation experiments on MEC dataset.

Setting		SSIM	NCC
TwinAxialFormer	Progressive Training		
✗	✗	0.497	0.452
✓	✗	0.533	0.484
✗	✓	0.536	0.483
✓	✓	**0.584**	**0.542**

Table 2. Comparative Performance on CREMI and MEC Datasets with a fitter parameter set to 0.2.

Method	CREMI			MEC	
	Dice	SSIM	NCC	SSIM	NCC
Elastic [6]	0.735	0.629	0.411	0.419	0.378
bUnwarpJ [5]	0.746	0.631	0.429	0.428	0.390
ssEMnet [13]	0.780	0.668	0.501	0.471	0.443
EMReg [15]	0.887	0.773	0.668	0.549	0.505
Ours	**0.926**	**0.803**	**0.741**	**0.584**	**0.542**

3.4 Quantitative Comparisons with Baselines

To evaluate the efficacy of our proposed image registration framework, we performed quantitative comparisons against established baseline methods, namely Elastic, bUnwarpJ [5], ssEMnet [13], and EMReg [15]. Table 2 presents a comprehensive overview of the performance metrics across both datasets for all evaluated methods. Utilizing the Structural Similarity Index Measure (SSIM) [22] and Normalized Cross-Correlation (NCC) as our primary metrics, supplemented by the Dice coefficient (Dice Score) for the CREMI dataset due to its available segmentation information.

Figure 2 provides a direct visual comparison of registration results obtained by various methods. Our approach exhibits superior conformity to the ground truth, successfully avoiding the edge stretching issues observed with EMReg's registrations,and provide more accurate local textures and edges compared to other methods

Figure 3 showcases a comparative analysis of image registration methods across varied levels of dataset distortions, including CREMI0.2, CREMI0.25, and CREMI0.3. "CREMI0.2" signifies the application of a thin plate spline transformation with a flitter parameter set to 0.2. Similarly, the labels "CREMI0.25" and "CREMI0.3" refer to datasets created by employing different flitter parameters, 0.25 and 0.3, respectively. The bar graph illustrates our method's superiority in maintaining higher Dice Scores, SSIM, and NCC values, regardless of distortion degree, signaling our method's robustness and precision in capturing the true alignment even under challenging deformation scenarios.

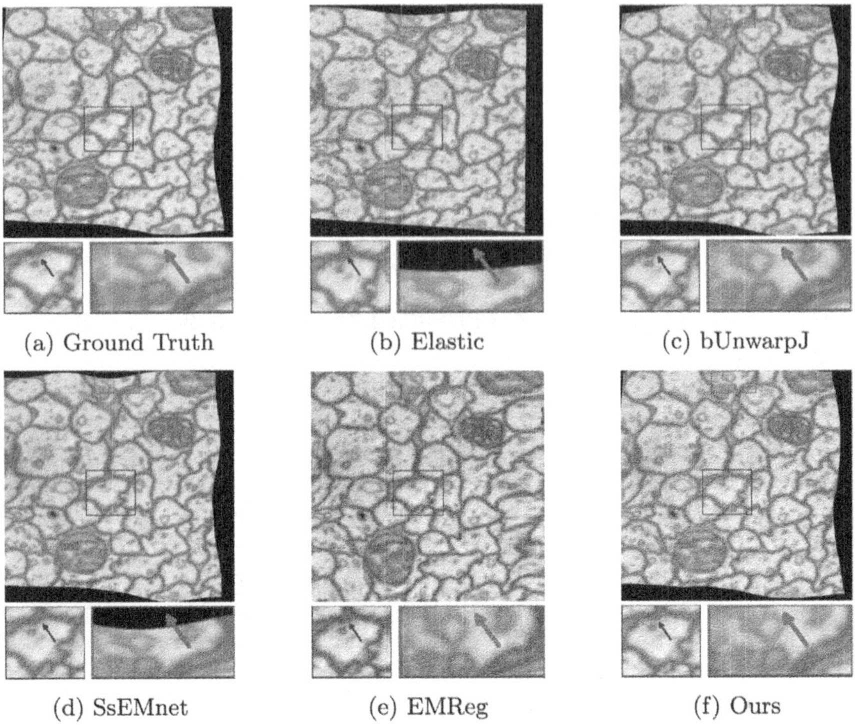

Fig. 2. Visualization of registration results on CREMI-A. Our approach achieves a registration that is notably closer to the ground truth, delivering superior results both in terms of fine detail and at the edges.

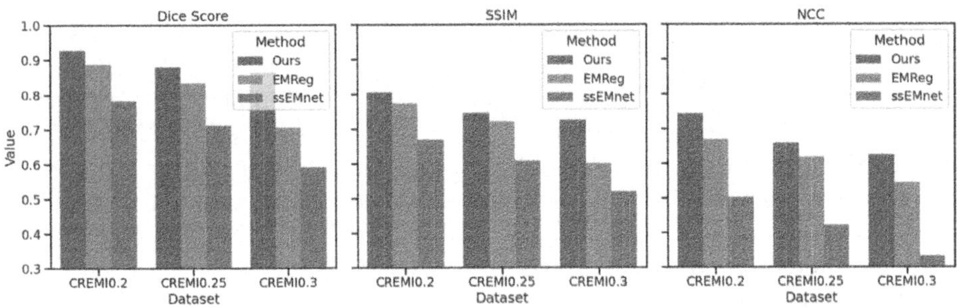

Fig. 3. Comparative Analysis of Image Registration Methods Across Varied Dataset Distortions

4 Conclusion

Our study introduces a deep learning framework designed for EM image registration. It takes a multi-scale approach and integrates seamlessly with the Vision Transformer architecture. By employing dual feature encoders, our system builds a hierarchical feature pyramid and performs residual optical flow calculations, achieving recursive prediction of optical flow fields from coarse to fine. In addition, the progressive training method we adopt effectively overcomes the training difficulties of large deformation data. Evaluation results on the CREMI and MEC datasets show that our method has good registration effects for large deformations.

Acknowledgements. This study was funded by the Institute of Artificial Intelligence, Hefei Comprehensive National Science Center (2YGXT003), and the National Natural Science Fundation of China (32271430).

References

1. David G Lowe. Distinctive image features from scale-invariant keypoints. *International journal of computer vision*, 60:91–110, 2004.
2. Ethan Rublee, Vincent Rabaud, Kurt Konolige, and Gary Bradski. Orb: An efficient alternative to sift or surf. In *2011 International conference on computer vision*, pages 2564–2571. Ieee, 2011.
3. Martin A Fischler and Robert C Bolles. Random sample consensus: a paradigm for model fitting with applications to image analysis and automated cartography. *Communications of the ACM*, 24(6):381–395, 1981.
4. Stefan Klein, Marius Staring, Keelin Murphy, Max A Viergever, and Josien PW Pluim. Elastix: a toolbox for intensity-based medical image registration. *IEEE transactions on medical imaging*, 29(1):196–205, 2009.
5. Ignacio Arganda-Carreras, Carlos OS Sorzano, Roberto Marabini, José María Carazo, Carlos Ortiz-de Solorzano, and Jan Kybic. Consistent and elastic registration of histological sections using vector-spline regularization. In *Computer Vision Approaches to Medical Image Analysis: Second International ECCV Workshop, CVAMIA 2006 Graz, Austria, May 12, 2006 Revised Papers 2*, pages 85–95. Springer, 2006.
6. Stephan Saalfeld, Richard Fetter, Albert Cardona, and Pavel Tomancak. Elastic volume reconstruction from series of ultra-thin microscopy sections. *Nature methods*, 9(7):717–720, 2012.
7. Philippe Weinzaepfel, Jerome Revaud, Zaid Harchaoui, and Cordelia Schmid. Deepflow: Large displacement optical flow with deep matching. In *Proceedings of the IEEE international conference on computer vision*, pages 1385–1392, 2013.
8. Alexey Dosovitskiy, Philipp Fischer, Eddy Ilg, Philip Hausser, Caner Hazirbas, Vladimir Golkov, Patrick Van Der Smagt, Daniel Cremers, and Thomas Brox. Flownet: Learning optical flow with convolutional networks. In *Proceedings of the IEEE international conference on computer vision*, pages 2758–2766, 2015.
9. Anurag Ranjan and Michael J Black. Optical flow estimation using a spatial pyramid network. In *Proceedings of the IEEE conference on computer vision and pattern recognition*, pages 4161–4170, 2017.
10. Guha Balakrishnan, Amy Zhao, Mert R Sabuncu, John Guttag, and Adrian V Dalca. Voxelmorph: a learning framework for deformable medical image registration. *IEEE transactions on medical imaging*, 38(8):1788–1800, 2019.
11. Guha Balakrishnan, Amy Zhao, Mert R Sabuncu, John Guttag, and Adrian V Dalca. An unsupervised learning model for deformable medical image registration. In *Proceedings of the IEEE conference on computer vision and pattern recognition*, pages 9252–9260, 2018.

12. Shenglong Zhou, Bo Hu, Zhiwei Xiong, and Feng Wu. Self-distilled hierarchical network for unsupervised deformable image registration. *IEEE Transactions on Medical Imaging*, 2023.
13. Inwan Yoo, David GC Hildebrand, Willie F Tobin, Wei-Chung Allen Lee, and Won-Ki Jeong. ssemnet: Serial-section electron microscopy image registration using a spatial transformer network with learned features. In *Deep Learning in Medical Image Analysis and Multimodal Learning for Clinical Decision Support: Third International Workshop, DLMIA 2017, and 7th International Workshop, ML-CDS 2017, Held in Conjunction with MICCAI 2017, Québec City, QC, Canada, September 14, Proceedings 3*, pages 249–257. Springer, 2017.
14. Shenglong Zhou, Zhiwei Xiong, Chang Chen, Xuejin Chen, Dong Liu, Yueyi Zhang, Zheng-Jun Zha, and Feng Wu. Fast and accurate electron microscopy image registration with 3d convolution. In *Medical Image Computing and Computer Assisted Intervention–MICCAI 2019: 22nd International Conference, Shenzhen, China, October 13–17, 2019, Proceedings, Part I 22*, pages 478–486. Springer, 2019.
15. Xinzhao Liu, Yueyi Zhang, Shenglong Zhou, Zhiwei Xiong, and Xiaoyan Sun. Electron microscopy image registration using correlation volume. In *2023 IEEE 20th International Symposium on Biomedical Imaging (ISBI)*, pages 1–5. IEEE, 2023.
16. Koen AJ Eppenhof, Maxime W Lafarge, Mitko Veta, and Josien PW Pluim. Progressively trained convolutional neural networks for deformable image registration. *IEEE transactions on medical imaging*, 39(5):1594–1604, 2019.
17. Jonathan Ho, Nal Kalchbrenner, Dirk Weissenborn, and Tim Salimans. Axial attention in multidimensional transformers. arXiv preprint arXiv:1912.12180, 2019.
18. J Funke, S Saalfeld, DD Bock, SC Turaga, and E Perlman. Miccai challenge on circuit reconstruction from electron microscopy images, 2016.
19. Adrien Bartoli, Mathieu Perriollat, and Sylvie Chambon. Generalized thin-plate spline warps. *International Journal of Computer Vision*, 88:85–110, 2010.
20. Adam Paszke, Sam Gross, Francisco Massa, Adam Lerer, James Bradbury, Gregory Chanan, Trevor Killeen, Zeming Lin, Natalia Gimelshein, Luca Antiga, et al. Pytorch: An imperative style, high-performance deep learning library. *Advances in neural information processing systems*, 32, 2019.
21. Diederik P Kingma and Jimmy Ba. Adam: A method for stochastic optimization. arXiv preprint arXiv:1412.6980, 2014.
22. Zhou Wang, Alan C Bovik, Hamid R Sheikh, and Eero P Simoncelli. Image quality assessment: from error visibility to structural similarity. *IEEE transactions on image processing*, 13(4):600–612, 2004.

General Vision Encoder Features as Guidance in Medical Image Registration

Fryderyk Kögl[1,2,3](✉), Anna Reithmeir[2,3,4], Vasiliki Sideri-Lampretsa[2], Ines Machado[5,6], Rickmer Braren[1,7], Daniel Rueckert[1,2,9], Julia A. Schnabel[2,3,8], and Veronika A. Zimmer[1,2]

[1] School of Medicine and Health, Klinikum Rechts der Isar, Technical University of Munich, Munich, Germany
fryderyk.koegl@tum.de
[2] School of Computation, Information and Technology, Technical University of Munich,Munich, Germany
[3] Institute of Machine Learning in Biomedical Imaging, Helmholtz, Munich, Germany
[4] Munich Center for Machine Learning, Munich, Germany
[5] Cancer Research UK Cambridge Institute, University of Cambridge, Cambridge, UK
[6] Department of Oncology, University of Cambridge, Cambridge, UK
[7] German Cancer Consortium (DKTK), Partner Site Munich, Heidelberg, Germany
[8] School of Biomedical Engineering and Imaging Sciences, King's College London, London, UK
[9] Department of Computing,Imperial College London, London, UK

Abstract. General vision encoders like DINOv2 and SAM have recently transformed computer vision. Even though they are trained on natural images, such encoder models have excelled in medical imaging, e.g., in classification, segmentation, and registration. However, no in-depth comparison of different state-of-the-art general vision encoders for medical registration is available. In this work, we investigate how well general vision encoder features can be used in the dissimilarity metrics for medical image registration. We explore two encoders that were trained on natural images as well as one that was fine-tuned on medical data. We apply the features within the well-established B-spline FFD registration framework. In extensive experiments on cardiac cine MRI data, we find that using features as additional guidance for conventional metrics improves the registration quality. The code is available at github.com/compai-lab/2024-miccai-koegl.

Keywords: Foundation models · feature-based distance measures

F. Kögl, A. Reithmeir—Equal contribution.

M. Modat et al. (Eds.): WBIR 2024, LNCS 15249, pp. 265–279, 2024.
https://doi.org/10.1007/978-3-031-73480-9_21

1 Introduction

Image registration is a crucial step in medical image analysis that enables the alignment of corresponding anatomical or functional regions across multiple images, regardless of modality or differences between time points. This process creates a unified and accurate representation of patient data, facilitating enhanced diagnostic accuracy, improved treatment planning, and disease monitoring [17,35].

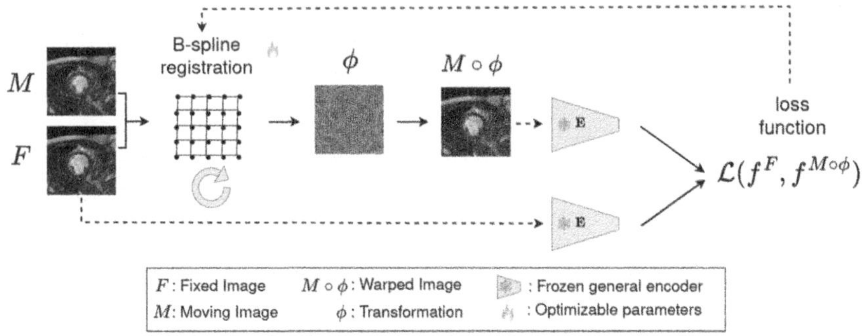

Fig. 1. Overall framework: We explore how different pre-trained natural image encoders can be used within the registration objective function. The fixed image and the warped moving image are encoded with the frozen encoders, and the distance between the features is measured. This distance measure term is then used in the objective function of the iterative B-splines registration.

Effective registration requires the assessment of the dissimilarity between images. This is not trivial due to the highly complex, non-linear relationships arising from anatomical or functional differences. As a result, choosing a distance measure is crucial and strongly influences the result. Consequently, several measures have been tested in the context of registration. Pixel-based distance measures such as the Euclidian distance (mean squared error - MSE) between pixel intensities and the local or non-local normalized cross-correlation (NCC) are robust and common choices used in conventional [1,2,5,30,37,39] and learning-based registration [4,9,12,13,28]. However, because registration is a non-convex optimization problem, these measures might be susceptible to local minima. To account for that, hand-crafted local descriptor and histogram-based losses, e.g., modality independent neighborhood descriptor (MIND) [19], normalized gradient fields (NGFs) [16], and mutual information (MI) and its variants [23,36,41], have also been widely adopted, especially in the case of images with different contrasts. However, Mutual Information and its variants operate on intensity histograms; therefore, they are agnostic to the underlying geometry of the image structures. Moreover, handcrafted features, derived from intensity and gradient information, may be sensitive to local details but lack global context, rendering them susceptible to local minima.

To account for the shortcomings of voxel-intensity measures [42], several works have studied the effects of data-driven measures in the context of registration [11,18,27,32,33]. However, each *learned* measure comes with certain shortcomings. Haskins et al. [18] require ground truth transformations. Czolbe et al. [11] propose to pre-train a feature extractor on a supplementary segmentation task and then use the semantic features to drive the optimization of a learning-based registration model for mono-modal registration. In contrast, [27,32] are tested only in the context of rigid/affine registration. However, all the above methods need retraining to apply them to a new modality or image domain, rendering them impractical for most clinical scenarios.

Building upon the success of feature-based measures, we explore the potential of using features learned by large vision encoders to drive deformable image registration. General vision encoders are deep learning models designed to learn semantic information from visual input such as images [8,21,40,44] and videos [25], and they can be applied to various downstream tasks once they have been trained. These general vision encoders provide versatile, rich feature embeddings, leveraging large, unlabeled datasets to enable task-agnostic performance, eliminating the need for task-specific data. Additionally, they allow for seamless adaptation to new tasks without retraining, which is very important in the medical image domain, where data can be scarce. For this reason, general vision encoders are being investigated for a wide range of applications [3,7,10,14,20,29,43]. Nonetheless, these encoders have not been extensively explored in the context of medical image registration, with DINO-Reg [34] to be the only work that explores the registration of DINO-v2 [26] features. In this work, instead of estimating the deformation field between feature maps as in [26], we propose to utilize these semantically rich encoders as a measure for image distance in the feature domain. Although most general vision encoders are trained using natural image cohorts, which vastly differ in appearance and properties from medical images, we demonstrate that the rich embedding spaces could be utilized to drive image registration. An overview of our approach is shown in Fig. 1. To our knowledge, this is the first attempt to investigate whether the features of these modes are applicable as part of distance measures in image registration.

Our **main contributions** can be summarized as follows:

- Explore whether the features of various general vision encoders are suitable for assessing image distance in deformable image registration
- Benchmark the performance of pure feature-based losses vs combined with voxel-intensity-based measures.
- Explore the feasibility of integrating feature-based losses in conventional iterative registration.

2 Methods

Given two n-dimensional images, a fixed image F and a moving image M with $F, M : \Omega \subset \mathbb{R}^n \rightarrow \mathbb{R}$ and $n \in \{2, 3, 4\}$, image registration aims to find an optimal spatial transformation $\phi : \mathbb{R}^n \to \mathbb{R}^n$ such that the transformed moving image is most similar to the fixed image. Typically, this is formulated as an optimization problem $\phi^* = \arg\min_\phi \mathcal{J}(F, M, \phi)$ where the distance between the images is minimized with constraints on the transformation. We denote the objective function as

$$\mathcal{J}(F, M, \phi) = \mathcal{D}_\mathcal{I}(F, M \circ \phi) + \lambda \mathcal{R}(\phi), \tag{1}$$

where $\mathcal{D}_\mathcal{I}$ is an image-based intensity dissimilarity measure (e.g. MI or NCC) and $\mathcal{R}$, is the regularization on the transformation, which is controlled by the regularization parameter λ. A common choice is the diffusion regularizer $\mathcal{R}(\phi) = ||\nabla\phi||^2$, used to assure smoothness of the deformation.

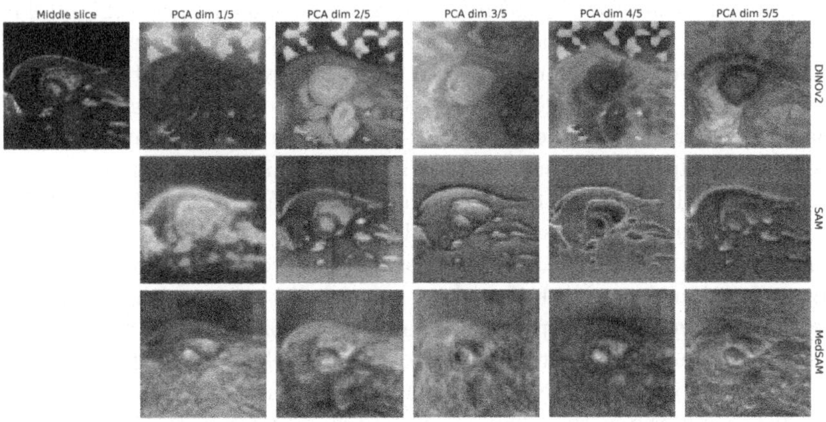

Fig. 2. Qualitative comparison of the first 5 principal components of DINOv2, SAM, and MedSAM features for a test image. DINOv2 captures coarser features, while SAM focuses on edges and MedSAM on the underlying texture.

In this work, we explore the use of general vision encoder features for medical image registration. In Sect. 2.1, we describe the encoders used for the extraction of high-dimensional features, and in Sect. 2.2, we present the dissimilarity measures used for feature comparison. Finally, in Sect. 2.3 we describe two variants of incorporating feature dissimilarity in the registration process.

2.1 General Vision Encoders for Feature Extraction

We compare three different pre-trained encoders: DINOv2 [26], SAM (segment-anything) [22] and MedSAM (segment-anything in medical images) [24]. All three are based on the vision transformer (ViT) architecture [15]. DINOv2 was developed using 142M images from 27 classification, segmentation, and depth estimation datasets in a self-supervised manner by self-distillation without labels. SAM was trained on a large curated segmentation dataset of 11M images. In contrast to DINOv2, SAM employed a supervised learning approach with ground truth segmentations. While the two models above were trained on natural images, i.e., photographs of real-world objects, the recently published MedSAM model specifically aims to segment medical images. It is a version of SAM fine-tuned with a dataset containing 1.6M medical images that covers 10 different modalities. Similarly to SAM, it was trained in a supervised manner.

Importantly, none of the three models have been trained on registration tasks. The features of one test image, extracted with the three encoders, are shown in Fig. 2. Since DINOv2 and SAM were trained on 2D RGB images and require such inputs, we use 2D image slices extracted from 3D medical images for all three encoders. We use these encoders in the registration process by individually mapping the moving and the fixed images to their respective feature spaces. The feature dimension $N_{\mathcal{G}}$ differ between the models - feature maps extracted from DINOv2 have 1024 dimensions, while SAM and MedSAM's have 256 dimensions.

2.2 Feature-Based Distance Measures

To use the extracted features $f^F, f^M \in \mathbb{R}^{N_{\mathcal{G}}}$ of the fixed and warped images for the registration process, we measure their dissimilarity $\mathcal{D}_{\mathcal{F}}$. Here, $N_{\mathcal{G}}$ denotes the feature dimension of the respective encoder model. To this end, we explore two different dissimilarity measures: (i) the negative cosine similarity $cos(f^F, f^M) = \frac{f^F.f^M}{||f^F||||f^M||}$ which measures the cosine of the angle between two vectors and which we denote as cosine dissimilarity in the following, and (ii) the mean absolute error (MAE)/L1 distance $L1(f^F, f^M) = \frac{1}{N_{\mathcal{G}}}\sum_i^{N_{\mathcal{G}}} |f_i^f - f_i^M|$.

2.3 General Vision Encoder Feature Distance in Image Registration

General vision encoders will be denoted as $\mathcal{G}$ and the extracted features as $f^I = \mathcal{G}(I)$ for an image I. We explore two ways of enhancing the registration objective function with a feature-based distance $\mathcal{D}_{\mathcal{F}}$:

Variant 1: Feature distance as the main loss term Here, we replace the voxel intensity-based image distance measure term $\mathcal{D}(M \circ \phi, F)$ in 1 with a feature-based distance $\mathcal{D}_{\mathcal{F}}(f^{M\circ\phi}, f^F)$:

$$\mathcal{J}(F, M, \phi) = \mathcal{D}_{\mathcal{F}}(f^F, f^{M\circ\phi}) + \lambda\mathcal{R}(\phi). \tag{2}$$

Variant 2: Feature distance as additional loss term Here, rather than exchanging the voxel intensity-based dissimilarity for the feature-based one, we use both terms in the objective function:

$$\mathcal{J}(F, M, \phi) = (1-\alpha)\mathcal{D}_{\mathcal{I}}(F, M \circ \phi) + \lambda\mathcal{R}(\phi) + \alpha\mathcal{D}_{\mathcal{F}}(f^F, f^{M\circ\phi}). \tag{3}$$

The parameter $\alpha \in \mathbb{R}$, with $0 \leq \alpha \leq 1$, controls the strength of the feature-based distance term. Note, that α and λ have to be chosen carefully.

3 Experiments and Results

3.1 Dataset and Implementation Details

Dataset and Data Preprocessing. We use the ACDC dataset [6], which consists of cardiac MRI images (multi-slice 2-D cine MRI) acquired from 150 patients. Ground truth segmentations of the left ventricular cavity (LV), myocardium (Myo), and right ventricular cavity (RV) are available for both phases. From the 3D volumes, we extract one middle slice per image after cropping around the heart and use the end-diastolic (ED) slices as the fixed and end-systolic (ES) slices as the moving images. The cropped slices have a size of 128×128. Additionally, we resample them to uniform spacing of $1.8\,\text{mm} \times 1.8\,\text{mm}$ and normalize the intensity values to the range $[0, 1]$.

Evaluation Metrics. Due to the lack of ground truth deformation fields, we evaluate the registration performance using surrogate measures. To assess the registration accuracy, we compare the Dice score (DSC) and the 95th percentile Hausdorff distance (HD95) between the fixed and warped moving segmentation maps. We consider the class-wise DSC of the three classes LV, RV, Myo, and the mean DSC over those. To assess the plausibility of the displacement field, we evaluate the standard deviation of the logarithm of the Jacobian determinant (SDlogJ) of the displacement field.

Implementation Details. To perform the registration, we use the Airlab [31] registration framework. This framework not only supports GPU-based computation that accelerates the registration process but also utilizes the Pytorch native autograd for the optimization process. The latter is particularly useful since it eliminates the need for a manual implementation of the backpropagation function for the encoders. Additionally, we choose B-Spline free-form deformation [30] as a transformation parametrization with a 24×24 control point lattice, a learning rate of 5e−4, a regularizer weight of 120, and 1500 iterations. The hyperparameters were found empirically by maximizing the DSC while maintaining plausible deformations. All computations have been performed with Python 3.10 and PyTorch 2.2.2 on an NVIDIA RTX A6000 GPU. The code is available at github.com/compai-lab/2024-miccai-koegl.

For a fair comparison, we select openly available encoder instances with approximately similar numbers of trainable parameters, where DINOv2 has $87M$, and SAM and MedSAM have $94M$ parameters.[1]. We repeat the images three times along the channel axis to fit the greyscale images to the three-channel input the encoders require. Furthermore, DINOv2 and SAM/MedSAM are trained with different patch sizes, resulting in the resolution of the feature maps being smaller than the input image and differing across the encoders. Thus,

[1] For DINOv2, we chose the base model *dinov2_vitb14_reg*. For SAM and MedSAM, we selected the base models *sam_vit_b_01ec64* and *medsam_vit_b*.

we upsample the images before extracting the features to obtain a consistent spatial resolution for the feature maps across all encoders. The feature maps have a spatial resolution of 64×64. Note that the MedSAM model can handle both 2D and 3D images. However, since the other encoders require 2D inputs, we restrict ourselves to 2D images.

3.2 Results

Comparison of Feature-Based Distance Measures' Curves for Rotations and Translations. In this experiment, we investigate the effectivity of the L1 and cosine dissimilarity measures between features when recovering rigid transformations. We conduct two experiments, one where we rotate the image by steps of $\frac{\pi}{32} rad$ in the range of $[-\pi/2, \pi/2]$ and another one where we translate the image along one axis by steps of $1.2mm$ in the range $[-57.6, 57.6]$ mm. To this end, we choose one test image and gradually transform it. At each step, we encode the transformed and original images, normalize the features to the range $[0, 1]$, and measure the dissimilarity between the two feature maps.

Figure 3 shows the curves for the measured feature distances with L1 and cosine dissimilarity for all three encoders for rotation and translation. In all cases, there is a clear global minimum at zero transformation, which is a desired behavior for a dissimilarity measure in the context of registration. Moreover, the cosine dissimilarity curves demonstrate a wider capture range for both rotation and translation while being smoother than the L1 curves. The latter appear less steep and not so smooth, demonstrating more local minima, making them more difficult to optimize.

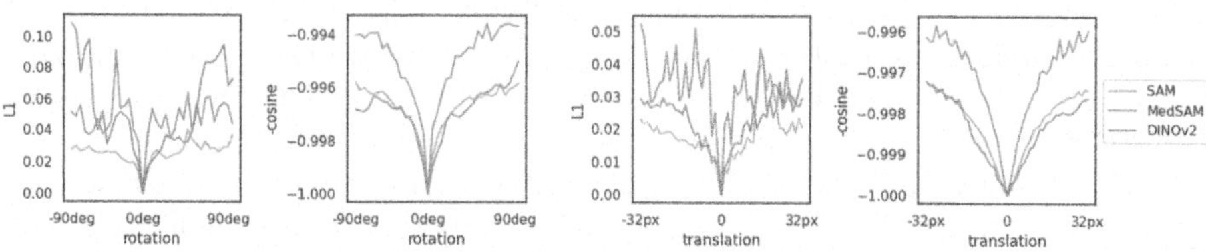

Fig. 3. Comparison of the L1 and cosine dissimilarity measures on the features of the three encoder models for rotation and translation of one test image. Rotation given in degree (deg) and translations given in pixels (px). For both rotation and translation tasks, the cosine dissimilarity curves appear smoother and with a wider capture range making it a better choice as a dissimilarity measure.

Feature-Based Distance Measures in Medical Image Registration. In this experiment, we investigate whether the general vision encoders' features can effectively be used within a distance measure for image registration. To evaluate this we conduct three comparisons: (i) using the feature distance as the sole dissimilarity measure (*variant 1*) versus using it as additional guidance alongside the image intensity-based distance (*variant 2*), (ii) using the L1 versus the cosine dissimilarity as distance measures between features, and (iii) using each of the three general vision encoders. As the baseline, we use the standard registration objective of Eq. 1 with NCC as an image dissimilarity measure and a weighted diffusion regularization term.

Variant 1 vs. Variant 2: Table 1 and Fig. 4 demonstrate that incorporating general vision encoded features as an auxiliary dissimilarity term (*variant 2*) enhances registration performance compared to the baseline. This improvement is observed for both dissimilarity measures and all encoder models, as indicated by the mean DSC and mean HD95. Specifically, a mean DSC of 0.845 and mean HD95 of 1.765 is achieved using the MedSAM encoder with cosine dissimilarity, surpassing the baseline mean DSC of 0.836 and mean HD95 of 1.916. This trend is also evident in class-wise DSC (see Fig. 4). On the other hand, *variant 1*, which does not utilize encoded features, results in decreased registration performance compared to the baseline.

L1 Distance vs. Cosine Dissimilarity: Firstly, the cosine dissimilarity generally improves the HD95 across most experiments for both variants. For *variant 2*, it consistently achieves a slightly higher mean DSC compared to the L1 distance, as shown in 1. Conversely, for *variant 1*, this trend holds true for two out of three encoder models (DINOv2 and SAM).

DINOv2 vs. SAM vs. MedSAM: The results indicate that MedSAM outperforms SAM in terms of both DSC and HD95, particularly in *variant 1*. In *variant 2*, the DSC and HD95 show less variation between the models compared to *variant 1*. Here, MedSAM achieves the best DSC while DINOv2 achieves the best HD95. Last but not least, SAM achieves the smoothest deformations with *variant 1*. Finally, there is no observable folding across all models and variants.

As expected, the MedSAM encoder achieves better results than SAM since it was fine-tuned on medical data and thus should be able to capture anatomical structures better than the general vision encoders. Nevertheless, it is surprising that the DINOv2 encoder, despite not seeing medical images during training, improves the results for *variant 2*. A reason that its features achieve such low performance when used in *variant 1* could be connected to the feature artifacts in the background regions (see Fig. 2. This hypothesis is supported by Fig. 5: It shows the three Jacobian determinant maps that have been obtained with *variant 1* for the DINOv2 and SAM encoders as well as the baseline registration. In the DINOv2-based deformation, local deformations are also visible in the background, while for the NCC-based registration, the deformation concentrates around the relevant regions of the heart. The general lack of folding in the qualitative results and Fig. 5 is most likely achieved with the relatively high regularization weight of the diffusion regularizer. Additionally, we find that the

Table 1. Quantitavie results: class-wise and mean DSC, HD95 and SDlogJ before and after registration. We see that *variant 2* consistently outperforms the baseline in terms of DSC and HD95.

Encoder $\mathcal{G}$	Variant	DSC↑	HD95↓	SDlogJ↓
no encoder	initial	0.651 ± 0.127	4.125 ± 1.816	-
no encoder	NCC	0.836 ± 0.051	1.916 ± 1.154	0.385 ± 0.111
DINOv2	L1	0.730 ± 0.113	3.095 ± 1.518	0.264 ± 0.058
	Cos	0.742 ± 0.108	2.936 ± 1.465	0.244 ± 0.053
	L1 + NCC	0.833 ± 0.075	1.859 ± 1.127	0.330 ± 0.087
	Cos + NCC	0.844 ± 0.063	**1.711 ± 0.918**	0.321 ± 0.082
SAM	L1	0.758 ± 0.106	3.052 ± 1.579	**0.161 ± 0.030**
	Cos	0.761 ± 0.106	2.980 ± 1.614	0.226 ± 0.040
	L1 + NCC	0.841 ± 0.059	1.862 ± 1.184	0.331 ± 0.100
	Cos + NCC	0.842 ± 0.063	1.900 ± 1.204	0.348 ± 0.095
MedSAM	L1	0.798 ± 0.100	2.363 ± 1.422	0.243 ± 0.055
	Cos	0.791 ± 0.105	2.534 ± 1.572	0.333 ± 0.061
	L1 + NCC	0.845 ± 0.055	1.819 ± 1.133	0.363 ± 0.104
	Cos + NCC	**0.845 ± 0.054**	1.765 ± 1.077	0.395 ± 0.096

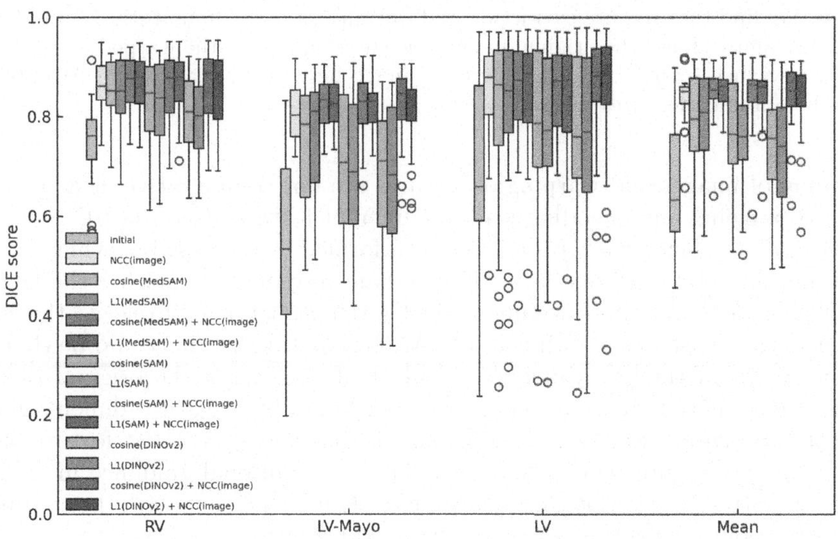

Fig. 4. Quantitative comparison: Class-wise and mean DSC scores before and after registration. Methods using the same encoder share the same color with varying hues. The boxplots are grouped horizontally by class: right ventricular cavity (RV), left ventricular myocardium (LV-Mayo), left ventricular cavity (LV), and the mean over all classes. We see that *variant 2* outperforms the baseline in both class-wise and mean DSC.

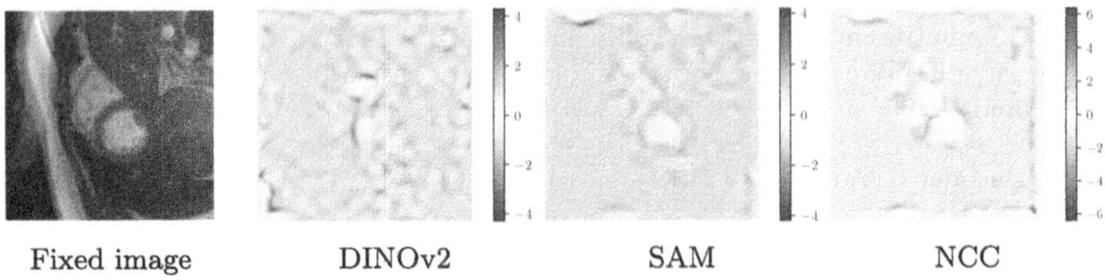

Fig. 5. Qualitative comparison of the Jacobian determinant maps from *variant 1* with the DINOv2 and SAM encoders, as well as the baseline. From left to right, we see that the Jacobian determinant map of DINOv2 has many local deformations in the background regions, while the one of SAM has fewer. The baseline transformation appears smoother outside of the heart. The fixed and moving images differ only in the heart, so we would expect most of the deformation to occur there. White areas show shrinkage, and blue areas expansion.

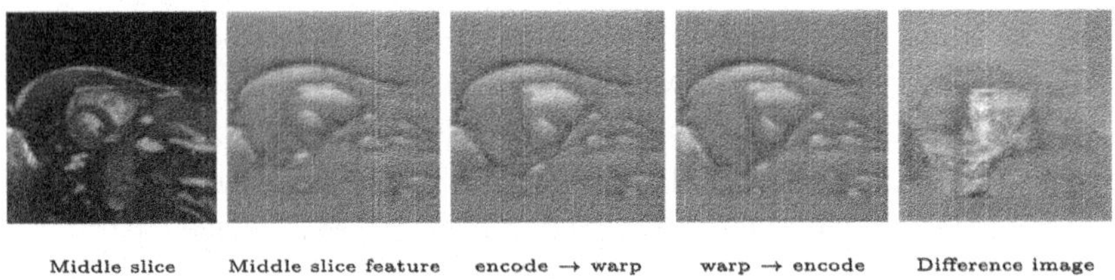

Fig. 6. We find that feature extraction and warping are not commutative. From left to right, the figure shows the original image, the encoded image, the warped features, the encoded warped image, and the difference image between the latter two. We observe that the obtained features differ.

encoding of images and warping of features are not commutative: changing the order of warping and encoding leads to different features (see Fig. 6).

Overall, the fact that *variant 1* is outperformed by the baseline, which in turn is outperformed by *variant 2* on the one hand reveals that neither the general vision encoders nor the fine-tuned MedSAM encoder is suitable as the main driver of the registration optimization. An explanation for this could be that the feature maps in the experiments are only half the size of the original images. Thus, some spatial information is lost, which the high feature dimensionality cannot compensate for. On the other hand, this also shows that feature-based dissimilarity measures can successfully serve as additional guidance in medical image registration, even if the encoder models have not explicitly been trained on medical data, as is the case with the DINOv2 encoder.

Qualitative Comparison of General Vision Encoder Features. We qualitatively compare the extracted features of one randomly chosen image between the three encoder models. Figure 2 shows the major principal components of the features. The DINOv2-encoded features capture large structures with coarse outlines. SAM-encoded features appear to focus more on edges. Also, small struc-

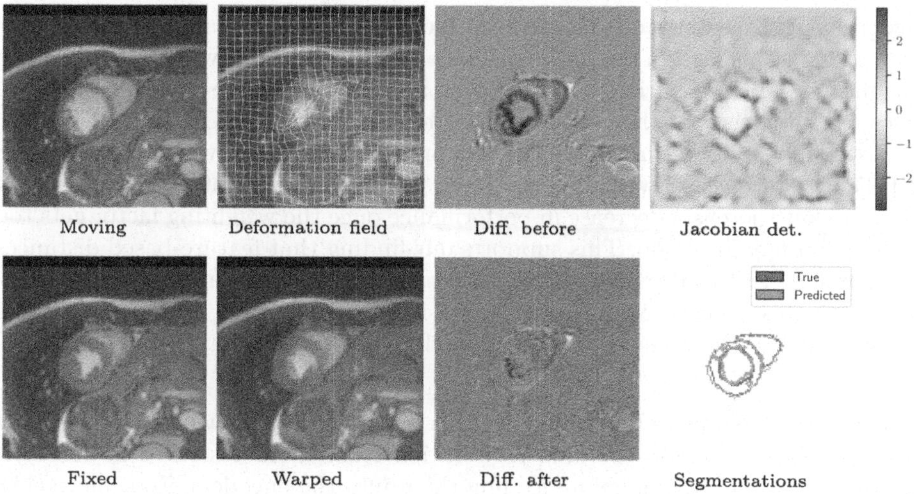

Fig. 7. Qualitative registration results with *variant 2* with $\alpha = 0.3$ for one image pair. The difference images before and after show a clear improvement and the Jacobian determinant map shows shrinkage inside the ventricles, as expected. The segmentation overlap improves from a DSC of 0.622 and HD95 of 7.13 to 0.860 and 1.28, respectively.

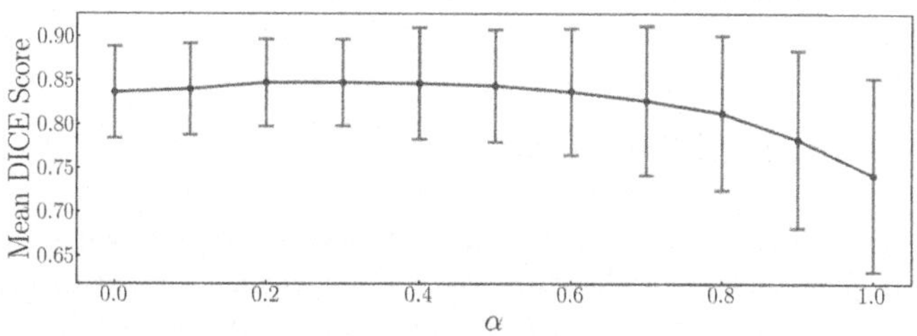

Fig. 8. We evaluate how the weight of the feature-based distance loss term influences the registration results. $\alpha = 0$ corresponds to the baseline and $\alpha = 1$ corresponds to *variant 2*. Here, the DINOv2 encoder with the cosine dissimilarity. The peak DSC is 0.847 ± 0.06 at $\alpha = 0.3$ with 1.723 ± 1.018 HD95.

tures are visible, and the features appear to be more fine-grained compared to the features of the DINOv2 encoder. In contrast, the MedSAM-encoded features highlight the texture, but the edges are less apparent. Additionally, the DINOv2-encoded features of the background region are non-homogeneous, but this is not the case for the other two encoders.

Influence of the Feature Distance Weighting. Furthermore, we evaluate how the parameter α, which balances the image dissimilarity term $\mathcal{D}_{\mathcal{I}}$ and the feature-based distance term $\mathcal{D}_{\mathcal{F}}$, influences the registration results for *variant 2*. To this end, we vary α in steps of 0.1 in the range of $[0.0, 1.0]$. To maintain the overall weighting between the distance measurement terms and the regularization term, we also weigh the image distance measure term $\mathcal{D}_I$ with $1-\alpha$, i.e., both weightings sum up to 1 in every case. For this experiment, we chose the DINOv2 encoder, as it showed the best combined result in terms of DICE and HD95 in Fig. 4. The results, with a peak DICE score of 0.847 ± 0.06 with 1.723 ± 1.018 HD95 at $\alpha = 0.3$ is shown in Fig. 8. We do not see a significant peak. However, it is clear that there is a decrease in performance once the weighting factor gets too extreme in any direction. This supports our finding that feature-based distances are not sufficient to drive the deformable registration task. An example registration with this best configuration is shown in Fig. 7. Here, we can see a plausible deformation in the region of the heart and an improved difference image.

Influence of Image Upscaling. In this experiment, we evaluate the effect of the upscaling factor on the final registration results. As described in Sect. 3.1, we upsample the images before feeding them into the encoders. To achieve this, we perform registration with the DINOv2 encoder without prior upscaling of the images. This yields a mean DSC of 0.683 ± 0.122 and a mean HD95 of 3.664 ± 1.658. In comparison, with upscaling, a DSC of 0.742 ± 0.108 and a HD95 of 2.936 ± 1.465 is obtained. In both cases, the cosine dissimilarity and loss *variant 1* were used. This indicates that upscaling the input images and, thus, obtaining feature maps with higher resolution results in an improved registration, as expected.

4 Conclusion and Outlook

In this work, we compared two variants of incorporating features from three general vision encoders in medical image registration. Our extensive evaluation on a cardiac cine MR dataset shows that using feature-based distance as an additional guidance term improves the registration.

A limitation of the general vision encoders is that they require the input images to be 2D images. For 3D volumes, a possible solution could be to stack multiple encoded 2D slices. However, this requires very large computing and memory resources. A disadvantage of our approach is the increased runtime with the repeated inference of the encoders. This could be circumvented by encoding the images once and performing the optimization in feature space. However, the meaning of a spatial deformation of high-dimensional feature maps is unclear.

In future work, we plan to explore the use of feature maps from intermediate encoder levels and perform evaluation on a diverse range of datasets. Furthermore, we aim to extend our approach to 3D volumes and to compare them to 3D medical foundation models, such as UniGradICON [38].

Acknowledgments. This work was supported by the Federal Ministry of Education and Research (BMBF, Grant Nr. 01ZZ2315B and 01KX2021), the Bavarian Cancer Research Center (BZKF, Lighthouse AI and Bioinformatics), the German Cancer Consortium (DKTK, Joint Imaging Platform), the ERC grant Deep4MI (884622),

Cancer Research UK (A22905), Cancer Research UK Cambridge Centre [CTRQQR-2021-100012 and A25177] and The Mark Foundation for Cancer Research [RG95043].

References

1. Avants, B.B., Epstein, C.L., Grossman, M., Gee, J.C.: Symmetric diffeomorphic image registration with cross-correlation: evaluating automated labeling of elderly and neurodegenerative brain. Medical image analysis **12**(1), 26–41 (2008)
2. Avants, B.B., Tustison, N.J., Song, G., Cook, P.A., Klein, A., Gee, J.C.: A reproducible evaluation of ants similarity metric performance in brain image registration. Neuroimage **54**(3), 2033–2044 (2011)
3. Azad, B., Azad, R., Eskandari, S., Bozorgpour, A., Kazerouni, A., Rekik, I., Merhof, D.: Foundational models in medical imaging: A comprehensive survey and future vision (2023)
4. Balakrishnan, G., Zhao, A., Sabuncu, M.R., Guttag, J., Dalca, A.V.: Voxelmorph: a learning framework for deformable medical image registration. IEEE transactions on medical imaging **38**(8), 1788–1800 (2019)
5. Beg, M.F., Miller, M.I., Trouvé, A., Younes, L.: Computing large deformation metric mappings via geodesic flows of diffeomorphisms. International journal of computer vision **61**, 139–157 (2005)
6. Bernard, O., Lalande, A., Zotti, C., Cervenansky, F., Yang, X., Heng, P.A., Cetin, I., Lekadir, K., Camara, O., Ballester, M.A.G., et al.: Deep learning techniques for automatic mri cardiac multi-structures segmentation and diagnosis: is the problem solved? IEEE transactions on medical imaging **37**(11), 2514–2525 (2018)
7. Blankemeier, L., Cohen, J.P., Kumar, A., Van Veen, D., Gardezi, S.J.S., Paschali, M., Chen, Z., Delbrouck, J.B., Reis, E., Truyts, C., et al.: Merlin: A vision language foundation model for 3d computed tomography. arXiv preprint arXiv:2406.06512 (2024)
8. Bordes, F., Pang, R.Y., Ajay, A., Li, A.C., Bardes, A., Petryk, S., Mañas, O., Lin, Z., Mahmoud, A., Jayaraman, B., et al.: An introduction to vision-language modeling. arXiv preprint arXiv:2405.17247 (2024)
9. Chen, J., Frey, E.C., He, Y., Segars, W.P., Li, Y., Du, Y.: Transmorph: Transformer for unsupervised medical image registration. Medical image analysis **82**, 102615 (2022)
10. Cui, B., Islam, M., Bai, L., Ren, H.: Surgical-dino: Adapter learning of foundation models for depth estimation in endoscopic surgery (2024)
11. Czolbe, S., Krause, O., Feragen, A.: Deepsim: Semantic similarity metrics for learned image registration. ArXiv **abs/2011.05735** (2020)
12. Dalca, A., Rakic, M., Guttag, J., Sabuncu, M.: Learning conditional deformable templates with convolutional networks. Advances in neural information processing systems **32** (2019)
13. Dalca, A.V., Balakrishnan, G., Guttag, J., Sabuncu, M.R.: Unsupervised learning for fast probabilistic diffeomorphic registration. In: Medical Image Computing and Computer Assisted Intervention–MICCAI 2018: 21st International Conference, Granada, Spain, September 16-20, 2018, Proceedings, Part I. pp. 729–738. Springer (2018)
14. Denner, S., Zimmerer, D., Bounias, D., Bujotzek, M., Xiao, S., Kausch, L., Schader, P., Penzkofer, T., Jäger, P.F., Maier-Hein, K.: Leveraging foundation models for content-based medical image retrieval in radiology (2024)

15. Dosovitskiy, A., Beyer, L., Kolesnikov, A., Weissenborn, D., Zhai, X., Unterthiner, T., Dehghani, M., Minderer, M., Heigold, G., Gelly, S., et al.: An image is worth 16x16 words: Transformers for image recognition at scale. arXiv preprint arXiv:2010.11929 (2020)
16. Haber, E., Modersitzki, J.: Intensity gradient based registration and fusion of multimodal images. Medical image computing and computer-assisted intervention: International Conference on Medical Image Computing and Computer-Assisted Intervention **9 Pt 2**, 726–33 (2006)
17. Haskins, G., Kruger, U., Yan, P.: Deep learning in medical image registration: a survey. Machine Vision and Applications **31**, 1–18 (2020)
18. Haskins, G., Kruecker, J., Kruger, U., Xu, S., Pinto, P.A., Wood, B.J., Yan, P.: Learning deep similarity metric for 3d mr–trus image registration. International journal of computer assisted radiology and surgery **14**, 417–425 (2019)
19. Heinrich, M., Jenkinson, M., Bhushan, M., Matin, T., Gleeson, F., Brady, M., Schnabel, J.: Mind: Modality independent neighbourhood descriptor for multimodal deformable registration. Medical image analysis **16 7**, 1423–35 (2012)
20. Huix, J., Ganeshan, A., Haslum, J., Soderberg, M., Matsoukas, C., Smith, K.: Are natural domain foundation models useful for medical image classification? In: 2024 IEEE/CVF Winter Conference on Applications of Computer Vision (WACV). pp. 7619–7628. IEEE Computer Society, Los Alamitos, CA, USA (jan 2024)
21. Jin, Y., Li, J., Liu, Y., Gu, T., Wu, K., Jiang, Z., He, M., Zhao, B., Tan, X., Gan, Z., et al.: Efficient multimodal large language models: A survey. arXiv preprint arXiv:2405.10739 (2024)
22. Kirillov, A., Mintun, E., Ravi, N., Mao, H., Rolland, C., Gustafson, L., Xiao, T., Whitehead, S., Berg, A.C., Lo, W.Y., et al.: Segment anything. In: Proceedings of the IEEE/CVF International Conference on Computer Vision. pp. 4015–4026 (2023)
23. Loeckx, D., Slagmolen, P., Maes, F., Vandermeulen, D., Suetens, P.: Nonrigid image registration using conditional mutual information. IEEE transactions on medical imaging **29**(1), 19–29 (2009)
24. Ma, J., He, Y., Li, F., Han, L., You, C., Wang, B.: Segment anything in medical images. Nature Communications **15**, 1–9 (2024)
25. Madan, N., Møgelmose, A., Modi, R., Rawat, Y.S., Moeslund, T.B.: Foundation models for video understanding: A survey. arXiv preprint arXiv:2405.03770 (2024)
26. Oquab, M., Darcet, T., Moutakanni, T., Vo, H., Szafraniec, M., Khalidov, V., Fernandez, P., Haziza, D., Massa, F., El-Nouby, A., et al.: Dinov2: Learning robust visual features without supervision. arXiv preprint arXiv:2304.07193 (2023)
27. Pielawski, N., Wetzer, E., Öfverstedt, J., Lu, J., Wählby, C., Lindblad, J., Sladoje, N.: Comir: Contrastive multimodal image representation for registration. Advances in neural information processing systems **33**, 18433–18444 (2020)
28. Qiu, H., Qin, C., Schuh, A., Hammernik, K., Rueckert, D.: Learning diffeomorphic and modality-invariant registration using b-splines. In: Medical Imaging with Deep Learning (2021)
29. Rabbani, N., Bartoli, A.: Can surgical computer vision benefit from large-scale visual foundation models? International Journal of Computer Assisted Radiology and Surgery pp. 1–7 (2024)
30. Rueckert, D., Sonoda, L.I., Hayes, C., Hill, D.L., Leach, M.O., Hawkes, D.J.: Nonrigid registration using free-form deformations: application to breast mr images. IEEE transactions on medical imaging **18**(8), 712–721 (1999)
31. Sandkühler, R., Jud, C., Andermatt, S., Cattin, P.C.: Airlab: autograd image registration laboratory. arXiv preprint arXiv:1806.09907 (2018)

32. Sideri-Lampretsa, V., Zimmer, V.A., Qiu, H., Kaissis, G., Rueckert, D.: Mad: Modality agnostic distance measure for image registration. In: International Conference on Medical Image Computing and Computer-Assisted Intervention. pp. 147–156. Springer (2023)
33. Simonovsky, M., Gutiérrez-Becker, B., Mateus, D., Navab, N., Komodakis, N.: A deep metric for multimodal registration. In: International Conference on Medical Image Computing and Computer-Assisted Intervention (2016)
34. Song, X., Xu, X., Yan, P.: General purpose image encoder dinov2 for medical image registration. arXiv preprint arXiv:2402.15687 (2024)
35. Sotiras, A., Davatzikos, C., Paragios, N.: Deformable medical image registration: A survey. IEEE Transactions on Medical Imaging **32**, 1153–1190 (2013)
36. Studholme, C., Hill, D., Hawkes, D.: An overlap invariant entropy measure of 3d medical image alignment. Pattern Recognition **32**, 71–86 (1999)
37. Thirion, J.P.: Image matching as a diffusion process: an analogy with maxwell's demons. Medical image analysis **2**(3), 243–260 (1998)
38. Tian, L., Greer, H., Kwitt, R., Vialard, F.X., Estepar, R.S.J., Bouix, S., Rushmore, R., Niethammer, M.: unigradicon: A foundation model for medical image registration. arXiv preprint arXiv:2403.05780 (2024)
39. Vercauteren, T., Pennec, X., Perchant, A., Ayache, N.: Non-parametric diffeomorphic image registration with the demons algorithm. In: International conference on medical image computing and computer-assisted intervention. pp. 319–326. Springer (2007)
40. Wadekar, S.N., Chaurasia, A., Chadha, A., Culurciello, E.: The evolution of multimodal model architectures. arXiv preprint arXiv:2405.17927 (2024)
41. Wells III, W.M., Viola, P., Atsumi, H., Nakajima, S., Kikinis, R.: Multi-modal volume registration by maximization of mutual information. Medical image analysis **1**(1), 35–51 (1996)
42. Zhang, R., Isola, P., Efros, A.A., Shechtman, E., Wang, O.: The unreasonable effectiveness of deep features as a perceptual metric. In: Proceedings of the IEEE conference on computer vision and pattern recognition. pp. 586–595 (2018)
43. Zhang, S., Metaxas, D.: On the challenges and perspectives of foundation models for medical image analysis. Medical Image Analysis **91**, 102996 (2024)
44. Zhou, Y., Feng, L., Ke, Y., Jiang, X., Yan, J., Yang, X., Zhang, W.: Towards vision-language geo-foundation model: A survey. arXiv e-prints pp. arXiv–2406 (2024)

Rigid Single-Slice-in-Volume Registration via Rotation-Equivariant 2D/3D Feature Matching

Stefan Brandstätter[1,3(✉)], Philipp Seeböck[1], Christoph Fürböck[1], Svitlana Pochepnia[2,3], Helmut Prosch[2,3], and Georg Langs[1,3]

[1] Computational Imaging Research Lab, Department of Biomedical Imaging and Image-guided Therapy, Medical University of Vienna, Vienna, Austria
{stefan.brandstatetter,georg.langs}@meduniwien.ac.at

[2] Division of General and Pediatric Radiology, Department of Biomedical Imaging and Image-guided therapy, Medical University Vienna, Vienna, Austria

[3] Christian Doppler Laboratory for Machine Learning Driven Precision Imaging, Department of Biomedical Imaging and Image-guided therapy, Medical University Vienna, Vienna, Austria
https://www.cir.meduniwien.ac.at

Abstract. 2D to 3D registration is essential in tasks such as diagnosis, surgical navigation, environmental understanding, navigation in robotics, autonomous systems, or augmented reality. In medical imaging, the aim is often to place a 2D image in a 3D volumetric observation to augment or compare information. Current approaches for rigid single slice in volume registration are limited by requirements such as pose initialization, stacks of adjacent slices, or reliable anatomical landmarks. Here, we propose a self-supervised 2D/3D registration approach to match a single 2D slice to the corresponding 3D volume. The method does not require anatomical priors or an initialization. It addresses the dimensionality disparity and establishes correspondences between 2D in-plane and 3D out-of-plane rotation-equivariant features by using group equivariant CNNs. These rotation-equivariant features are extracted from the 2D query slice and aligned with their 3D counterparts. Results demonstrate the robustness of the proposed slice-in-volume registration on the NSCLC-Radiomics CT and KIRBY21 MRI datasets, attaining an absolute median angle error of less than 2°C and a mean-matching feature accuracy of 89% at a tolerance of 3 pixels.

Keywords: 2D/3D · Slice-in-Volume · Group Equivariant CNNs

Aligning a 2D image within a 3D volume poses a fundamental challenge across various disciplines, including medical imaging [7], robotics [14], autonomous systems [26], and augmented reality [36]. In medical imaging, the ability to conduct reliable 2D/3D registration enables combining information across multiple imaging modalities and dimensions for diagnosis, treatment decisions, therapeutic interventions or surgery [7,10,25,29,33]. Most existing approaches estimate

M. Modat et al. (Eds.): WBIR 2024, LNCS 15249, pp. 280–294, 2024.
https://doi.org/10.1007/978-3-031-73480-9_22

the pose of a *projection* image, such as an X-Ray, in relation to a 3D volume relying on the fact that the 2D image contains information of the entire volume. Here, we propose a method to find the position of a 2D image slice in a 3D image volume. We do not rely on any anatomical priors to enable analysis of irregular structures such as tumors. The approach is based on group rotation-equivariant features [37] that enable the matching from 2D to 3D, even-though the image slice only contains a small fraction of the image volume information.

1 Related Work

Current methods that address 2D/3D-registration are optimization- [22,30], learning- [10,34,38], or landmark-based [33], or a combination of these techniques [28].

Optimization-based methods are the most commonly used approach and iteratively adjust a 3D pose to match the similarity of the 2D data [22] starting from an initial pose estimate [33]. Iterative approaches require a considerable number of optimization steps often entailing simulating 2D projections, leading to high computational costs [22,29], and are sensitive to poor initialization [29].

Landmark and key-point based approaches aim at reducing computational costs [11] by relying on either manual landmark annotations, or automated candidate point selection by approaches such as Canny edge detectors [5]. Deep learning has made strides in automatically selecting landmarks with the assistance of labeled ground truth [28]. Algorithms match key points either in 2D-to-2D [9,24] or in 3D-to-3D [2,39] settings. However, it is still an open research question to accurately determine 2D key point positions in a 3D volume.

Machine learning-based 2D/3D-registration approaches can embed both a 2D projection and the 3D volume into a shared feature space to directly predict the transformation parameters [10,34]. They enhance the conventional similarity measure by utilizing a neural network as their similarity function [7]. They often rely on large, paired datasets [21] including multiple examples of the same anatomical structure for training. Optimization based approaches perform iterative refinement and rely on initialization [7]. Generating digitally reconstructed radiographs (DRR) [31] from the 3D volume and pose parameters enables comparison with the observed 2D projection image to guide registration. Recent work includes auto-differentiable DRRs to facilitate rapid matching [16] during surgery. Projection images contain information of a large part of the 3D volume, and their change is relatively well behaved with changing camera pose. In contrast, slice in volume registration aims at determining the position and rotation of a 2D slice extracted from within the volume. The slice only contains a small local part of the volume information. This is relevant in fetal imaging, where slice stack to volume registration approaches have been proposed for reconstruction [32,38] and motion tracking [3]. In [6] stacks of histology slices and brain MR data were registered for multi-modal high-resolution atlasing.

Equivariant representations exploit symmetries and equivariances present in a registration task [11]. Recent methods address this issue by utilizing rotation-equivariance [23,35]. However, their applicability is confined to point clouds and consistent dimensionality such as 2D-to-2D [20] or 3D-to-3D [3,27]. For 2D/3D registration, existing methods are still limited by the need for manual initialization, a restricted search space or iterative refinement, and prior knowledge about the appearance and the geometry of the anatomical structures that are registered. This limits their utility in registering individual 2D slices of irregular structures such as tumors in 3D imaging data.

Contribution. Here, we propose *SLIV-Reg*, a method for reliable 2D slice in 3D volume rigid registration to address two challenges. First, the method matches the comparably sparse in-plane 2D slice image information to 3D volume data. Second, it does not rely on image priors to enable registration of irregular structures such as tumors. Group-equivariant Convolutional Neural Networks (CNNs) extract in-plane-rotation-equivariant features for candidate points in both the slice and the volume. To cope with out-of-plane rotations we efficiently sample 2D features in different orientations from the 3D volume. We then match a large number of corresponding 2D to 3D candidates, and predict the position of the 2D slice within the 3D volume using robust pose estimation. To reduce computational costs, we restrict candidate points to Canny edge detector points. Compared to previous methods, we overcome the limitations of (1) manual landmark selection, (2) confining the search space, (3) pose initialization, (4) the need for iterative refinement steps, (5) the reliance on a stack of adjacent slices, (6) prior knowledge about the anatomical geometry of the imaged objects, and (7) the need of a 2D projection image. Our experiments demonstrate the robustness and accuracy of the proposed approach in CT data of lung tumors.

2 Method

We propose a method to achieve precise alignment between a 2D query slice $\mathbf{Q} \in \mathbb{R}^{H \times W}$ and a 3D search volume $\mathbf{S} \in \mathbb{R}^{H \times W \times D}$ through the utilization of rotation-equivariant feature matching (Fig. 1). For training, we conduct self-supervised pre-training of a rotation-equivariant encoder. During inference, we identify candidate points $\mathbf{c}_Q^i \in \mathbb{R}^2$, $i = 1, \ldots, I$ on the query slice and $\mathbf{c}_S^j \in \mathbb{R}^3$, $j = 1, \ldots, J$ in the search volume with the help of Canny edge detector. For readability we omit the indices of candidate points through out the remaining manuscript, where not necessary. We extract patches $\mathbf{p}_Q$ at the candidate positions $\mathbf{c}_Q$ in $\mathbf{Q}$. For each candidate point in the search volume $\mathbf{S}$, we extract R patches $\mathbf{p}_S^{j,r}$ with different orientations centered at each $\mathbf{c}_S^j$ with the Plane extractor (PE) a cube centered around each $\mathbf{c}_S^j$ in $\mathbf{S}$ and then extract . The pre-trained encoder is used to extract the feature vectors $\mathbf{f}_Q^i$ and $\mathbf{f}_S^{j,r}$ for all $\mathbf{p}_Q^i$ and $\mathbf{p}_S^{j,r}$. After computing the pairwise distances between $\mathbf{f}_Q$ and $\mathbf{f}_S$, the features with the lowest distance create pairs between $\mathbf{c}_Q$ and $\mathbf{c}_S$. These pairs are our

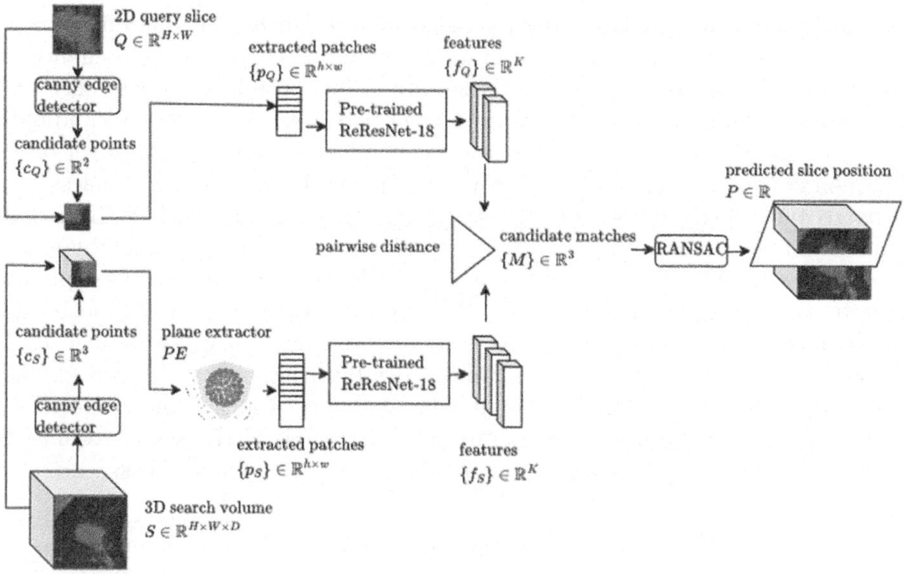

Fig. 1. SLIV-Reg model: a 2D query slice $\mathbf{Q}$ and a 3D search volume $\mathbf{S}$ are processed to identify candidate points $\mathbf{c}_Q$ and $\mathbf{c}_S$. At each point in $\mathbf{Q}$ one patch is extracted, and in $\mathbf{S}$ a set of R co-centered 2D patches in an equidistante sampling of orientations are extracted with the plane extractor (PE). A pre-trained ReResNet-18 encodes each patch to features $\mathbf{f}_Q$ and $\mathbf{f}_S$. For each $\mathbf{f}_Q$ we find the closest feature vector $\mathbf{f}_S$, resulting in a set of candidate matches $\mathbf{M}$. These matches serve as basis for a RANSAC estimate of the 2D slice pose in the 3D volume.

candidate matches $\mathbf{M} \in \mathbb{R}^3$ and serve as basis for robust pose P estimation of the slice in the volume with the RANSAC algorithm.

Steerable CNNs. Steerable CNNs implement E(2)-equivariant convolutions, meaning they are equivariant to rotations and reflections on the image plane $\mathbb{R}^2$ [37]. In these networks, the feature spaces consist of feature fields, which are characterized by a group representation that dictates how they transform in response to transformations of the input [37].

Rotation-Equivariant Pre-training. We use a rotation-equivariant ResNet-18 [20] as feature extractor. The model is based on the ResNet-18 architecture but enhanced with rotation-equivariant convolutional layers from the E(2)-CNN framework [37] and maps an image to a feature vector $\mathbf{N} : \mathbf{p} \mapsto \mathbf{f}$. In the pre-training phase, we adopt the self-supervised setup employed by [20]. The primary goal of pretraining is to enable the encoder to learn matching features in rotated images through extensive data augmentation [20]. We trained on the non-small cell lung cancer (NSCLC) [1] dataset and on the Kirby21 dataset [19]. This ensures that the extracted features are distinguishable not only by their spatial location but also by their in-plane rotation.

Candidate Points Extraction in 2D and 3D Data. For matching sets of 2D to 3D points we first identify candidates at locations that have a chance to correspond. We assume that object edges are visible both in the query slice and the search volume, and use the Canny edge detector [5] to extract candidate points in the 2D- and 3D data. We selected the Canny edge detector over other common keypoint detectors because our initial experiments showed it is less sensitive to out-of-plane image rotations compared to algorithms such as SIFT [24], SuperPoint [9], and the Harris corner detector [18]. Additionally, the hyperparameters of the Canny edge detector allow to reduce the number of keypoints, thereby decreasing computation time. To use the Canny edge detection in 3D, we analyzed each z-index of the 3D volume and applied the Canny algorithm to each 2D slice. The query slice $\mathbf{Q}$ undergoes blurring with a Gaussian kernel and is then processed with the Canny edge detector, resulting in the positions of edge points $\mathbf{c}_Q$. Similarly, we iterate over the z-axis of the search volume $\mathbf{S}$ and find the candidate points $\mathbf{c}_S$ accumulated over the volume slices.

Patch Extraction. For each candidate point $\mathbf{c}_Q$ in the query slice $\mathbf{Q}$ we extract a patch $\mathbf{p}_Q \in \mathbb{R}^{h \times w}$ centered at the point. For each candidate point $\mathbf{c}_S$ of the search volume we use a *plane extractor (PE)* to extract R 2D planes in different orientations, centered at the candidate point from the 3D volume. We implement this by extracting a local cube, and define orientations by normal vectors sampled on the unit sphere. This results in R distinct patches $\mathbf{p}_S^{j,r} \in \mathbb{R}^{h \times w}$ for R orientations, all centered at the candidate point $\mathbf{c}_S^j$.

Rotation-Equivariant Feature Extraction. The pre-trained 2D ReResNet-18 encodes each patch $\mathbf{p}_Q$ and $\mathbf{p}_S$ to a feature vector $\mathbf{f}_Q = \mathbf{N}(\mathbf{p}_Q) \in \mathbb{R}^K$ and $\mathbf{f}_S = \mathbf{N}(\mathbf{p}_S) \in \mathbb{R}^K$. In contrast to pre-training, we are not utilizing the group-alignment feature and multi-scale forward of [20] during this step.

Feature Matching. For each query slice feature $\mathbf{f}_Q$ we identify the closest search volume feature $\mathbf{f}_S$ based on the L2 distance between feature vectors. Each 2D candidate point $\mathbf{c}_Q$ is matched to one 3D candidate point $\mathbf{c}_S$, resulting in a set of candidate matches $\mathbf{M}$.

Slice Position Prediction. To predict the final query slice position in the search volume, we use all candidate matches M (i.e. corresponding $\mathbf{c}_S$ coordinates) as input to the RANSAC algorithm [13]. RANSAC predicts the slice position $P = \langle \mathbf{n}, \mathbf{t} \rangle$ (normal vector, plane offset) that has the strongest support among the matched points (Fig. 2).

3 Experimental Setup

Data: We used two medical imaging data sets of different modalities and anatomical structures for evaluation. The first dataset comprises 422 computed tomography (CT) volumes of subjects with non-small cell lung cancer (NSCLC) tumors [1]. We partitioned this data into training (335), validation (62), and test (25) sets. Each CT volume contains one tumor, and corresponding manual pixel-wise annotations. The second dataset - Kirby21 [19] - contains 41 T1 weighted

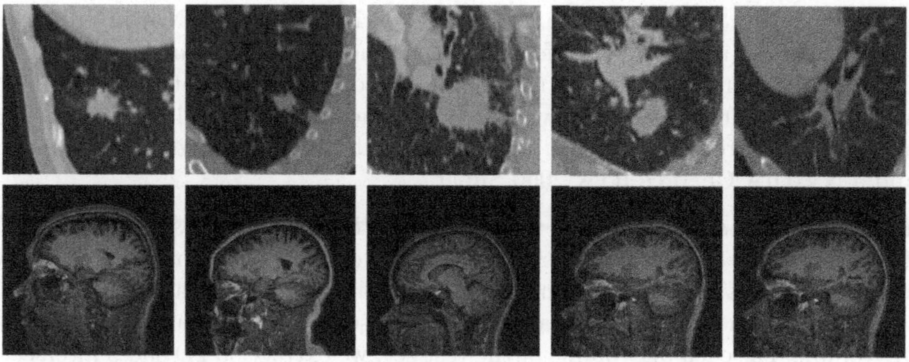

Fig. 2. Data used for method evaluation. First row: tumors in CT volumes of NSCLC patients illustrating the variability of tumor shapes; second row: T1-weighted MRI scans of the brain in the Kirby21 dataset.

magnetic resonance imaging (T1w MRI) volumes of the brain. For training and validation we used volumes KKI06 to KKI42, for testing, we used volumes KKI01 to KKI05. Experiments A, B, and C are conducted using the NSCLC dataset, while Experiment D is performed using both datasets.

Implementation Details: The ReResNet-18 [17] is pre-trained with 512×512 slices extracted from each volume. We follow the strategy of Lee et al. [20] to enhance the model's geometric and photometric robustness by augmenting the slices with random affine homographies, gaussian and speckle noise addition, and motion blurring. We use a cyclic group order of 16, a batch size of 16, a learning rate of 10-4, weight decay of 0.1 and 12 epochs with 322 iterations. The extracted patches $\mathbf{p}_Q$ and $\mathbf{p}_S$ at the location of the candidate points $\mathbf{c}_Q$ and $\mathbf{c}_S$ have a size of 21×21. Features $\mathbf{f}_Q$ and $\mathbf{f}_S$ have a feature size K of 512. The parameter of the RANSAC [13][1] algorithm are a treshold of 1 pixel, which defines the distance to be counted as a inlier, maximum iterations equals 1000, and minPoints was set to 40. The pretraining for the Rigid Transformation Parameter Initialization (RTPI) network is done based on the RTPInetV3[2] as the model [8] with a change of the PRO-ST[3] module [15], to the standard pytorch grid sampler to extract the centre slice instead of the resulting 2D projective image. We use a batch size of 16, a learning rate of 10-3, stepLR scheduler with gamma of 0.5 every 30 epochs with 100 epochs in total. Pretraining has been conducted separately on both datasets using the same setup and configurations. For the CMA-ES[4] there is no pretraining needed. The hyperparameters specify that the initial rotation matrix should not include any rotation of the volume, a sigma of 10-3, no lr

[1] https://github.com/leomariga/pyRANSAC-3D.

[2] https://github.com/m1nhengChen/SOPI.

[3] https://github.com/gaocong13/Projective-Spatial-Transformers.

[4] https://github.com/m1nhengChen/CMAES-reg.

adaptation and population size. For the experiments on the NSCLC dataset, subvolumes are selected to perform the inference on image parts that contain tumor tissue. For both datasets, the volume is cropped after rotation to eliminate any trivial features correlating with rotation, such as existing black borders. All experiments were conducted on a machine with an AMD Ryzen 5 3600 CPU and a NVIDIA GeForce RTX 2060 Super.

Experiment A: Accuracy of 2D Slice in 3D Volume Pose Estimation. To evaluate the registration accuracy of the proposed algorithm, we randomly sample locations and orientations of 2D slices in a 3D volume, and test whether the model can accurately estimate the pose based on the 2D query slice, and a 3D volume. For each 3D volume of the test set, we extract 90 2D slices intersecting the bounding box of the tumor. First, 30 slices are sampled in 30 random approximately equidistant directions. Then, slices with the same orientations, but a random translation off-set and distance of -6 and +6 voxels are extracted resulting in overall 90 slices varying in orientation and location. We use the proposed algorithm to perform registration, estimating the pose of each 2D query slice in the corresponding 3D search volume. Registration accuracy is evaluated by comparing the angle difference of the orthogonal vectors of predicted and the ground-truth slice position. We vary the number R of directions used be the plane extractor from 10 to 700 to assess the influence of this hyper parameter on accuracy.

Experiment B: Accuracy of 2D/3D Image Feature Matching. In addition to the accuracy of the estimated pose, we evaluate the mean matching accuracy (MMA) of feature points in the correct position of the 2D query slice, and its estimated position with pixel thresholds of 3/5/10 before and after RANSAC matching [13]. MMA evaluates the fraction of correctly matched feature points being located within a certain radius or "pixel threshold".

Experiment C: Sensitivity of the Feature Distance to Angle Differences. The model yields a feature distance between points in the 2D query slice and slices extracted from the 3D search volume. To facilitate correct point matching, the feature distance should be relatively small for the same point and different angles, and large for different points in the 3D volume. To investigate this behavior, we compare the feature distance, when gradually changing orientation (intra-point feature distance) with the feature distance between slices sampled at different positions (inter-point feature distance). For each candidate point, we extract 400 features $\mathbf{f}_S$ resulting from different orientations ($R = 400$). We then sample pairs of random orientations at the same point (reference feature A, intra-point feature B) with an angle difference 10 to 90°C and compute their feature distance (β). To generate inter-point feature distances, we sample with the same angle difference but at a different random candidate point (inter-point feature C) and compute the feature distance to A (γ). β should increase with increasing angle difference, β should be lower compared to γ, and the difference $\beta - \gamma$ should decrease with increasing angle.

Additionally, we extract 360 in-plane features $\mathbf{f}_S$ for each candidate point in the volume, by conducting the plane extraction only within the xz-plane using 360 equidistant directions. We use the feature at the 0-degree angle as reference and calculate all intra-and inter-point feature distances. We calculate intra-point feature distances as a function of the angle difference, and compare intra- and inter-point feature distances. The metric used to assess the distance between features is the L2 norm.

Experiment D: Comparison with State-of-the-Art Methods. To our knowledge, there is no prior algorithm addressing the exact same problem. However, 2D projection image to 3D volume registration approaches exist, and the comparison is instructive. We compare the proposed algorithm with two existing approaches. First, we apply an optimization based algorithm called Covariance matrix adaptation evolution strategy (CMA-ES) as proposed by [7], with normalized cross-correlation as the cost function [12], second, we apply the Rigid Transformation Parameter Initialization (RTPI) module suggested by [8]. We evaluate both algorithms for the median absolute angle error without any translation on 30 slices against SLIV-Reg with $R = 300$ and $R = 700$ directions on the sphere.

4 Results

Figure 3(a) shows the median absolute angle error between true and estimated pose of the query slice as a function of directional sampling density R. SLIV-Reg yields a median absolute error of $\leq 3°$ for $R \geq 150$, and the error converges after that. Figure 3(b) provides an example visualization of the estimated and true pose of the query slice, and the corresponding images extracted from the 3D volume (GT at the correct pose, PRED at the estimated pose). The example illustrates the level of similarity after successful matching.

Figure 4(a) shows the 2D/3D feature matching evaluation results as MMA as a function of directional sampling density of the plane extractor. Starting from $R \geq 300$, $\geq$80% of feature points are successfully matched (@3px) after RANSAC. This is also reflected in qualitative results illustrated in Fig. 4(b), where the candidate points of the query slice on the right are connected to their corresponding matches among all 3D candidate points, highlighted in blue. Results show a modest enhancement of 1.64% from 97.42% to 99.06% before and after RANSAC at an error rate of 10 (MMA@10px) and $R = 300$. A notable improvement before and after RANSAC is observed for MMA@3px, with a substantial increase of 66.26%, rising from 14.39% to 80.664% for $R = 200$.

Figure 5(a) shows the feature distance for slices extracted at the same location with different orientations (orange) and slices extracted at different locations and the corresponding orientations (blue) in an interval of one degree. Intra-point feature distances are markedly below the feature distances to other points up to an angle difference of appr. $\pm 40°$. This is consistent with the observation that a sampling density of 150 directions on the sphere achieves a median angle error of $\leq 5°$C. The sampling directions have an approximately 8° offset, which is

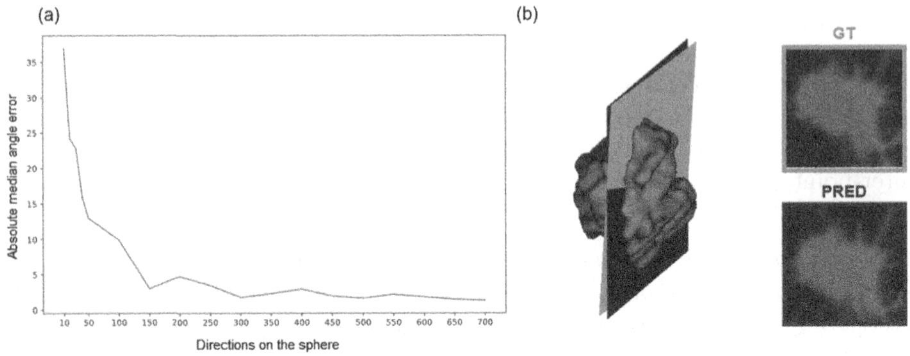

Fig. 3. Experiment A: (a) The median absolute angle error for varying directional sampling density of the plane extractor. (b) Qualitative 3D visualization of a registration result, showing the predicted and ground truth slice pose, together with their 2D projections on the right hand side.

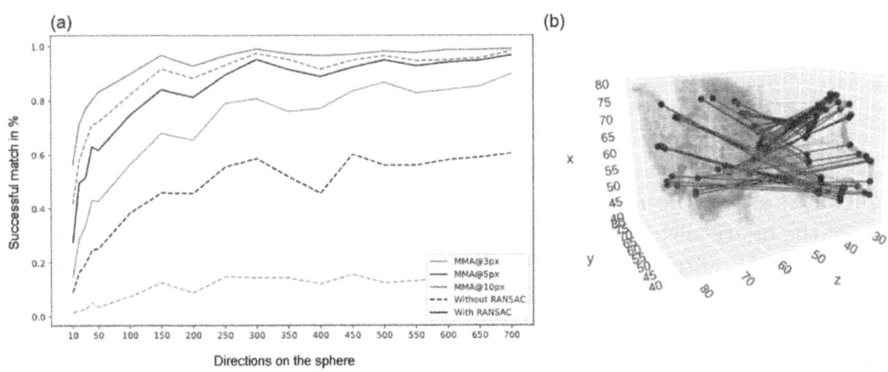

Fig. 4. Experiment B: (a) The MMA with pixel thresholds 3/5/10 before (dashed line) and after RANSAC (solid line) for varying directional sampling densities R. (b) 3D visualization of a matching result: $\mathbf{c}_S$ candidate points are shown in blue in the 3D space, $\mathbf{c}_Q$ points are visualized in the yz-plane on the right hand side, and accurate/inaccurate matches are highlighted in green/red (MMA pixel threshold of 3). (Color figure online)

within the acceptable range. In line with these results, the median difference between the intra- and inter-point feature distances ranges from −0.15 to 0.01, with increasing difference for decreasing angles (Fig. 5(b)).

The quantitative comparison of SLIV-reg with two state-of-the-art approaches is shown in Fig. 6. We evaluated two sampling density values for SLIV-reg ($R = 300$ and $R = 700$) with CMA-ES and RTPI. Experiments were performed on two different data sets (CT tumor, MRI brain). We compare the mean absolute angle error of the estimated query slice pose, and the correct pose. Results demonstrate a median absolute angle error of ≤ 2°C on both NSCLC and Kirby21 datasets for the proposed SLIV-reg with $R = 700$. The CMA-ES algo-

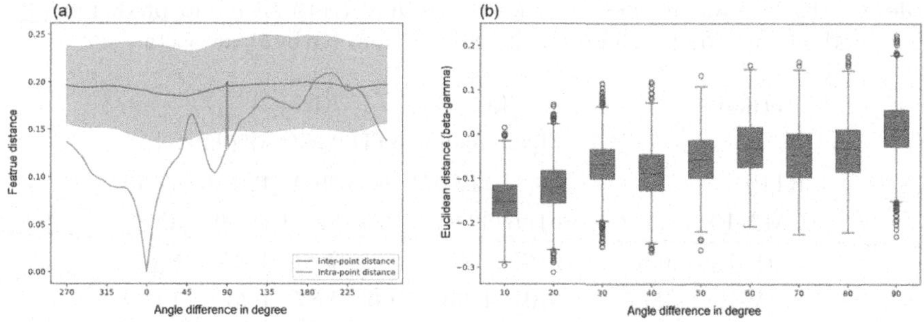

Fig. 5. Experiment C: (a) For varying angles, intra-point feature distances are illustrated in orange, while inter-point distances are highlighted in blue, accompanied by standard deviation. (b) Median difference between intra- and inter-point feature distances at given directions, shown for absolute angle differences of 10 to 90°C (Color figure online).

rithm depends heavily on its starting point. It can find the correct position with an angle error of ≤ 0.5°C if the initial rotation matrix is close to the true one. However, if the initial matrix is substantially different to the correct orientation, the algorithm cannot converge as seen in Fig. 6 where initialization was chosen randomly. The Rigid Transformation Parameter Initialization Module (RTPI) was trained unsuccessful and could not converge. Note that this is not surprising, since both RTPI, and CMA-ES are developed to match projection images to 3D volumes for estimating the correct rotation. Thus, they do not perform well in the different slice in volume matching problem addressed in this paper. Projection images contain more information about the entire volume, compared to the relatively minimal information of an extracted slice (Table 1).

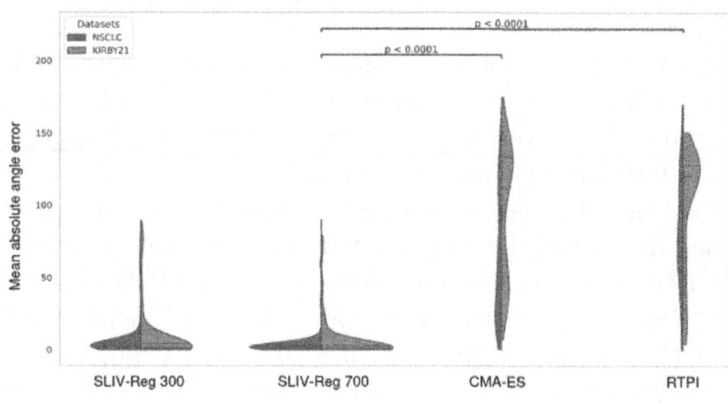

Fig. 6. Experiment D: Median absolute angle error between the ground truth slice and the predicted slice. Each method was evaluated using both the NSCLC and Kirby21 datasets. P-values indicate the statistical significance of our method.

Table 1. Absolute angle error between the ground truth slice and predicted slice of our method against the baselines on the NSCLC and Kirby21 datasets.

Method	NSCLC			Kirby21		
	Mean	Median	STD	Mean	Median	STD
RTPI	96.43	112.07	45.34	99.05	120.26	42.46
CMA-ES	61.97	52.81	43.36	82.14	78.29	39.88
SLIV-Reg 300 (ours)	6.37	2.71	13.48	**2.85**	**1.23**	**8.83**
SLIV-Reg 700 (ours)	**4.01**	**1.93**	**9.63**	5.87	1.33	13.80

5 Discussion and Conclusion

We propose *SLIV-Reg*, a slice-in-volume registration approach that reliably predicts the pose of a 2D query slice in a 3D search volume based on rotation-equivariant 2D-3D feature matching and robust pose estimation. Results on CT data of lung tumors and MRI data of the brain show that it can accurately identify the location and rotation of a slice, independent of prior knowledge of anatomy. This makes it useful for matching imaging data of irregular structures such as tumors. In contrast to existing algorithms, the method does not require initialization, iterative refinement, constraints of the search space, stacks of neighboring slices, or an image that is a projection of the 3D volume. Our method is inspired by recent 2D-to-2D registration approaches [20] that attain an MMA@3px of 91.69% for the Roto-360 dataset. We introduce a 2D-to-3D matching approach that can cope with in-plane and out-of-plane rotations. It achieves comparable quantitative results in a 2D-to-3D task, reporting a MMA@3px of 89.82% for the NSCLC-Radiomics dataset. This indicates that SLIV-Reg successfully addresses the challenge of appropriately matching features across 2D and 3D domains for rigid registration by sampling multiple rotation-equivariant 2D features for a single 3D position. Accuracy benefits from increased directional sampling density R of the 3D image descriptors, but makes matching feasible already with low directional sampling density. This is due to the feature distances being relatively well behaved with orientation changes compared to features extracted at different positions (Fig. 5a).

The lack of 2D-in-3D registration baselines leads us to compare our algorithm with approaches developed for the related but different problem of X-ray to CT volume registration. These approaches rely on a DDR image summarizing the 3D volume depending on the projection direction, instead of holding only a small part of the image information in a 2D slice. Comparative evaluation, demonstrates that these approaches are not applicable to the present problem as they are sensitive to good initialization.

6 Limitations and Future Work

While future work will evaluate translation to different modalities and non-medical datasets, rotation-equivariant networks have shown to generalize well with little risk of overfitting [4,37]. As no other comparable 2D-3D single slice registration approach has been proposed, we rely on the comparable 2D-2D registration approach [20] and similar learning tasks as 2D X-ray to 3D CT volume as reference when comparing to the state of the art.

The proposed method offers a means to perform 2D-3D slice-in-volume registration accurately and reliably. Comparably to existing 3D-3D machine learning based approaches it offers the perspective of integrating multi-modal data such as in-vivo and ex-vivo imaging data. A wider impact is expected by using the proposed algorithm for the alignment of histology images and CT scans to improve information available for treatment response prediction. At this point the method is validated against two medical datasets but we plan to extend our validation efforts on industrial CT scans and geological scans to include non-medical datasets in future research.

Acknowledgement. The financial support by the Austrian Federal Ministry for Digital and Economic Affairs, the National Foundation for Research, Technology and Development and the Christian Doppler Research Association, Siemens Healthineers, the Austrian Science Fund (FWF, P 35189-B - ONSET), and the Vienna Science and Technology Fund (WWTF, PREDICTOME [10.47379/LS20065]) is gratefully acknowledged.

Disclosure of Interests. The authors have no competing interests to declare that are relevant to the content of this article.

References

1. Aerts, H.J.W.L., Wee, L., Rios Velazquez, E., Leijenaar, R.T.H., Parmar, C., Grossmann, P., Carvalho, S., Bussink, J., Monshouwer, R., Haibe-Kains, B., Rietveld, D., Hoebers, F., Rietbergen, M.M., Leemans, C.R., Dekker, A., Quackenbush, J., Gillies, R.J., Lambin, P.: Data from NSCLC-Radiomics (2019)
2. Balakrishnan, G., Zhao, A., Sabuncu, M.R., Guttag, J., Dalca, A.V.: Voxelmorph: a learning framework for deformable medical image registration. IEEE transactions on medical imaging **38**(8), 1788–1800 (2019)
3. Billot, B., Moyer, D., Dey, N., Hoffmann, M., Turk, E.A., Gagoski, B., Grant, E., Golland, P.: Se (3)-equivariant and noise-invariant 3d motion tracking in medical images. arXiv preprint arXiv:2312.13534 (2023)
4. Bylander, K.: More efficient training using equivariant neural networks (2023)
5. Canny, J.: A computational approach to edge detection. IEEE Transactions on pattern analysis and machine intelligence (6), 679–698 (1986)
6. Casamitjana, A., Lorenzi, M., Ferraris, S., Peter, L., Modat, M., Stevens, A., Fischl, B., Vercauteren, T., Iglesias, J.E.: Robust joint registration of multiple stains and mri for multimodal 3d histology reconstruction: Application to the allen human brain atlas. Medical image analysis **75**, 102265 (2022)

7. Chen, M., Li, T., Zhang, Z., Kong, Y.: An optimization-based baseline for rigid 2d/3d registration applied to spine surgical navigation using cma-es. arXiv preprint arXiv:2402.05642 (2024)
8. Chen, M., Zhang, Z., Gu, S., Kong, Y.: Embedded feature similarity optimization with specific parameter initialization for 2d/3d medical image registration. In: ICASSP 2024-2024 IEEE International Conference on Acoustics, Speech and Signal Processing (ICASSP). pp. 1521–1525. IEEE (2024)
9. DeTone, D., Malisiewicz, T., Rabinovich, A.: Superpoint: Self-supervised interest point detection and description. In: Proceedings of the IEEE conference on computer vision and pattern recognition workshops. pp. 224–236 (2018)
10. Dong, G., Dai, J., Li, N., Zhang, C., He, W., Liu, L., Chan, Y., Li, Y., Xie, Y., Liang, X.: 2d/3d non-rigid image registration via two orthogonal x-ray projection images for lung tumor tracking. Bioengineering **10**(2), 144 (2023)
11. Evan, M.Y., Wang, A.Q., Dalca, A.V., Sabuncu, M.R.: Keymorph: Robust multi-modal affine registration via unsupervised keypoint detection. In: Medical Imaging with Deep Learning (2021)
12. Ferrante, E., Paragios, N.: Slice-to-volume medical image registration: A survey. Medical image analysis **39**, 101–123 (2017)
13. Fischler, M.A., Bolles, R.C.: Random sample consensus: a paradigm for model fitting with applications to image analysis and automated cartography. Communications of the ACM **24**(6), 381–395 (1981)
14. Gao, C., Farvardin, A., Grupp, R.B., Bakhtiarinejad, M., Ma, L., Thies, M., Unberath, M., Taylor, R.H., Armand, M.: Fiducial-free 2d/3d registration for robot-assisted femoroplasty. IEEE transactions on medical robotics and bionics **2**(3), 437–446 (2020)
15. Gao, C., Liu, X., Gu, W., Killeen, B., Armand, M., Taylor, R., Unberath, M.: Generalizing spatial transformers to projective geometry with applications to 2d/3d registration. In: Medical Image Computing and Computer Assisted Intervention–MICCAI 2020: 23rd International Conference, Lima, Peru, October 4–8, 2020, Proceedings, Part III 23. pp. 329–339. Springer (2020)
16. Gopalakrishnan, V., Golland, P.: Fast auto-differentiable digitally reconstructed radiographs for solving inverse problems in intraoperative imaging. In: Workshop on Clinical Image-Based Procedures. pp. 1–11. Springer (2022)
17. Han, J., Ding, J., Xue, N., Xia, G.S.: Redet: A rotation-equivariant detector for aerial object detection. In: Proceedings of the IEEE/CVF Conference on Computer Vision and Pattern Recognition. pp. 2786–2795 (2021)
18. Harris, C., Stephens, M., et al.: A combined corner and edge detector. In: Alvey vision conference. vol. 15, pp. 10–5244. Citeseer (1988)
19. Landman, B.A., Huang, A.J., Gifford, A., Vikram, D.S., Lim, I.A.L., Farrell, J.A., Bogovic, J.A., Hua, J., Chen, M., Jarso, S., et al.: Multi-parametric neuroimaging reproducibility: a 3-t resource study. Neuroimage **54**(4), 2854–2866 (2011)
20. Lee, J., Kim, B., Kim, S., Cho, M.: Learning rotation-equivariant features for visual correspondence. In: Proceedings of the IEEE/CVF Conference on Computer Vision and Pattern Recognition. pp. 21887–21897 (2023)
21. Li, P., Pei, Y., Guo, Y., Ma, G., Xu, T., Zha, H.: Non-rigid 2d-3d registration using convolutional autoencoders. In: 2020 IEEE 17th International Symposium on Biomedical Imaging (ISBI). pp. 700–704. IEEE (2020)
22. Liao, H., Lin, W.A., Zhang, J., Zhang, J., Luo, J., Zhou, S.K.: Multiview 2d/3d rigid registration via a point-of-interest network for tracking and triangulation. In: Proceedings of the IEEE/CVF Conference on Computer Vision and Pattern Recognition (CVPR) (June 2019)

23. Lin, C.W., Chen, T.I., Lee, H.Y., Chen, W.C., Hsu, W.H.: Coarse-to-fine point cloud registration with se (3)-equivariant representations. In: 2023 IEEE International Conference on Robotics and Automation (ICRA). pp. 2833–2840. IEEE (2023)
24. Lowe, D.G.: Distinctive image features from scale-invariant keypoints. International journal of computer vision **60**, 91–110 (2004)
25. Markelj, P., Tomaževič, D., Likar, B., Pernuš, F.: A review of 3d/2d registration methods for image-guided interventions. Medical image analysis **16**(3), 642–661 (2012)
26. Miao, H., Lu, F., Liu, Z., Zhang, L., Manocha, D., Zhou, B.: Robust 2d/3d vehicle parsing in cvis. arXiv preprint arXiv:2103.06432 (2021)
27. Moyer, D., Abaci Turk, E., Grant, P.E., Wells, W.M., Golland, P.: Equivariant filters for efficient tracking in 3d imaging. In: Medical Image Computing and Computer Assisted Intervention–MICCAI 2021: 24th International Conference, Strasbourg, France, September 27–October 1, 2021, Proceedings, Part IV 24. pp. 193–202. Springer (2021)
28. Nath, T., Mathis, A., Chen, A.C., Patel, A., Bethge, M., Mathis, M.W.: Using deeplabcut for 3d markerless pose estimation across species and behaviors. Nature protocols **14**(7), 2152–2176 (2019)
29. Nguyen, V., Alves Pereira, L.F., Liang, Z., Mielke, F., Van Houtte, J., Sijbers, J., De Beenhouwer, J.: Automatic landmark detection and mapping for 2D/3D registration with BoneNet. Front Vet Sci **9**, 923449 (2022)
30. Otake, Y., Wang, A.S., Stayman, J.W., Uneri, A., Kleinszig, G., Vogt, S., Khanna, A.J., Gokaslan, Z.L., Siewerdsen, J.H.: Robust 3d–2d image registration: application to spine interventions and vertebral labeling in the presence of anatomical deformation. Physics in Medicine & Biology **58**(23), 8535 (2013)
31. Staub, D., Murphy, M.J.: A digitally reconstructed radiograph algorithm calculated from first principles. Medical physics **40**(1), 011902 (2013)
32. Taymourtash, A., Kebiri, H., Schwartz, E., Nenning, K.H., Tourbier, S., Kasprian, G., Prayer, D., Bach Cuadra, M., Langs, G.: Spatio-temporal motion correction and iterative reconstruction of in-utero fetal fmri. In: International Conference on Medical Image Computing and Computer-Assisted Intervention. pp. 603–612. Springer (2022)
33. Unberath, M., Gao, C., Hu, Y., Judish, M., Taylor, R.H., Armand, M., Grupp, R.: The impact of machine learning on 2d/3d registration for image-guided interventions: A systematic review and perspective. Frontiers in Robotics and AI **8**, 716007 (2021)
34. Van Houtte, J., Audenaert, E., Zheng, G., Sijbers, J.: Deep learning-based 2d/3d registration of an atlas to biplanar x-ray images. International Journal of Computer Assisted Radiology and Surgery **17**(7), 1333–1342 (2022)
35. Wang, A.Q., Evan, M.Y., Dalca, A.V., Sabuncu, M.R.: A robust and interpretable deep learning framework for multi-modal registration via keypoints. Medical Image Analysis **90**, 102962 (2023)
36. Wang, K., Liu, D., Liu, Z., Duan, G., Hu, L., Tan, J.: A fast object registration method for augmented reality assembly with simultaneous determination of multiple 2d-3d correspondences. Robotics and computer-integrated manufacturing **63**, 101890 (2020)
37. Weiler, M., Cesa, G.: General e (2)-equivariant steerable cnns. Advances in neural information processing systems **32** (2019)

38. Xu, J., Moyer, D., Grant, P.E., Golland, P., Iglesias, J.E., Adalsteinsson, E.: Svort: iterative transformer for slice-to-volume registration in fetal brain mri. In: International Conference on Medical Image Computing and Computer-Assisted Intervention. pp. 3–13. Springer (2022)
39. Zhao, S., Lau, T., Luo, J., Eric, I., Chang, C., Xu, Y.: Unsupervised 3d end-to-end medical image registration with volume tweening network. IEEE journal of biomedical and health informatics **24**(5), 1394–1404 (2019)

A Self-supervised Image Registration Approach for Measuring Local Response Patterns in Metastatic Ovarian Cancer

Inês P. Machado[1,2,3(✉)], Anna Reithmeir[4,5], Fryderyk Kogl[4,5], Leonardo Rundo[6], Gabriel Funingana[1,2,3], Marika Reinius[1,2,3], Gift Mungmeeprued[1,2,3], Zeyu Gao[1,2,3], Cathal McCague[1,2], Eric Kerfoot[7], Ramona Woitek[8], Evis Sala[9], Yangming Ou[10], James Brenton[1,2], Julia Schnabel[4,5,7], and Mireia Crispin[1,2,3]

[1] Department of Oncology, University of Cambridge, Cambridge, UK
im549@cam.ac.uk
[2] Cancer Research UK, Cambridge Institute, University of Cambridge, Cambridge, UK
[3] Early Cancer Institute, University of Cambridge, Cambridge, UK
[4] School of Computation, Information and Technology, Technical University of Munich, Munich, Germany
[5] Institute of Machine Learning in Biomedical Imaging, Helmholtz, Munich, Germany
[6] Department of Information and Electrical Engineering, University of Salerno, Fisciano, Italy
[7] School of Biomedical Engineering and Imaging Sciences, King's College London, London, UK
[8] Research Center for Medical Image Analysis and AI, Danube University, Krems an der Donau, Austria
[9] Department of Radiologic Sciences, Università Cattolica del Sacro Cuore, Milan, Italy
[10] Department of Radiology, Boston Children's Hospital, Harvard Medical School, Boston, MA, USA

Abstract. High-grade serous ovarian carcinoma (HGSOC) is characterised by significant spatial and temporal heterogeneity, typically manifesting at an advanced metastatic stage. A major challenge in treating advanced HGSOC is effectively monitoring localised change in tumour burden across multiple sites during neoadjuvant chemotherapy (NACT) and predicting long-term pathological response and overall patient survival. In this work, we propose a self-supervised deformable image registration algorithm that utilises a general-purpose image encoder for image feature extraction to co-register contrast-enhanced computerised tomography scan images acquired before and after neoadjuvant chemotherapy. This approach addresses challenges posed by highly complex tumour deformations and longitudinal lesion matching during treatment. Localised tumour changes are calculated using the Jacobian determinant maps of the registration deformation at multiple disease sites and their macroscopic areas, including hypo-dense (i.e., cystic/necrotic), hyper-dense (i.e., calcified), and intermediate density (i.e., soft tissue)

M. Modat et al. (Eds.): WBIR 2024, LNCS 15249, pp. 295–307, 2024.
https://doi.org/10.1007/978-3-031-73480-9_23

portions. A series of experiments is conducted to understand the role of a general-purpose image encoder and its application in quantifying change in tumour burden during neoadjuvant chemotherapy in HGSOC. This work is the first to demonstrate the feasibility of a self-supervised image registration approach in quantifying NACT-induced localised tumour changes across the whole disease burden of patients with complex multi-site HGSOC, which could be used as a potential marker for ovarian cancer patient's long-term pathological response and survival.

Keywords: Foundation Models · Deformable Image Registration · Medical Imaging · Cancer Research · Treatment Response

1 Introduction

Despite notable advances in the treatment of ovarian cancer, it continues to be one of the leading causes of gynecological death among women [1]. High-grade serous ovarian carcinoma (HGSOC) is the most common type and is a highly heterogeneous disease that typically presents with advanced, multi-site metastatic disease. Neoadjuvant chemotherapy (NACT) followed by delayed primary surgery is becoming the most frequent treatment strategy for advanced HGSOC [2]. HGSOC often exhibits lesion-wise heterogeneous response to NACT with some lesions disappearing or shrinking, some remaining stable, some growing, and new lesions appearing despite ongoing treatment. The response of individual lesions to treatment has been shown to drive progression and provides predictive information for the effectiveness of delayed primary surgery and survival [3]. Consequently, treatment response assessment following NACT is critical to properly plan subsequent stages of the patient's clinical pathway.

In current clinical practice, tumour response assessment is usually based on two-dimensional measurements to approximate change in tumour size on serial contrast-enhanced computerised tomography (CE-CT) scan images. Although Response Evaluation Criteria in Solid Tumours (RECIST 1.1) is the current clinical guideline to assess size change of solid tumours after therapeutic treatment, it is based on measurements of lesion diameters, and therefore is not able to capture the three-dimensional complexity of tumours. Additionally, radiologists have to choose which target lesions to track and measure, which adds subjectivity to the clinical evaluation process.

Future clinical trials for ovarian cancer require stronger, more sensitive response metrics that can provide more robust discrimination between different response patterns. Here, a first challenge is how to effectively monitor volumetric tumour changes across multiple disease sites during the course of NACT and, more importantly, how to effectively reveal heterogeneous changes in various sub-regions within tumours rather than tumours as a whole, since the heterogeneity may be more indicative of long-term pathological response. A second challenge is how to match multiple lesions between longitudinal scans. Lesions may be numerous and densely packed within a single organ or tissue. For example, metastatic ovarian cancer patients with a high disease burden can exhibit

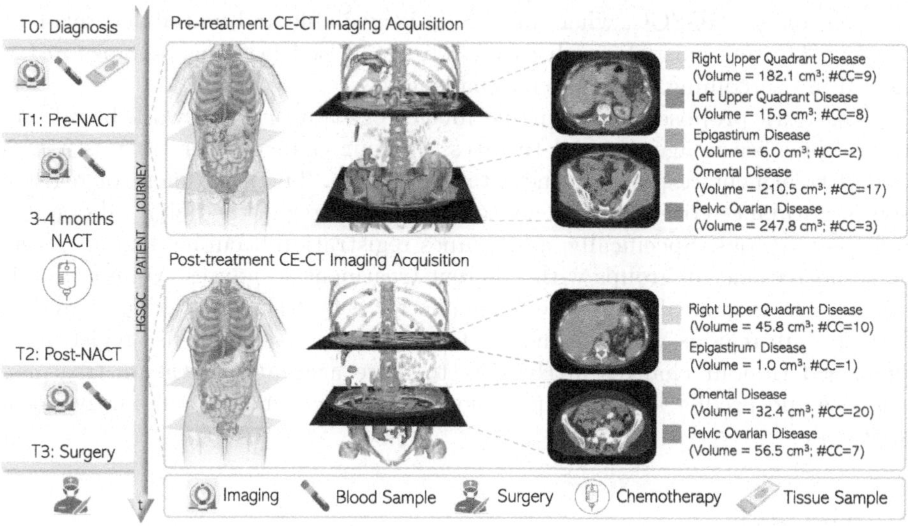

Fig. 1. Overview of the treatment schedule and characteristics of an advanced-stage high-grade serous carcinoma. Initial staging and diagnosis were based on CE-CT imaging and biopsy. Surgery was performed after two cycles of NACT. According to RECIST 1.1 criteria, the patient was classified as a partial responder. Pre- and post-treatment CE-CT scans revealed the metastatic nature of the disease, identifying 16 sites of disease burden with varying numbers of connected components (CC) and highlighting heterogeneity in lesion response.

dozens of disease sites with hundreds of small connected components [4]. Additionally, lesions may grow, shrink, split, or merge over time. Thus, matching lesions between longitudinal images is critical to ensure that heterogeneous response patterns can be identified, and imaging data can be best utilised to inform treatment decisions. Figure 1 illustrates pre- and post-treatment CE-CT scans that highlight the metastatic nature of HGSOC and the lesion-wise heterogeneity in response to chemotherapy, with some lesions disappearing or shrinking, some remaining stable, some growing, and new lesions appearing.

Registration of longitudinal CE-CT images acquired during the course of treatment has received increasing interest as a useful tool to enable localised lesion tracking during therapy. Despite its importance, longitudinal registration of ovarian cancer CE-CT scans has been less studied, and the registration performance is less understood compared to other organs (e.g., lung [5,6], liver [7,8], brain [9], prostate [10], cervix [11] and head and neck [12]).

With the emergence of deep learning, there has been a notable shift in image registration techniques. These approaches directly predict a displacement field by leveraging self-learned hierarchical features extracted from both the input moving and reference images. In particular, the introduction of foundation models in computer vision marks a significant paradigm shift in the approach to complex visual tasks. Self-supervised vision foundation models demonstrate this evolution by providing pre-trained models that have learned rich feature representations from large-scale unlabeled datasets. This is particularly advantageous in scenarios like HGSOC, where little data is available and no publicly available imaging datasets exist. In such cases, self-supervised learning becomes instrumental in overcoming data scarcity challenges.

This work proposes a deformable image registration framework based on a general-purpose image encoder to assess tumour changes over time as a measure of treatment response in metastatic HGSOC. The assessment of registration performance covers various anatomical structures in the pelvic, abdominal, and chest regions. Specifically, it examines registration accuracy among disease sites and patient subgroups with different treatment responses. Volumetric disease change is quantified for the first time for each lesion individually, and for different sub-regions within tumours. Our ultimate goal is to use registration-quantified tumour changes in HGSOC to more accurately measure treatment response and predict long-term pathological response and patient survival after treatment.

2 Methods

Registration Framework. The source and target images in the registration problem are denoted by $I_A : \Omega \to \mathbb{R}$ and $I_B : \Omega \to \mathbb{R}$, respectively. The transformation map is denoted by $\Phi_{AB} : \mathbb{R}^d \to \mathbb{R}^d$ with the intention that $I_A \circ \Phi_{AB} \sim I_B$. We adopt the GradICON [13] registration network which uses a two-step registration process. In the first step, images are passed through a three-level multiresolution registration network. At each level, a U-Net accepts the warped image from the previous level and the target image. Additionally, the input images are downsampled to $\frac{1}{4}$ and $\frac{1}{2}$ for each level, respectively. In the second step, images are passed through a single U-Net, which takes the warped image from the first step and the target image at full resolution. All four U-Nets share the same architecture as illustrated in Fig. 2.

Volumetric Feature Encoder and Training. We adopt the publicly-available uniGradICON [14] feature encoder, a foundation model trained and evaluated on 12 public medical image datasets from the Learn2Reg Grand Challenge[1] covering various anatomical regions, imaging modalities, and deformation patterns. Given an ordered image pair (I_A, I_B), the registration network outputs the transformation map Φ_{AB} which maps I_A to the space of I_B. By swapping the input pair (I_B, I_A), we obtain the estimated inverse map Φ_{BA}. The similarity loss L_{sim} is computed between the warped image $I_A \circ \Phi_{AB}$ and the target image I_B, and vice versa following the formulation:

$$L = L_{\text{sim}}(I_A \circ \Phi_{AB}, I_B) + L_{\text{sim}}(I_B \circ \Phi_{BA}, I_A) + \lambda \|\nabla(\Phi_{AB} \circ \Phi_{BA}) - I\|_F^2$$

We use localised normalised cross correlation (1 - LNCC) as the similarity measure following the implementation described in [13]. The third term is the gradient inverse consistency regularizer, which penalizes differences between the Jacobian of the composition of Φ_{AB} with Φ_{BA} and the identity matrix I; $\|\cdot\|_F^2$

[1] https://learn2reg.grand-challenge.org/,.

is the Frobenius norm, $\lambda > 0$. The first and second stages of the registration network were trained for 800 and 200 epochs, respectively, with a learning rate of 5×10^{-5}.

Quantification of Volumetric Tumour Progression. The obtained deformation vector fields were further used to calculate the Jacobian maps. The Jacobian map was computed as the determinant of the gradient of the deformation vector field (DVF). The Jacobian determinants indicate the volumetric change ratio at each voxel, where a value greater than 1 indicates voxel expansion, a value less than 1 indicates shrinkage, and a value of exactly 1 indicates volume preservation.

Intra- and Inter-tumoural Heterogeneity. HGSOC is characterised by high levels of macroscopic heterogeneity, with frequent hyper-dense (calcified), hypo-dense (cystic/necrotic), or intermediate dense (soft tissue) tumour regions, each of which can have different prognostic significance and influence therapy response. To understand how different tumour regions respond to treatment, an automated sub-segmentation of these regions was performed using a tissue-specific sub-segmentation framework based on an unsupervised fuzzy clustering technique previously described in [15]. Jacobian maps were computed as the determinant of the gradient of the deformation vector field to quantify local changes for each tissue-specific sub-segmentation.

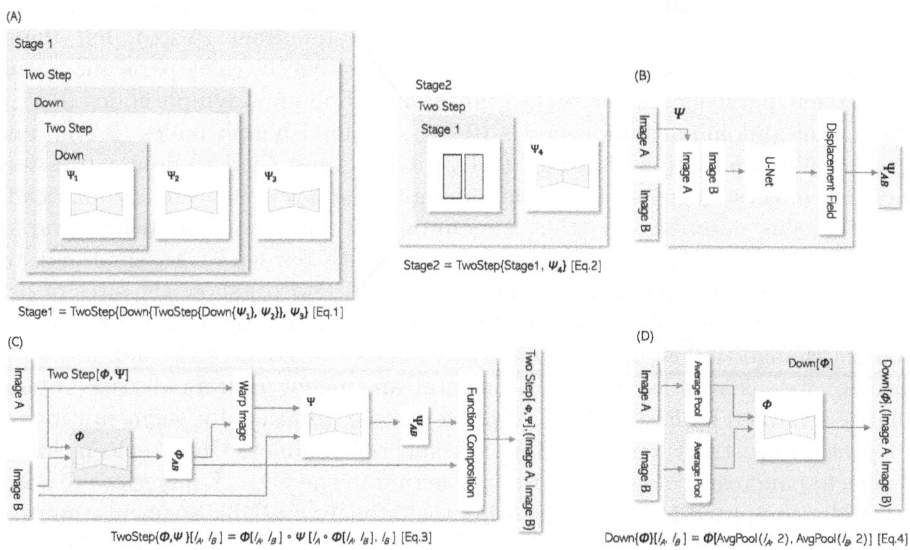

Fig. 2. Image registration follows a multi-step, multi-resolution approach trained with a two-stage process (A) via the Two Step (C) and the Downsample (D) operators. The basic component Ψ of the registration network is represented by a U-Net instance, taking as input two images and returning a displacement field (B). The downsample operator predicts the warp between two high-resolution images using a network that operates on low-resolution images, and the two-step operator predicts the warp between two images in two steps, first capturing the coarse transform via Φ and then the residual transform via Ψ.

3 Experiments

Patient Cohort. A total of 99 patients with a confirmed histopathological diagnosis of HGSOC, who underwent neoadjuvant chemotherapy before delayed primary surgery, were included in this study. All patients were treated at Cambridge University Hospitals NHS Foundation Trust and were recruited into a clinical observational study approved by the local research ethics committee (REC reference number: 08/H0306/61). Clinical CE-CT scans covering the abdomen and pelvis were acquired. Figure 3 (a) shows the median (min - max) age of the patient cohort in years, the total number of NACT cycles administered, and the FIGO stage (I to IV), which describes how far the cancer has spread. The radiological response is measured based on the RECIST 1.1 criteria and is categorised into four groups: complete response (CR), partial response (PR), stable disease (SD), or progressive disease (PD). Chemotherapy Response Scores (CRS) describe the histopathologic response to NACT in the omentum (fatty tissue layer in the abdominal cavity) and are categorized into I to III.

Image Segmentation and Labelling. On CE-CT axial images reconstructed with a slice thickness of typically 5 mm, all cancer lesions were segmented semi-automatically by a board-certified radiologist with ten years of experience in clinical imaging, using Microsoft Radiomics (project InnerEye; Microsoft, Redmond, WA, USA). The volumes of interest were annotated for their anatomic location with the following disease sites and volume percentages: ovaries and pelvis (66.4%), omentum (24.4%), right upper quadrant (5.1%), left upper quadrant (1.1%), epigastrium (0.1%), mesentery (0.3%), right paracolic gutter (0.8%), left paracolic gutter (0.3%), infrarenal abdominal lymph nodes (0.8%), suprarenal abdominal lymph nodes (0.2%), inguinal lymph nodes (0.2%), and supradiaphragmatic lymph nodes (0.1%). Cystic and solid tumour parts were included in these segmentations. Intra-tumoural sub-regions were automatically segmented as described in [15]. Tumour in the omentum and pelvic/ovarian locations accounts for the majority of the disease burden at baseline and are the most frequent tumour locations. Figure 3 (b) and (c) show the sites of primary and metastatic disease in HGSOC and the distribution of tumour volumes, volume change (follow-up/ baseline) and number of connected components by disease site, as well as the total, omental and pelvic/ovarian disease volume change stratified by RECIST 1.1 response status. Additionally, segmentations of 10 anatomical structures, including left and right kidneys, liver, spleen, lungs, stomach, pancreas, colon, vertebra and sacrum in the CE-CT images were automatically computed for each scan using the publicly-available segmentation tool Total Segmentator v2 [16] and used for registration evaluation. Baseline and follow-up CE-CT scans were evaluated according to RECIST 1.1 for treatment response assessment.

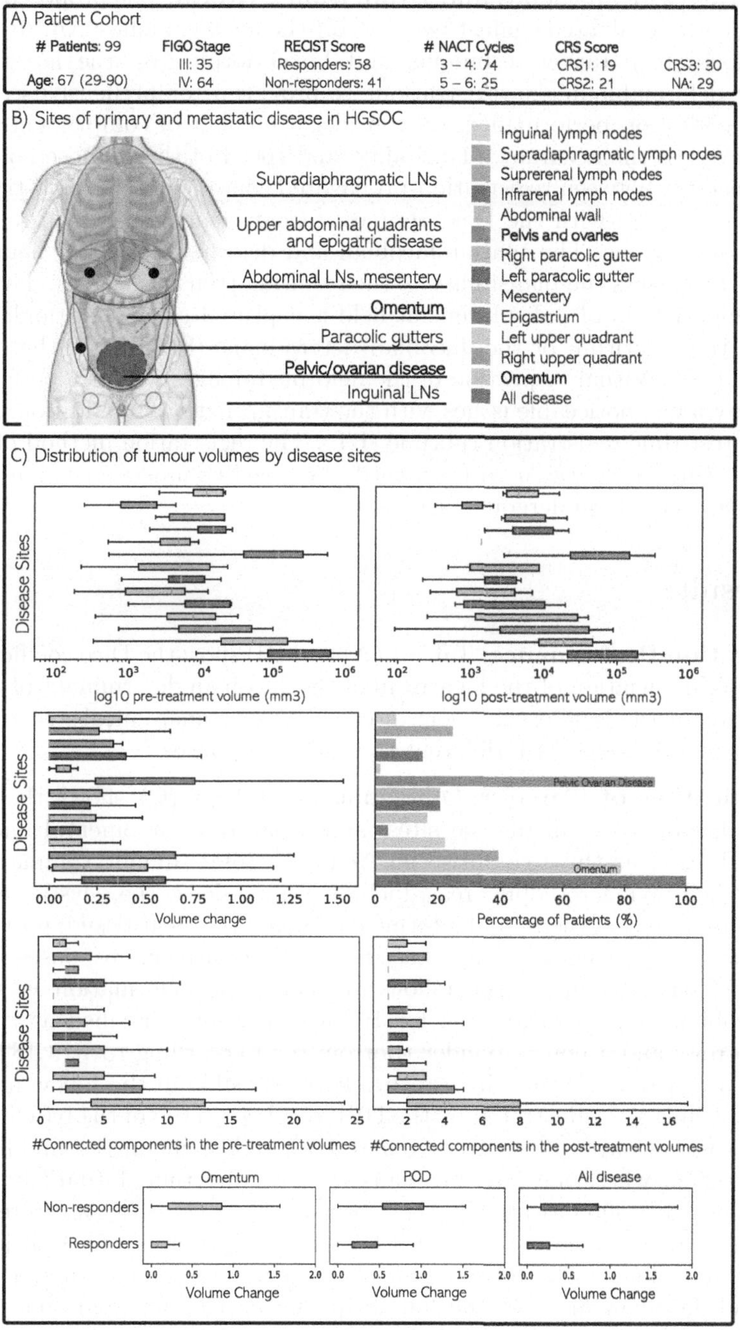

Fig. 3. (A) Main characteristics of the patient cohort. (B) Sites of primary and metastatic HGSOC disease. (C) Distribution of tumour volumes by disease site.

Comparative Approaches and Evaluation Criteria. The proposed foundation model was evaluated against two state-of-the-art techniques: NiftyReg [17], a conventional registration technique, and VoxelMorph [18], the latest benchmark for deformable image registration. To provide a comprehensive evaluation of the registration performance, we consider a number of complementary metrics that assess the accuracy, plausibility, and speed of the proposed approach. The agreement between registration-propagated contours and ground-truth contours was assessed with the Dice Similarity Coefficient (DSC) for tumour and non-tumour regions. The spatial Jacobian and deformation vector field images were used to assess the plausibility of the resulting transformations. The plausibility (smoothness) of a displacement field is captured using the standard deviation of the logarithm of the Jacobian determinant (SDlogJ) of the displacement field [19]. Visual inspection of the deformed images was also performed to detect any large, noticeable issues with the transformation. In addition, we measure the test-time registration runtime (RT). The registration method obtaining the highest accuracy was used to quantify localised volumetric tumour changes in HGSOC patients undergoing NACT.

4 Results

Registration Performance. Table 1 (A) and (B) show the Dice coefficient and the standard deviation of the logarithm of the Jacobian determinant of different registration methods across different disease sites (A), anatomical structures (B), and patient subgroups with different treatment responses.

Quantification of Tumour Deformation. Table 1 (C) shows the average median Jacobian for all disease sites, and separately for omental and pelvic ovarian disease, as these comprise 90.8% of the total tumour volume. Results are stratified by macroscopic areas, including hypo-dense (i.e., cystic/necrotic), hyper-dense (i.e., calcified), and intermediate dense (i.e., soft tissue) portions and categorised by patient subgroups based on their treatment responses. Figure 4 shows the baseline image, the follow-up image, the deformation vector field and Jacobian map in axial view for HGSOC patients classified as a responder (top row) and a non-responder (bottom row) according to RECIST 1.1. In the top row, pelvic ovarian disease shrinkage is visible in the follow-up image. Tumour shrinkage is illustrated by the DVF overlaid on top of the baseline image (red contour). Converging vectors generated a sink point in the middle of the tumour region where median Jacobians were smaller than 1 (dark blue) indicating shrinkage. Quantitatively, the majority of voxels (97.53%) were shrinking with average Jacobian of 0.46 ± 0.15, indicating overall a 54% shrinkage in the tumour volume. The bottom row shows a non-responder tumour with its DVF and Jacobian map. In the follow-up image, a large expansion appeared on the left distal side of the pelvic ovarian lesion. DVF overlaid on the baseline image illustrates that vectors are diverging toward the right distal side of tumour (red contour) where median Jacobians were larger than 1 (dark red), indicating expansion. In this case, majority of voxels (98.26%) were expanding with Jacobian greater than 1. The mean Jacobian was 1.34 ± 0.18, indicating overall a 34% expansion in the tumour volume. Figure 5 shows examples of tissue-specific sub-segmentations. The whole tumour, hyper-dense and hypo-dense region contours are displayed as light blue, yellow and magenta solid lines, respectively.

5 Discussion

We benchmarked state-of-the-art medical image registration algorithms, including traditional methods (NiftyReg), deep learning-based methods (VoxelMorph), and foundation models (uniGradICON), for the first time in the context of longitudinal CE-CT registration for HGSOC. Table 1 shows that the proposed foundation model outperforms the traditional and DL-based methods in terms of registration accuracy in both tumour and non-tumour regions and speed. We also show that registration accuracy varies across different regions (disease sites vs. anatomical structures) and among patient subgroups with different responses to chemotherapy. Omental disease poses a more challenging registration task compared to the pelvic ovarian disease, as omental lesions are usually numerous and densely packed within a single organ.

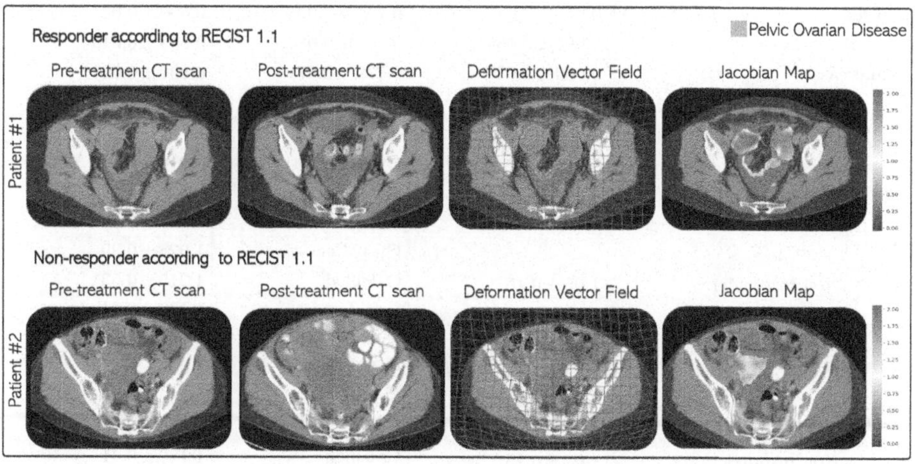

Fig. 4. Baseline, follow-up, DVF, and Jacobian maps in axial view for HGSOC patients classified as a responder (top row) and a non-responder (bottom row) according to RECIST 1.1. Tumour segmentations are overlaid on top of both the DVFs and the Jacobian maps, showing longitudinal volume changes. Red and blue colours in Jacobian maps represent expansion and shrinkage respectively, whereas white colour shows volume preservation. (Color figure online)

These lesions split or merge over time more frequently than those in the pelvic region, where the number of connected components between pre- and post-treatment scans is generally more stable. Additionally, patients with complete

Table 1. Mean and standard deviation of Dice coefficient and the standard deviation of the logarithm of the Jacobian determinant as DSC ± SD(SDlogJ), and running time of registration methods across disease sites (A), anatomical structures (B), and patient subgroups. Initial DSC is 0.376 ± 0.45 for tumour regions and 0.546 ± 0.71 for non-tumour regions. (C) Median Jacobian determinants stratified by disease sites - omental disease (OD) and pelvic ovarian disease (POD) - and their macroscopic areas, including hypo-dense (i.e., cystic/necrotic), hyper-dense (i.e., calcified), and intermediate dense (i.e., soft tissue) portions.

Location / Registration Method		NiftyReg [14min]	VoxelMorph [8s]	GradICON [8s]
(A) DSC ± SD (SDlogJ) in Tumour Regions				
POD	Responder	0.575 ± 0.089 (0.061)	0.645 ± 0.042 (0.052)	0.796 ± 0.261 (0.025)
	Non-responder	0.599 ± 0.071 (0.059)	0.751 ± 0.212 (0.035)	0.802 ± 0.145 (0.018)
OD	Responder	0.501 ± 0.073 (0.036)	0.602 ± 0.056 (0.061)	0.701 ± 0.383 (0.071)
	Non-responder	0.572 ± 0.076 (0.083)	0.700 ± 0.021 (0.041)	0.725 ± 0.178 (0.020)
(B) DSC ± SD (SDlogJ) in Healthy Anatomical Structures				
Kidneys		0.824 ± 0.048	0.815 ± 0.067	0.913 ± 0.052
Liver		0.801 ± 0.059	0.807 ± 0.031	0.937 ± 0.076
Sacrum		0.889 ± 0.051	0.845 ± 0.078	0.901 ± 0.033
Vertebra		0.811 ± 0.086	0.978 ± 0.062	0.912 ± 0.220
Colon		0.510 ± 0.052	0.525 ± 0.043	0.720 ± 0.068
Stomach		0.522 ± 0.076	0.560 ± 0.035	0.779 ± 0.040
Spleen		0.645 ± 0.083	0.712 ± 0.028	0.800 ± 0.031
Lungs		0.661 ± 0.084	0.724 ± 0.084	0.889 ± 0.010
Pancreas		0.609 ± 0.035	0.739 ± 0.071	0.891 ± 0.091

(C) Median Jacobians stratified by Tumour Sub-Regions				
POD	Responder	0.80 ± 0.10	Cyst/Necrotic	0.918 ± 0.55
			Calcified	1.129 ± 0.36
			Soft Tissue	0.804 ± 0.39
	Non-responder	1.05 ± 0.15	Cyst/Necrotic	1.028 ± 0.29
			Calcified	1.215 ± 0.53
			Soft Tissue	1.106 ± 0.31
OD	Responder	0.64 ± 0.14	Cyst/Necrotic	0.962 ± 0.46
			Calcified	0.606 ± 0.21
			Soft Tissue	0.640 ± 0.43
	Non-responder	1.02 ± 0.15	Cyst/Necrotic	1.004 ± 0.39
			Calcified	1.182 ± 0.41
			Soft Tissue	1.116 ± 0.39

responses exhibit greater tumour changes, posing a more significant registration challenge, as shown by the lowest DSC in Table 1 compared to non-responders.

Due to the higher accuracy of the proposed foundation model compared to other registration methods, we further evaluated its ability to quantify longitudinal tumour changes induced by NACT. The average median Jacobian for all disease sites was 0.82 ± 0.11 and 1.06 ± 0.15 for responders and non-responders, respectively. These indicated that on average responders had 18% median volume

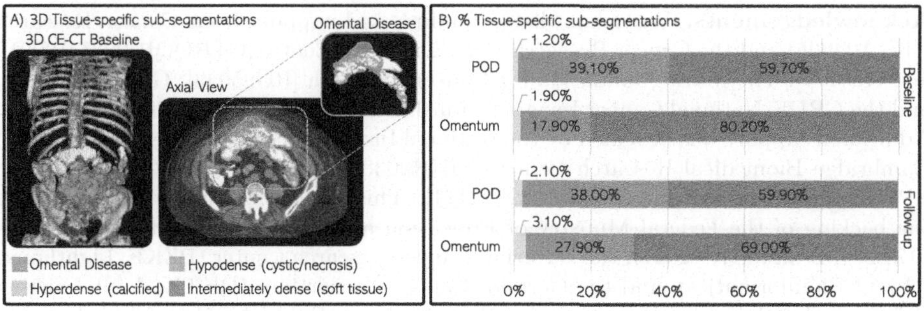

Fig. 5. (A) 3D and axial CT reconstructions showing segmented ROIs: hyper-dense, hypo-dense and soft tissue components relative to the whole omental tumour. (B) Total percentage of each tissue type for two disease sites at baseline and follow-up for the cohort.

shrinkage while non-responders had 6% median volume expansion. For pelvic ovarian disease, the average median Jacobian was 0.80 ± 0.10 for responders and 1.05 ± 0.15 for non-responders, respectively. This indicates that, on average, responders experienced a 20% median pelvic tumor volume shrinkage, while non-responders experienced a 5% median pelvic tumor volume expansion. For omental disease, the average median Jacobian was 0.64 ± 0.14 for responders and 1.02 ± 0.15 for non-responders, indicating that, on average, responders had a 36% median volume shrinkage, while non-responders had a 2% median volume expansion. Omental disease showed a significantly better response than pelvic ovarian disease. We also quantify, for the first time, localised volumetric deformation in different sub-regions (soft tissue, calcified, and necrotic regions) within tumors. We show that cysts are more prevalent in the POD compared to omental disease, and that calcified regions are typically more prominent in follow-up scans, representing areas of calcified fibrosis or calcified deposits resulting from treatment or disease progression. Table 1 (C) shows that soft tissue shrinks in responders and expands in non-responders, while cyst regions preserve their volume and calcified areas generally increase during NACT.

Upon further evaluation of the performance of registering longitudinal CE-CT images to quantify localised tumour change in HGSOC, our future work includes (a) quantitative evaluation of registration accuracies on longitudinal images with more than one follow-up time point; (b) evaluation of the performance of registration in larger datasets and across a broader range of registration algorithms; and (c) the use of registration-quantified changes and registration-revealed heterogeneity of changes to predict long-term pathological responses to treatment. In conclusion, this work quantifies, for the first time, localised heterogeneous tumour changes across different disease sites in ovarian cancer and demonstrates the ability of registration to characterise spatially heterogeneous effects induced by treatment, which may ultimately be used as more sensitive markers for evaluating treatment response and predicting patient outcomes.

Acknowledgements. We acknowledge funding and support from Cancer Research UK (A22905) and the Cancer Research UK Cambridge Centre [CTRQQR-2021-100012 and A25177], The Mark Foundation for Cancer Research [RG95043], GE HealthCare, and the CRUK National Cancer Imaging Translational Accelerator (NCITA) [A27066]. Additional support was provided by the National Institute for Health Research (NIHR) Cambridge Biomedical Research Centre [NIHR203312 and BRC-1215-20014] and the EPSRC Tier-2 capital grant [EP/T022221/1]. This work was further supported by the backing of the Federal Ministry of Education and Research (BMBF, Grant Nos. 01ZZ2315B and 01KX2021), the Bavarian Cancer Research Center (BZKF, Lighthouse AI and Bioinformatics), and the German Cancer Consortium (DKTK, Joint Imaging Platform). Further support was received through the St. Baldrick's Career Development Grant, NIH R61NS126792, NIH R21TR004265, and NIH R21NS21735. Additionally, we recognize the contribution of the Helmholtz Information and Data Science Academy. The authors would like to acknowledge the support of the Scientific Computing team from the Cancer Research UK Cambridge Institute.

Disclosure of Interests. The authors have no competing interests to declare that are relevant to the content of this article.

References

1. Vergote, Ignace, et al. "Neoadjuvant chemotherapy or primary surgery in stage IIIC or IV ovarian cancer." New England Journal of Medicine 363.10 (2010): 943-953.
2. Kehoe, Sean, et al. "Primary chemotherapy versus primary surgery for newly diagnosed advanced ovarian cancer (CHORUS): an open-label, randomised, controlled, non-inferiority trial." The Lancet 386.9990 (2015): 249-257.
3. Morgan, Robert D., et al. "Objective responses to first-line neoadjuvant carboplatin-paclitaxel regimens for ovarian, fallopian tube, or primary peritoneal carcinoma (ICON8): post-hoc exploratory analysis of a randomised, phase 3 trial." The Lancet Oncology 22.2 (2021): 277-288.
4. Crispin-Ortuzar, Mireia, et al. "Integrated radiogenomics models predict response to neoadjuvant chemotherapy in high grade serous ovarian cancer." Nature communications 14.1 (2023): 6756.
5. Zhang, Hang, et al. "An unsupervised convolutional LSTM network (C-LSTMNet) for lung 4D-CT registration." IEEE Access (2024).
6. Tahmasebi, Nazanin, et al. "Real-time lung tumor tracking using a CUDA enabled nonrigid registration algorithm for MRI." IEEE journal of translational engineering in health and medicine 8 (2020): 1-8.
7. Xue, Peng, et al. "Structure-aware registration network for liver DCE-CT images." IEEE Journal of Biomedical and Health Informatics (2024).
8. Pham, Xuan Loc, et al. "Multi-resolution Coarse-to-fine Registration Approach for Liver Computed Tomography Image Analysis." 2022 26th International Computer Science and Engineering Conference (ICSEC). IEEE, 2022.
9. van Garderen, Karin A., et al. "Evaluating the predictive value of glioma growth models for low-grade glioma after tumor resection." IEEE Transactions on Medical Imaging (2023).
10. Dupuy, Tamara, et al. "2D/3D deep registration along trajectories with spatiotemporal context: Application to prostate biopsy navigation." IEEE Transactions on Biomedical Engineering (2023).

11. Dupuy, Tamara, et al. "2D/3D deep registration along trajectories with spatiotemporal context: Application to prostate biopsy navigation." IEEE Transactions on Biomedical Engineering (2023).
12. Huang, Bin, et al. "3D lightweight network for simultaneous registration and segmentation of organs-at-risk in CT images of head and neck cancer." IEEE Transactions on Medical Imaging 41.4 (2021): 951-964.
13. Tian, Lin, et al. "GradICON: Approximate diffeomorphisms via gradient inverse consistency." Proceedings of the IEEE/CVF Conference on Computer Vision and Pattern Recognition. 2023.
14. Tian, Lin, et al. "uniGradICON: A Foundation Model for Medical Image Registration." arXiv preprint arXiv:2403.05780 (2024).
15. Rundo, Leonardo, et al. "Tissue-specific and interpretable sub-segmentation of whole tumour burden on CT images by unsupervised fuzzy clustering." Computers in biology and medicine 120 (2020): 103751.
16. Wasserthal, Jakob, et al. "Totalsegmentator: Robust segmentation of 104 anatomic structures in ct images." Radiology: Artificial Intelligence 5.5 (2023).
17. Modat, M., Ridgway, G.R., Taylor, Z.A., Lehmann, M., Barnes, J., Hawkes, D.J., et al.: Fast free-form deformation using graphics processing units. Computer methods and programs in biomedicine 98(3), 278-284 (2010)
18. Balakrishnan, Guha, et al. "Voxelmorph: a learning framework for deformable medical image registration." IEEE transactions on medical imaging 38.8 (2019): 1788-1800.
19. Leow, Alex D., et al. "Statistical properties of Jacobian maps and the realization of unbiased large-deformation nonlinear image registration." IEEE transactions on medical imaging 26.6 (2007): 822-832.

CAR: Contrast-Agnostic Deformable Medical Image Registration with Contrast-Invariant Latent Regularization

Yinsong Wang, Siyi Du, Shaoming Zheng, Xinzhe Luo, and Chen Qin(✉)

Department of Electrical and Electronic Engineering & I-X, Imperial College London, London, UK
{y.wang23,s.du23,s.zheng22,x.luo,c.qin15}@imperial.ac.uk

Abstract. Multi-contrast image registration is a challenging task due to the complex intensity relationships between different imaging contrasts. Conventional image registration methods are typically based on iterative optimizations for each input image pair, which is time-consuming and sensitive to contrast variations. While learning-based approaches are much faster during the inference stage, due to generalizability issues, they typically can only be applied to the fixed contrasts observed during the training stage. In this work, we propose a novel contrast-agnostic deformable image registration framework that can be generalized to arbitrary contrast images, without observing them during training. Particularly, we propose a random convolution-based contrast augmentation scheme, which simulates arbitrary contrasts of images over a single image contrast while preserving their inherent structural information. To ensure that the network can learn contrast-invariant representations for facilitating contrast-agnostic registration, we further introduce contrast-invariant latent regularization (CLR) that regularizes representation in latent space through a contrast invariance loss. Experiments show that CAR outperforms the baseline approaches regarding registration accuracy and also possesses better generalization ability to unseen imaging contrasts. Code is available at https://github.com/Yinsong0510/CAR.

Keywords: Multi-contrast Image Registration · Magnetic Resonance Imaging (MRI) · Latent Space Regularization · Random Convolution

1 Introduction

Multi-contrast images in magnetic resonance imaging (MRI) provide complementary information for characterizing tissues of human bodies, which are widely

Supplementary Information The online version contains supplementary material available at https://doi.org/10.1007/978-3-031-73480-9_24.

M. Modat et al. (Eds.): WBIR 2024, LNCS 15249, pp. 308–318, 2024.
https://doi.org/10.1007/978-3-031-73480-9_24

used in both qualitative and quantitative imaging in clinical diagnosis. However, due to variations of acquisition protocols as well as possible patient movement during the acquisition process, the images are often spatially misaligned, which entails image registration for accurate downstream multi-contrast analysis and interpretation. Registering images between different contrasts can be challenging due to the complex relationship between their intensity profiles. Conventional approaches typically tackle this problem by optimizing information-theoretic similarity measures such as mutual information over the misaligned image pairs [12,13]. Nevertheless, the iterative-based optimization framework is usually time-consuming and therefore can be limited in real-world applications.

Recently, deep learning-based registration approaches have shown their great potential for fast and accurate medical image registration. Several learning-based methods have also attempted to solve the multi-contrast image registration task. For instance, Qiu et al. [17] have proposed to leverage the conventional mutual information as the similarity metric and embed it into an end-to-end learning-based registration framework. An alternative way proposed by Qin et al. [16] reduced the multi-contrast registration task to a mono-contrast one by disentangling images into shape and appearance latent spaces and aligning the image contrasts with style transfer approaches. Dey et al. [6] also proposed to use contrastive representation learning to learn a similarity loss between the warped and fixed multi-contrast images based on multi-scale PatchNCE using an autoencoder. Nevertheless, due to the network generalizability issues, these approaches can only deal with fixed imaging contrasts, which require the contrast of the image pairs to be the same during both training and inference stages.

To address the contrast variations between training and inference, there have also been some recent efforts in achieving contrast agnostic registration. Hoffmann et al. proposed to synthesize arbitrary contrast of images based on segmentation or randomly generated labels [10]. They used these generated images as the input to the registration network and optimized that based on original segmentation labels using soft dice loss [7]. However, this approach relies on the availability of segmentation maps and may not register images well in terms of fine-grained details, as the employed soft dice loss mainly focuses on the structure overlaps rather than pixel-level or volume-level differences. Several recent works proposed training a neural network to learn a certain distance metric for multi-modal registration. Sideri et al. [20] relied on modality and affine geometric augmentation for learning a contrast-agnostic distance measure based on image patch centers to tackle multi-contrast rigid registration. Ronchetti et al. [18] proposed to use a small CNN to learn a feature-level distance that approximates the conventional LC^2 similarity. However, both approaches focused on similarity approximation, without explicitly enforcing contrast-agnostic learning.

In this work, we propose a contrast-agnostic registration framework (CAR) for deformable image registration that only relies on single contrast images for training but can be generalized to other unseen contrasts. Specifically, we propose a novel random convolution-based contrast-augmentation scheme to simulate arbitrary contrasts based on single contrast images. To encourage contrast-agnostic learning, we further enforce the network to learn contrast-invariant features by regularizing the feature representation in latent spaces using a proposed contrast invariance loss. Unlike existing learning-based multi-contrast registration frameworks [6,14,16,17], where each model can only register between fixed contrast images, CAR can train a single model that registers arbitrary contrasts of images without seeing the contrasts during training, which can significantly improve the generalizability and applicability for multi-contrast registration. In addition, CAR does not rely on segmentation maps for synthesis [10] or carefully designed networks for learning similarity losses [18,20]. Instead, we can adopt simple mono-contrast similarity losses such as negative local normalized cross-correlation (LNCC) based on pre-augmented single contrast images, which enables the training of the network to be more lightweight and efficient. Experiments are performed on both brain MRI of T1 weighting of T2 weighting data and cardiac T1 mapping data with varied T1 weightings, and our results demonstrate the superior performance of CAR against baseline methods across various contrasts, even though the proposed model has not seen those contrasts during training.

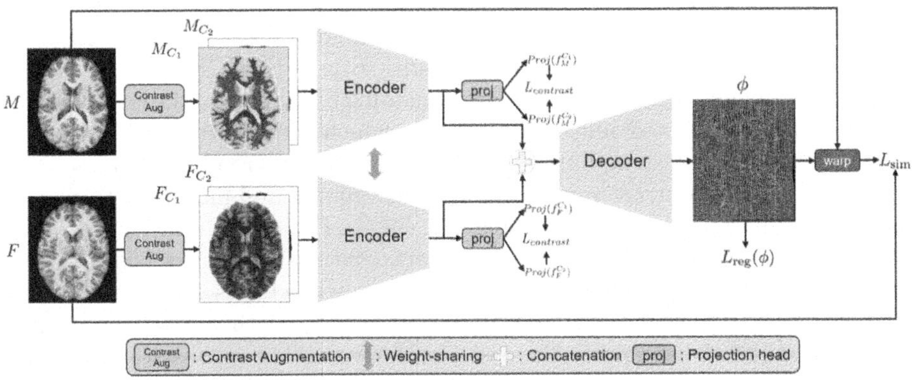

Fig. 1. Overview of the contrast-agnostic registration framework. The input fixed and moving image pair is fed to two Siamese encoders to produce contrast-invariant latent representations, which are taken by the decoder to predict the deformation field.

2 Methodology

Multi-contrast deformable image registration establishes dense and non-linear deformation field ϕ between a multi-contrast image pair, i.e. a moving image M and a fixed image F. Learning-based approaches estimate the deformation field through a neural network f_θ, $\phi = f_\theta(M, F)$, with loss functions typically consisting of a dissimilarity term L_{sim} and a regularization term L_{reg}, i.e.,

$$L_{\text{total}} = L_{\text{sim}}(M \circ \phi, F) + \lambda L_{\text{reg}}(\phi), \quad (1)$$

where λ is a regularization hyperparameter. Conventional learning-based methods typically forecast the desired spatial transformation directly from image appearances, which not only renders the learned model contrast-dependent but disregards the fact that registration aims to find the intrinsic structural correspondence usually invariant to image contrasts. To achieve contrast-agnostic registration that can deal with even unobserved contrasts as discussed in Sect. 1, in our method, we propose a random convolution-based contrast-augmentation scheme for contrast simulation (Sect. 2.1) and a contrast invariance loss for contrast-invariant representation learning (Sect. 2.2). Our overall framework is shown in Fig. 1.

2.1 Contrast Augmentation via Random Convolution

To enable a learning-based registration framework that can be contrast-agnostic, we first propose to simulate the variations of image contrasts by leveraging random convolution (RC) [25] for contrast augmentation. RC has been shown effective for learning robust representations for domain generalization in both the general computer vision field and medical imaging domain for tackling classification and segmentation tasks [15,20]. Specifically, RC leverages randomly initialized convolution kernels to convolve with input images to adjust their appearances while preserving inherent shape information. In earlier works [15,25], it is shown that RC with large kernel size would introduce blurring effect, which can adversely impact the registration problem as it heavily relies on fine-grained details of images for learning pixel-level correspondences. To avoid this, we propose to set the kernel size to 1×1 to minimize the potential effects of the induced artifacts on structural details. This allows the contrast of similar tissues to vary homogeneously in the process, and this synchronizes with the rendering of multi-contrast images determined by imaging physics. Additionally, to ensure that the augmented images can maximally capture the variations of image contrasts, we further propose to stack multiple RC layers with LeakyReLu activation as proposed in [25] to model more complex and diverse contrast mappings, as well as simulating the non-linear intensity relationships as among different MRI contrasts. In each RC layer, the parameters of the convolution kernel are sampled independently from certain uniform distributions to enable the mapping to be diverse. The generated augmented images are then used as inputs to train our contrast-agnostic registration network. Illustrations of this contrast augmentation scheme and augmented images can be found in supplementary materials.

2.2 Contrast-Invariant Latent Regularization

To equip the network with the ability to learn contrast-invariant feature representations and stabilize the training process, we propose to regularize the feature representation in latent space using our proposed contrast invariance loss. Specifically, we assume that the latent representation of images that share the same shape information but with different contrasts should be close in the latent space. Therefore, we aim to pull the feature representations of the images with the same shape information to be close in latent space. Inspired by the contrastive learning-based methods [5,24], we also leverage a projection head to map the representation to a low dimensional space before computing the loss. Instead of using contrastive-based approaches, we propose to use a mean squared difference loss to calculate the difference of projected representations, which can provide a more direct constraint than the InfoNCE loss [5]. In detail, for each optimization step, we perform the network forward passes with two contrast-augmented moving-fixed image pairs $\{M_{C_1}, F_{C_1}\}$ and $\{M_{C_2}, F_{C_2}\}$. Through the encoder and the projection head, the latent representations of the two image pairs can then be obtained, i.e., $f_M^{C_1}, f_F^{C_1}, f_M^{C_2}, f_F^{C_2}$, with dimension (C, N_x, N_y), where C is the channel dimension and (N_x, N_y) is the spatial dimension of the representation. Our contrast invariance loss function can be defined by

$$L_{\text{contrast}} = \frac{1}{N_x N_y} \sum_{i=1}^{N_x} \sum_{j=1}^{N_y} \Big[(\text{Proj}(f_M^{C_1}(i,j)) - \text{Proj}(f_M^{C_2}(i,j)))^2 + (\text{Proj}(f_F^{C_1}(i,j)) - \text{Proj}(f_F^{C_2}(i,j)))^2 \Big], \tag{2}$$

where Proj represents the projection head. Therefore, the contrast invariance loss essentially enforces the encoder to capture the intrinsic structural features that are invariant to various image contrasts or appearances [1].

2.3 Overall Network Architecture and Loss Function

The overall registration framework is illustrated in Fig. 1. The registration network is a U-shape network [3] except we duplicate the encoders for moving images and fixed images respectively. Skip connections [8] are adopted between the two Siamese encoders and the decoder. The moving image and the fixed image are taken as the input of their respective encoder and transformed to their corresponding latent representations f_M and f_F, with 1/16 of the original image size. The two latent representations are concatenated and fed into the decoder to output the final deformation field.

To train the registration network, our loss function is comprised of three terms: a dissimilarity loss L_{sim}, a regularization loss L_{reg}, and our proposed contrast invariance loss L_{contrast}. Since the input contrast-augmented image pair $\{M_{C_i}, F_{C_i}\}$ originated from the mono-contrast image pair $\{M, F\}$, we propose to use the pre-augmented mono-contrast image pairs for loss supervision with a mono-contrast dissimilarity measure. This therefore alleviates the need to fit any multi-contrast similarity loss through a neural network. In our work, we adopt the negative local normalized cross-correlation (LNCC) [2] as the dissimilarity, and a regularization loss formulated in the form of the L2-norm of the deformation field gradient. The overall loss function can be formulated as

$$L_{\text{total}} = -\text{LNCC}(M \circ \phi, F) + \lambda_1 ||\nabla \phi||_2^2 + \lambda_2 L_{\text{contrast}}, \tag{3}$$

where λ_1 and λ_2 are the respective hyperparameters that control the strength of the regularization loss and contrast invariance loss. In this way, we will be able to achieve contrast-agnostic registration by jointly optimizing the deformation field and enforcing the network to learn contrast-invariant feature representations by our contrast-invariant latent regularization.

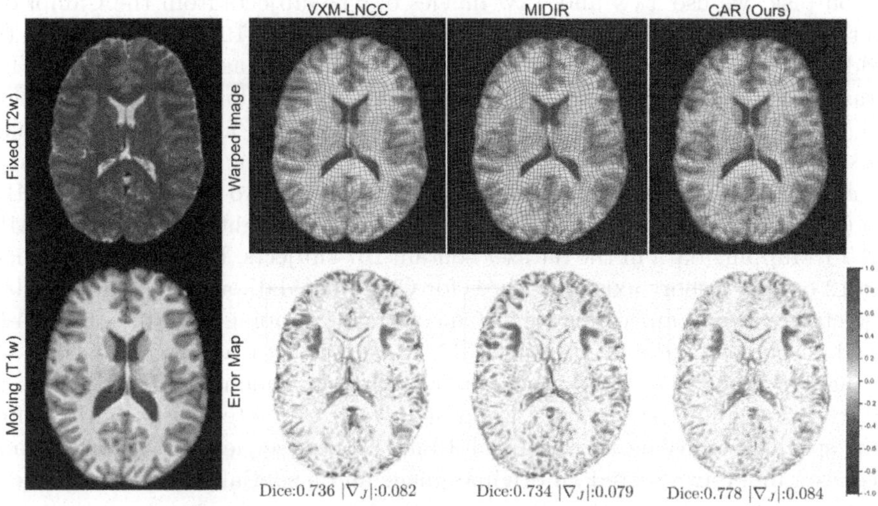

Fig. 2. Qualitative results on CamCAN dataset.

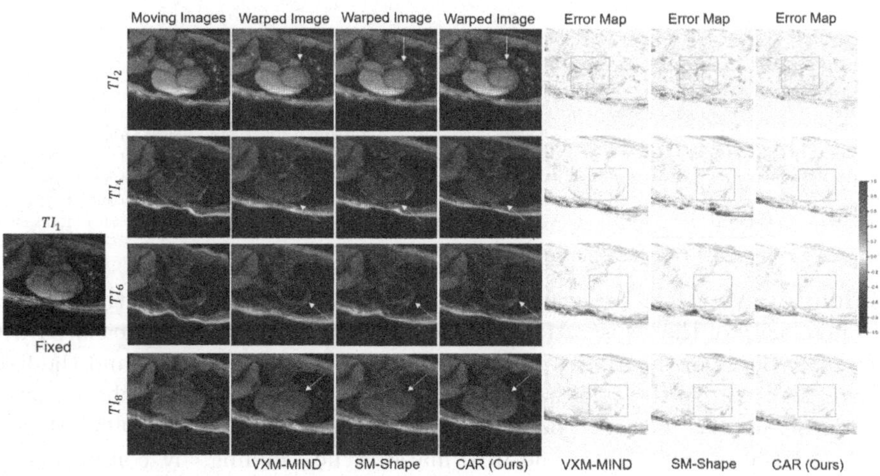

Fig. 3. Qualitative results on CMRxRecon dataset using T1 Mapping data for training.

3 Experiments and Results

Datasets. We evaluated CAR on two applications: (1) inter-subject 2D brain MRI registration of T1 weighted (T1w) - T2 weighted (T2w) images, and (2) intra-subject registration of multiple contrasts of T1 mapping acquisition in cardiac MRI. The 2D experiments on brain MRI registration mainly served as a proof-of-concept study to validate the method's feasibility, and the cardiac datasets only contain 2D stacks of data for 2D experiments. For the brain registration task, we use T1w and T2w images of 652 subjects from the Cambridge Centre for Ageing and Neuroscience (CamCAN) project [19,21]. We sampled the middle slice along the z-axis and cropped the sampled image into 160×192. The dataset was randomly split into 600, 10, and 42 volumes for training, validation, and testing, where subjects were randomly selected to form T1w-T2w image pairs. For the cardiac registration task, we used the public CMRxRecon cardiac datasets [22,23]. The CMRxRecon dataset consists of two parts: (1) Cine MRI, and (2) T1 mapping with images of nine different T1 weightings. Both Cine MRI and T1 Mapping data in the dataset contain 167 subjects. Each subject includes 7 to 12 or 4 to 5 short-axis view slices for Cine MRI data and T1 Mapping data respectively. We cropped images in Cine and T1 Mapping data into 128×128. The dataset was randomly split into 137, 10, and 20 for training, validation, and testing. For the cardiac data, we simulated different random deformations to distort each image in the sequence using artificially generated FFDs with various mesh spacings following the approach from [11]. For sequences within the same slice, we randomly selected an original image and a spatially distorted image as an image pair.

Table 1. Quantitative results of on both CamCAN and CMRxRecon datasets. * represents for values that CAR significantly outperformed with p-value < 0.01 in a paired t-test.

Methods	CamCAN				CMRxRecon			
	Dice ↑	$\lvert\nabla_J\rvert$ ↓	$J_{<0}\%$ ↓	HD95 ↓	Dice ↑	$\lvert\nabla_J\rvert$ ↓	$J_{<0}\%$ ↓	HD95 ↓
Unregistered	0.519 ± 0.109	–	–	5.418 ± 1.829	0.815 ± 0.066	–	–	2.445 ± 0.562
SyN	0.707* ± 0.066	**0.053** ± 0.009	**0.461** ± 0.432%	3.368* ± 1.028	0.608* ± 0.295	**0.046** ± 0.010	0.417* ± 0.478%	6.488* ± 5.238
VXM-LNCC	0.724* ± 0.056	0.085 ± 0.009	0.472 ± 0.163%	3.108* ± 0.685	0.859* ± 0.090	0.087* ± 0.018	0.108* ± 0.078%	2.114* ± 1.108
VXM-MIND	0.688* ± 0.061	0.108* ± 0.012	0.768* ± 0.288%	3.179* ± 0.703	0.863* ± 0.080	0.088* ± 0.012	0.166* ± 0.137%	1.888* ± 0.642
MIDIR	0.706* ± 0.045	0.068 ± 0.006	0.480 ± 0.216%	3.269* ± 0.599	0.763* ± 0.092	0.115* ± 0.015	0.491* ± 0.012%	3.286* ± 1.664
SM-Shape	0.715* ± 0.054	0.086 ± 0.010	0.515* ± 0.208%	3.212* ± 0.785	0.868* ± 0.080	0.075* ± 0.007	0.110* ± 0.001%	1.854* ± 0.594
CAR (Ours)	**0.743** ± 0.055	0.083 ± 0.009	0.477 ± 0.172%	**3.040** ± 0.681	**0.898** ± 0.068	0.047 ± 0.005	**0.002** ± 0.001%	**1.514** ± 0.505

Evaluation Metrics. We evaluated the registration accuracy and the deformation regularity of the registration. The registration accuracy was measured with Dice score between the segmentation of the warped moving image and the fixed image. To quantify the registration accuracy of the boundary of the structure of interest, we also measure the 95% of the Hausdorff distance (HD95). We also evaluated the extent of extreme deformations, i.e., folding, by computing the ratio of points with negative Jacobian determinant ($J_{<0}\%$), and the smoothness of the deformation by computing the magnitude of the spatial gradient of Jacobian determinant $|\nabla_J|$ [17].

Implementation Details. The backbone network has two Siamese encoders and a decoder. The channel dimension of each convolutional layer in the two encoders is 128, and 256 for the decoder. The projection head for contrast-invariant latent regularization consists of a 1×1 convolution layer with the output channel set to 32. The random convolution-based contrast augmentation consists of 4 RC layers, where the kernel weights of each layer in Sect. 2.1 are sampled from a uniform distribution $U(0, 10)$. The kernel weights are reshifted to be zero-centered after sampling. A LeakyReLu activation with a negative slope of 0.2 is added after each RC layer. Experiments were conducted using Adam Optimizer on an NVIDIA A5000 GPU with a batch size of 16. The initial learning rate was set to be 1×10^{-4} and started to decay to 1×10^{-5} from the 5th epoch for convergence.

Table 2. Quantitative results of the generalizability experiment. The models are trained on CMRxRecon Cine data while tested on CMRxRecon T1 Mapping data. * represents for values that CAR significantly outperformed with p-value < 0.01 in a paired t-test.

Methods	Dice ↑	$\lvert\nabla_J\rvert$ ↓	$J_{<0}\%$ ↓	HD95 ↓
Unregistered	0.815 ± 0.066	–	–	2.445 ± 0.562
VXM-LNCC	$0.808^* \pm 0.098$	0.080 ± 0.014	$0.950^* \pm 0.287\%$	$2.471^* \pm 0.890$
VXM-MIND	$0.730^* \pm 0.169$	$0.132^* \pm 0.030$	$1.678^* \pm 1.238$	$3.492^* \pm 1.903$
MIDIR	$0.721^* \pm 0.109$	$0.135^* \pm 0.027$	$0.918^* \pm 0.538\%$	$3.489^* \pm 1.889$
SM-Shape	$0.868^* \pm 0.080$	$\mathbf{0.075} \pm 0.007$	$0.110^* \pm 0.001\%$	$1.854^* \pm 0.594$
CAR (Ours)	$\mathbf{0.893} \pm 0.067$	0.080 ± 0.008	$\mathbf{0.043} \pm 0.001\%$	$\mathbf{1.567} \pm 0.485$

Comparison Study. CAR was first compared with a traditional iterative registration method SyN [2] with mutual information and a default Gaussian smoothing of 3 and three scales with 180, 80, 40 iterations, respectively following the setting of TransMorph [4]. We then compared CAR with SOTA learning-based methods, including VoxelMorph [3] using LNCC (VXM-LNCC) and MIND [9] (VXM-MIND) as the dissimilarity metrics respectively, a multi-contrast mutual information-based registration approach MIDIR [17], and a contrast-agnostic registration framework SynthMorph [10]. Note that the channel dimensions of all convolutional layers in VXM-LNCC, VXM-MIND, and SM-shape are set to 256 for a fair comparison with CAR. VoxelMorph and MIDIR were trained using all the available contrasts, i.e., both T1w and T2w images in brain MRI and all nine T1 weighting images in cardiac MRI. Synthetic images used for training SynthMorph were generated based on random label maps (SM-Shape). In contrast, CAR used only single contrast images for training, i.e., T1w images in brain MRI and TI_1 images in cardiac MRI.

A quantitative comparison result on both brain MRI and cardiac MRI is summarized in Table 1. For the cardiac registration task, the result is reported for registering misaligned images of $TI_{i=2:9}$ to TI_1. It can be observed that CAR achieves the best registration accuracy while possessing good deformation regularity for both tasks. For the brain MRI registration task, CAR demonstrates superior generalization ability as CAR only used T1w images for training but performed the best on the T1w-T2w registration task for all learning-based approaches in terms of registration accuracy and deformation regularity. This is similar to the cardiac registration task when our model was trained using only

TI_1 image but can be generalized to all other contrasts of images in T1 Mapping data. It can be noted that CAR has the lowest folding ratio. This is because we are dealing with nine different weightings of T1 images and the learned contrast-invariant representations may help the network get a more stable solution from images with different contrasts. Fig. 2 and Fig. 3 also show the qualitative comparison results of brain and cardiac registration tasks respectively, demonstrating that CAR can achieve low registration error and smooth deformation field. Both figures also show that CAR can register more fine-grained details than baseline approaches, which are shown in the highlighted regions.

To further evaluate the generalizability of CAR, we also trained our model and baseline methods on CMRxRecon Cine MRI data and evaluated model performance on the multi-contrast T1 Mapping data (also registering misaligned images of $TI_{i=2:9}$ to TI_1). The results are shown in Table 2. It can be shown all other methods except CAR and SM-shape failed to register the image with contrast they had not seen during training. In contrast, only CAR and SM-shape can still achieve good registration results. However, CAR has a significant increase in dice score and better deformation regularity than SM-shape, which demonstrates CAR's superior generalizability.

Table 3. Ablation Studies on the CamCAN dataset. * represents for values that CAR significantly outperformed with p-value < 0.01 in a paired t-test.

Methods	Dice ↑	$\|\nabla_J\|$ ↓	$J_{<0}\%$ ↓	HD95 ↓
CAR w/o CLR	0.733* ± 0.051	0.082 ± 0.008	0.504* ± 0.158%	3.089* ± 0.641
CAR with RC-3	0.714* ± 0.054	**0.082** ± 0.010	0.484 ± 0.186%	3.129* ± 0.659
CAR with InfoNCE	0.734* ± 0.057	0.084 ± 0.010	0.501* ± 0.191%	3.053 ± 0.642
CAR (Proposed)	**0.743** ± 0.055	0.083 ± 0.009	**0.477** ± 0.172%	**3.040** ± 0.681

Ablation Studies. To verify the effectiveness of CAR, we conducted ablation studies on the proposed random convolution-based contrast augmentation and CLR on the CamCAM datasets. We tested the effect of the random convolution kernel sizes on the model performance, where RC-3 represents that the kernel size is 3. We also replaced our proposed contrast invariance loss with InfoNCE loss [5] in contrastive learning to study its effectiveness. The results are summarized in Table 3. We can observe that any missing components of our model can lead to an inferior model performance from the results. Without CLR, the registration accuracy gets lower with a higher folding rate. Similar results can also be observed when replacing the proposed contrast invariance loss with InfoNCE loss in [5]. This means that CLR not only facilitates contrast-invariant representation learning but also improves the deformation regularity by producing a more plausible deformation field. Besides, a larger kernel size in the random convolution layer also leads to a performance drop. This is because a larger kernel size would lead to a blurring effect for the images, which can damage the voxel-to-voxel correspondences and therefore harm the registration task.

4 Conclusion

We present a novel contrast-agnostic deformable image registration framework (CAR) that can register arbitrary contrasts of images without observing them during training. To achieve this, we propose a random convolution-based approach to simulate variations of contrasts and develop a contrast invariance loss to regularize the feature representation in latent space for contrast-invariant feature representation learning. Experiments showed that on both tasks, CAR outperformed the state-of-the-art methods and demonstrated superior generalizability although CAR is contrast-agnostic. For future work, we will extend CAR to 3D applications as well as to more general multi-modal registration tasks.

Acknowledgement. This work was partially supported by the Engineering and Physical Sciences Research Council [grant number EP/Y002016/1]. X. Luo was supported by the Engineering and Physical Sciences Research Council [grant number EP/X039277/1].

References

1. Achille, A., Soatto, S.: Emergence of invariance and disentanglement in deep representations. The Journal of Machine Learning Research **19**(1), 1947–1980 (2018)
2. Avants, B.B., Epstein, C.L., Grossman, M., Gee, J.C.: Symmetric diffeomorphic image registration with cross-correlation: evaluating automated labeling of elderly and neurodegenerative brain. Medical image analysis **12**(1), 26–41 (2008)
3. Balakrishnan, G., Zhao, A., Sabuncu, M.R., Guttag, J., Dalca, A.V.: VoxelMorph: a learning framework for deformable medical image registration. IEEE Trans. Med. Imaging **38**(8), 1788–1800 (2019)
4. Chen, J., Frey, E.C., He, Y., Segars, W.P., Li, Y., Du, Y.: TransMorph: transformer for unsupervised medical image registration. Med. Image Anal. **82**, 102615 (2022)
5. Chen, T., Kornblith, S., Norouzi, M., Hinton, G.: A simple framework for contrastive learning of visual representations. In: International Conference on Machine Learning, pp. 1597–1607. PMLR (2020)
6. Dey, N., Schlemper, J., Salehi, S.S.M., Zhou, B., Gerig, G., Sofka, M.: ContraReg: contrastive learning of multi-modality unsupervised deformable image registration. In: Wang, L., Dou, Q., Fletcher, P.T., Speidel, S., Li, S. (eds.) Medical Image Computing and Computer Assisted Intervention - MICCAI 2022, MICCAI 2022, LNCS, vol. 13436, pp. 66–77. Springer, Cham (2022). https://doi.org/10.1007/978-3-031-16446-0_7
7. Dice, L.R.: Measures of the amount of ecologic association between species. Ecology **26**(3), 297–302 (1945)
8. He, K., Zhang, X., Ren, S., Sun, J.: Deep residual learning for image recognition. In: Proceedings of the IEEE Conference on Computer Vision and Pattern Recognition, pp. 770–778 (2016)
9. Heinrich, M.P., et al.: Mind: modality independent neighbourhood descriptor for multi-modal deformable registration. Med. Image Anal. **16**(7), 1423–1435 (2012)
10. Hoffmann, M., Billot, B., Greve, D.N., Iglesias, J.E., Fischl, B., Dalca, A.V.: SynthMorph: learning contrast-invariant registration without acquired images. IEEE Trans. Med. Imaging **41**(3), 543–558 (2021)

11. Luo, X., Zhuang, X.: $\mathcal{X}$-metric: an n-dimensional information-theoretic framework for groupwise registration and deep combined computing. IEEE Trans. Pattern Anal. Mach. Intell. **45**(7), 9206–9224 (2023)
12. Maes, F., Collignon, A., Vandermeulen, D., Marchal, G., Suetens, P.: Multimodality image registration by maximization of mutual information. IEEE transactions on Medical Imaging **16**(2), 187–198 (1997)
13. Maes, F., Vandermeulen, D., Suetens, P.: Medical image registration using mutual information. Proceedings of the IEEE **91**(10), 1699–1722 (2003)
14. Mahapatra, D., Antony, B., Sedai, S., Garnavi, R.: Deformable medical image registration using generative adversarial networks. In: 2018 IEEE 15th International Symposium on Biomedical Imaging (ISBI 2018), pp. 1449–1453. IEEE (2018)
15. Ouyang, C., Chen, C., Li, S., Li, Z., Qin, C., Bai, W., Rueckert, D.: Causality-inspired single-source domain generalization for medical image segmentation. IEEE Transactions on Medical Imaging **42**(4), 1095–1106 (2022)
16. Qin, C., Shi, B., Liao, R., Mansi, T., Rueckert, D., Kamen, A.: Unsupervised deformable registration for multi-modal images via disentangled representations. In: Chung, A., Gee, J., Yushkevich, P., Bao, S. (eds.) Information Processing in Medical Imaging, IPMI 2019, LNCS, vol. 11492, pp. 249–261. Springer, Cham (2019). https://doi.org/10.1007/978-3-030-20351-1_19
17. Qiu, H., Qin, C., Schuh, A., Hammernik, K., Rueckert, D.: Learning diffeomorphic and modality-invariant registration using b-splines. In: Medical Imaging with Deep Learning (2021)
18. Ronchetti, M., Wein, W., Navab, N., Zettinig, O., Prevost, R.: DISA: differentiable similarity approximation for universal multimodal registration. In: Greenspan, H., et al. Medical Image Computing and Computer Assisted Intervention - MICCAI 2023, MICCAI 2023, LNCS, vol. 14229, pp. 761–770. Springer, Cham (2023). https://doi.org/10.1007/978-3-031-43999-5_72
19. Shafto, M.A., Tyler, L.K., Dixon, M., Taylor, J.R., Rowe, J.B., Cusack, R., Calder, A.J., Marslen-Wilson, W.D., Duncan, J., Dalgleish, T., et al.: The cambridge centre for ageing and neuroscience (cam-can) study protocol: a cross-sectional, lifespan, multidisciplinary examination of healthy cognitive ageing. BMC neurology **14**, 1–25 (2014)
20. Sideri-Lampretsa, V., Zimmer, V.A., Qiu, H., Kaissis, G., Rueckert, D.: MAD: modality agnostic distance measure for image registration. In: Woo, J., et al. (ed.) Medical Image Computing and Computer Assisted Intervention - MICCAI 2023 Workshops, MICCAI 2023, LNCS, vol. 14394, pp. 147–156. Springer, Cham (2023). https://doi.org/10.1007/978-3-031-47425-5_14
21. Taylor, J.R., et al.: The cambridge centre for ageing and neuroscience (cam-can) data repository: structural and functional MRI, meg, and cognitive data from a cross-sectional adult lifespan sample. Neuroimage **144**, 262–269 (2017)
22. Wang, C., Li, Y., Lv, J., Jin, J., Hu, X., Kuang, X., Chen, W., Wang, H.: Recommendation for cardiac magnetic resonance imaging-based phenotypic study: imaging part. Phenomics **1**, 151–170 (2021)
23. Wang, C., et al.: Cmrxrecon: an open cardiac MRI dataset for the competition of accelerated image reconstruction. arXiv preprint arXiv:2309.10836 (2023)
24. Wang, X., Zhang, R., Shen, C., Kong, T., Li, L.: Dense contrastive learning for self-supervised visual pre-training. In: Proceedings of the IEEE/CVF Conference on Computer Vision and Pattern Recognition, pp. 3024–3033 (2021)
25. Xu, Z., Liu, D., Yang, J., Raffel, C., Niethammer, M.: Robust and generalizable visual representation learning via random convolutions. In: International Conference on Learning Representations (2021)

Efficiency

High Performance Groupwise Cortical Surface Registration with Multimodal Surface Matching

Renato Besenczi(✉), Yourong Guo, and Emma C. Robinson

Research Department of Biomedical Computing, School of Biomedical Engineering & Imaging Sciences, King's College London, London, UK
{renato.besenczi,yourong.guo,emma.robinson}@kcl.ac.uk

Abstract. Improving the accuracy of cortical surface registration while preserving individual characteristics is a difficult task, since cortical shape varies significantly across different individuals. In this paper we propose an extension of Multimodal Surface Matching (MSM) towards a hierarchical groupwise framework to address these challenges. Central to the success of this framework has been re-implementation of MSM to improve its efficiency. Experiments show that the proposed method supports unbiased and efficient alignment of large groups of up to 250 subjects, while significantly improving the alignment of cortical folds, relative to alignment to a single population average template. This implies significant advantages with regards to the search for subtle cortical signatures of cognition and disease. The code can be found at https://github.com/rbesenczi/newMSM.

Keywords: Cortical surface registration · Groupwise registration · Individual characteristics · High-performance · MSM

1 Introduction

Performing neuroimaging analyses on the surface of the human cerebral cortex has been proven imperative for the mapping of its morphological [7,16,25] and functional organisation [14,15,29], which then provides a framework for investigating how this is genetically encoded [3,8,37] and differentially expressed across neurological [5,39] and psychiatric [12,30] phenotypes. To directly compare cortical features across populations, surfaces must first be spatially normalised. During this process input surfaces are typically continuously deformed until they overlap, in some sense, with a reference image, which is historically generated from a global population average. As this problem is ill-posed, classic frameworks constrain deformations with biomechanically-inspired regularisation, to enforce diffeomorphisms, which prohibit folding or tearing of the deformation grid. Human brains, however, vary across individuals in ways that break these

R. Besenczi and Y. Guo—These authors contributed equally to this paper.

M. Modat et al. (Eds.): WBIR 2024, LNCS 15249, pp. 321–332, 2024.
https://doi.org/10.1007/978-3-031-73480-9_25

assumptions [2,14,36]; resulting in spatial misalignments, which reduce the sensitivity of studies to detect and model subtle signatures of cognition and disease.

One approach that has been popularly used to address the problem of cortical folding variation in neuroimage processing, has been to first cluster examples that share common shape variants. In this way, label fusion methods [20,40] have been used to improve cortical segmentation by propagating tissue labels from more similar brain atlases, and hierarchical groupwise image registration frameworks [1,6,22,28,38,42,44] have been used to align within and between clusters of brain images. Such frameworks build upon historical groupwise registration frameworks [4,23,34], which have repeatedly shown that simultaneous registration of inputs reduces the bias of templates to any single reference image, and improves the sharpness of the final population average image. Thus far the majority of groupwise registration frameworks have been implemented for brain volumes, despite strong evidence that alignment of cortical morphology and function can be improved through surface-driven alignment [10,11,14,32,33,35,43,45].

In this paper, we propose a fast and efficient groupwise cortical surface alignment framework, via extension and re-implementation of Multimodal Surface Matching (MSM) [32,33]. We validate this for alignment of the young adult Human Connectome Project (HCP) [13] to show that this framework significantly improves the accuracy and efficiency of alignment (relative to typical MSM) while also improving the sensitivity of downstream statistical analyses. The main contributions are the re-implementation and extension of MSM, development of a framework for fast and efficient groupwise and cross-group surface registration that levarage the advantages of the hierarchical clustering method, and fine-tuning the parameters of registration to achieve the best comparable results for validation of the framework.

2 Cortical Surface Registration

With the exception of a number of papers focused on spectral alignment of cortical shape [27,41] or function [19,26] cortical surface registration is typically performed following projection of all cortical anatomies to the sphere [10,32,33,43]. This offers specific advantages as it simplifies registration from that of mapping complex shapes in 3D, to moving points around on a smooth, 2D spherical surface, until features of cortical organisation better overlap between moving (M) and fixed (F) images.

Although shape remains the most commonly used feature for driving cortical alignment, spherical matching itself is agnostic to the choice of input, allowing for registration to be driven using a broader range of features. This motivated the development of MSM [32,33], which supports registration of any combination of multi-modal feature sets through leveraging a discrete optimisation framework:

$$E(G) = \min \sum_{\mathbf{p} \in G} \rho_p(l_p) + \lambda \sum_{\mathbf{p} \in G} \sum_{\mathbf{q},\mathbf{r} \in N(p)} W_{pqr}(l_p, l_q, l_r) \quad (1)$$

In this formulation, transformations ($T : M \rightarrow F$) are regularised through optimising displacements of low-resolution control-point (CP) grids (G), formed from regular tessellations of icosahedrons. At each iteration, each CP vertex ($\mathbf{p} \in G$) may only deform to one of a finite set of d target locations, indicated by labels ($l_p = \{1, 2....d\}$). Each label represents an equivalent deformation for each CP, implemented as a rotation of that point across the surface such that $T(\mathbf{p}) = \mathbf{R}^{l_p}\mathbf{R}'_p\mathbf{p}$, where, $\mathbf{R}'_p$ represents the aggregation of rotations across previous iterations.

The cost function (Eq. 1) balances a data similarity term, $\rho_p(l_p)$, which estimates the overlap of features ($N(p)$) in $T \circ M$ and F for that proposed move, with a regularisation penalty, $W_{pqr}(l_p, l_r, l_s)$, which penalises the hyper-elastic strain energy of the resulting deformation [32]. Calculation of $\rho_p(l_p)$ involves upsampling of the proposed warp to vertices ($\mathbf{v}$) of the moving mesh, after projecting $\mathbf{v}$ into the plane of the mesh face triangle defined by $\mathbf{p}, \mathbf{q}, \mathbf{r}$, implemented through barycentric interpolation:

$$T(\mathbf{v}) = \eta(\mathbf{v}, \mathbf{p}; \mathbf{q}, \mathbf{r})\mathbf{R}^{l_p}\mathbf{R}'_p\mathbf{p} + \eta(\mathbf{v}, \mathbf{q}; \mathbf{r}, \mathbf{p})\mathbf{R}^{l_q}\mathbf{R}'_q\mathbf{q} + \eta(\mathbf{v}, \mathbf{r}; \mathbf{p}, \mathbf{q})\mathbf{R}^{l_r}\mathbf{R}'_r\mathbf{r},$$

Here $\eta(\mathbf{v}, \mathbf{p}; \mathbf{q}, \mathbf{r}) = \frac{A_{vqr}}{A_{pqr}}$ and A_{vqr} is the area of triangle defined by $\mathbf{v}, \mathbf{q}, \mathbf{r}$. Most often Pearson's correlation is used to measure overlap and similarity is calculated between patches of corresponding vertices that surround each CP. The regularisation term $W_{pqr}(l_p, l_r, l_s)$ constraints the energy of warps as

$$W_{pqr} = \mu/2(R^k + R^{-k} - 2) + \kappa/2(J^k + J^{-k} - 2)$$

where $k \in \mathbb{N}$ is k-exponent, μ is shear modulus that penalizes changes in shape, κ is bulk modulus that penalises changes in size. $J = \lambda_1\lambda_2$ represents relative areal change, while $R = \lambda_1/\lambda_2$ is relative shape change, where λ_1 and λ_2 are principal in-plane stretches after projection of the deformation gradient of the mesh face to the tangent plane. Costs are first calculated separately for each proposed clique (unary for the data term with triplets for regularisation); then the optimal combination of labels is found through solving with fast primal-dual (FastPD) [24] optimisation of the Markov random field (MRF) energy (Eq. 1) - leveraging Higher-order Clique Reduction (HOCR) [21] to reduce triclique regularisation terms to paired cliques in order to support efficient optimisation. This process is repeated over multiple iterations for each CP grid, and registration is implemented in a coarse-to-fine fashion: propagating mappings from low-resolution icospheres (ico2 - 162 vertices) to high-resolution icospheres (to ico4 or ico5 with 2562 or 10,242 vertices) over a series of resolution levels.

More recently a number of learning-based approaches have been proposed for driving spherical alignment, including s3Reg [46], SUGAR [31], DDR [35]. These leverage a range of different surface convolutional frameworks to learn fast, unsupervised mappings of cortical surfaces. Of these deep-discrete registration (DDR) [35] is most similar to MSM in that it optimises discrete displacements through framing the mapping problem as a label classification problem, regularised with a deep conditional random field (CRF).

3 Method

Following [17,34], we extend the existing MSM framework to perform groupwise registration by simultaneously co-registering clusters of surfaces, which share common modes of cortical variation (in this case folding):

$$\begin{aligned} E(G_1,\ldots,G_n) = \min \sum_{i} \sum_{\mathbf{p}\in G_i} \sum_{\substack{\mathbf{q}\in \\ (N(\mathbf{p})\setminus G_i)}} \rho_{\mathbf{pq}}(l_{\mathbf{p}}, l_{\mathbf{q}}) \\ + \sum_{i} \sum_{\mathbf{p}\in G_i} \sum_{\mathbf{rs}\in N(p)} \lambda W_{\mathbf{pqr}}(l_{\mathbf{p}}, l_{\mathbf{r}}, l_{\mathbf{s}}) \end{aligned} \tag{2}$$

Let's say, there exist n moving meshes (M_i $(i = 1 \ldots n)$) for any given cluster; each input surface is therefore associated with its own low-resolution CP grid G_i. For each iteration, the groupwise registration must choose the optimal discrete displacement for all G_i by comparing the overlap of features, across all inputs, for the set of proposed moves $\{T_1(G_1), \ldots T_n(G_n)\}$. For each G_i, this involves calculating the similarity ($\rho_{\mathbf{pq}}(l_{\mathbf{p}}, l_{\mathbf{q}})$) between overlapping patches of features $N(\mathbf{p})$, defined around each of its CP vertices $\mathbf{p}$, and all overlapping patches defined for neighbouring vertices ($\mathbf{q}$) *in all other CP grids* ($\mathbf{q} \in (N(\mathbf{p}) \setminus G_i)$. These are defined by propagating the proposed mapping to the resolution of the input feature maps as $T_i(M_i)$, and resampling all data to a common reference frame (F) to calculate patch overlap. Here F is defined from a high resolution icospheric tessellation (ico6 - 40,962 vertices), and features are resampled using adaptive barycentric interpolation [13]. Regularisation is then implemented by calculating the hyper-elastic strain of every proposed move on all subjects. (Fig. 1)

After groupwise registration, all subjects are registered to the top level (root node) of the hierarchy to make population level analysis and comparison possible in the same space. This is performed by recursively registering all subject in a group to the group average template of the other group of the parent node using typical MSM. This process runs up the hierarchy until all subjects are registered the group average of the top level node (i.e. population average, therefore all examples are in the same space).

3.1 Implementation of Groupwise Mode of MSM

Several features of the original implementation of MSM made it impractical for extension to a groupwise registration, including inefficient code, slow nearest-point search, reduced efficiency of the MRF optimisation (due to the cost of splitting the higher-order MRF into a series of pairwise problems) and single-threaded implementation. This resulted in extremely long runtimes for typical MSM, necessitating the re-implementation of interpolation libraries and calculation of costs, together with re-factorisation of the entire code base. Interpolation was updated from a minimal distance search, which iterated through all vertices, to octree search [9], which recursively subdivides each mesh into eight octants, terminating at the point when each octant has 50 vertices or less. This defines

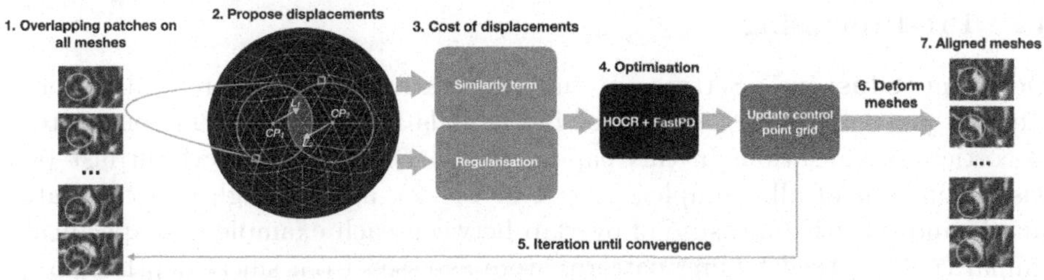

Fig. 1. Overview of the groupwise registration process. The input of the method is pairs of overlapping patches on all meshes across the group (step 1). These patches are moved around the labels (2), then the calculated costs of all displacements (3) are minimised by the optimisation method (4). Then all CP grids are updated. This iteration runs until convergence (5), then all meshes are deformed (6) according to the results of the optimisation. The aligned meshes (7) then resampled to the output space and saved. As an example, yellow circles highlight patches that are aligned during the process from step 1 to 7. Blue and red circles denote patches on different subjects, similarity costs are calculated on the overlapping area of patches. (Color figure online)

nearest neighbour search over a tree structure, significantly reducing time complexity of one search from $\mathcal{O}(n)$ to $\mathcal{O}(\log_8 n)$, where n is the number of vertices of the mesh. In the original implementation, similarity costs were calculated by separately rotating each control point to each label point; this was updated to rotate the whole mesh once, and performing interpolation once, for each choice of label. This resulted in an approximately 1GB on memory usage per subject per feature. Finally, major performance boosts were achieved through wider use of C++ Standard Template Library (STL) tools, and the introduction of multi-threading for interpolation and clique energy calculation (using the OpenMP library). The previous version of MSM has been reported to perform an experimental 10 subject groupwise registration in several days whilst newMSM can perform the same task within 30 min. This significant performance improvement has also made possible to perform groupwise registration with up to 250 examples in a single groupwise MSM process.

4 Experiments

4.1 Data

A total of 1110 subjects from young adult HCP were included in the current work (participant age between 22–35). The dataset underwent HCP processing pipelines for spatial artifact and distortion removal, alignment to standard space, cross-modal registration, and surface reconstruction. Full details of the HCP processing pipeline have been described in [13]. All left hemispheres are aligned to the right, doubling the examples. Sulcal depth and curvature metrics were used as features to drive the groupwise registration, with their contributions weighted equally during alignment.

4.2 Pre-Processing

Defining Clusters. Setting up the registration as a hierarchical problem, requires pre-definition of groups of individuals which share common patterns of cortical organisation. In this paper, we use DDR [35] to perform fast pairwise alignment of all examples (step 1 in Fig. 2), and through this calculate a simple and intuitive measure of overlap between each example (2). Specifically, similarity of cortical folding patterns were assessed from sulcal depth maps by calculating the correlation, and DICE overlap of gyral crowns. Next, agglomerative hierarchical clustering was used to group subjects with similar cortical folding patterns (3). The resulting hierarchy was thresholded at a level to return 50 clusters (4), cluster sizes between 3–71 examples. Clusters with less than 10 examples are excluded, resulting in a total of 2038 examples across clusters.

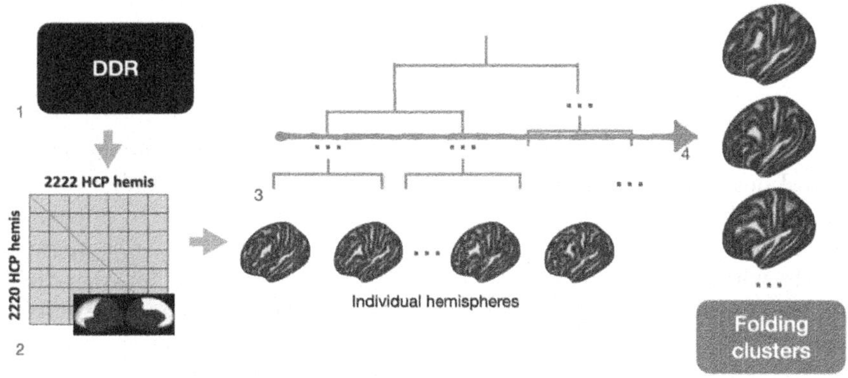

Fig. 2. Defining folding clusters. DDR [35] was used for fast pairwise alignment of all examples (1). Calculating the correlation, and DICE overlap of gyral crowns from sulcal depth maps (2). Hierarchical clustering (3) revealed 50 folding clusters (4). (See further explanation in Sect. 4.2.)

The sulcal depth and curvature metrics were resampled to icosphere of order 6 using adaptive barycentric interpolation. First, the rigid rotation part of DDR was trained between all HCP hemispheres to the HCP population sulcal depth template to estimate the 12 parameters of the rotation matrix. Then the deformation part was trained using 615,495 pairs of hemispheres in HCP, divided into 5 folds for cross-validation. A sum of MSE and cross-correlation loss optimised the network. Predictions from the 5 folds were averaged to generate the mean deformed meshes.

4.3 Validation

We validate the groupwise MSM against the typical registration to a population average space (typical MSM). In order to achieve a fair comparison, distortions were controlled at a similar level, and k-exponent was $k = 2$ in both cases.

To compare the ability to capture the individual variability, we used the average cross-correlation (CC) similarity and DICE overlap ratio of gyral crowns between all pairs of data within the clusters. The average CC and DICE before and after the between-group registration were also compared, intending to validate the ability to keep the topographic feature of specific clusters. The mean and standard deviation maps of deformed sulcal depth and curvature within folding clusters were compared before and after the between-group registration with the typical MSM; their sharpness and uncertainty were compared through visual inspection.

We also validated the recent newMSM version (version 0.6.3-BETA) by comparing its output with the output of the previous MSM implementation. We randomly selected 100 hemispheres from the HCP dataset, registered them to the HCP global average template (typical MSM) with the same configuration parameters and reported error by mean squared error (MSE). Also, we compared average runtimes of single and multithreaded newMSM and previous MSM (only single thread execution available) in the same environment.

4.4 Results

Evaluation of newMSM relative to the original implementation is shown in Table 1 and Fig. 3. This demonstrates that newMSM achieves comparable alignment (MSE=0.0069) in much less time than the original implementation.

Table 1. Runtime (minutes) comparison of different MSM versions (number of execution threads in parenthesis).

	Runtime (1)	Runtime (16)
newMSM	35.98±29.64	9.17±3.84
previous MSM	74.72±64.49	NA

Three representative groups with different sizes (identified as NODE2117 (with 71 examples), NODE1988 (31), and NODE2138 (14)) were chosen to evaluate the performance of groupwise MSM relative to registration to a single population average (Fig. 4). We also evaluate performance after cross-group registration on the top level of the hierarchy, meaning all subjects in the population of the study (identified as NODE2218 (2083)). Results show that groupwise MSM generates much sharper cluster averages with much reduced variance. Table 2 shows that this increases CC and DICE overlap across all individuals ($p<0.005$), without increasing warp distortion (Fig. 5).

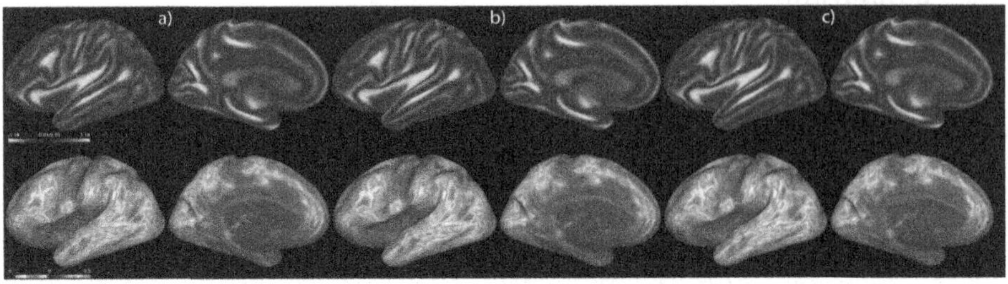

Fig. 3. Sulcal depth means (first row) and variance maps (second row) of output of a) newMSM (16 thread), b) newMSM (1 thread), and c) previous implementation of MSM showing negligible differences in registration between implementations with MSE=0.0069.

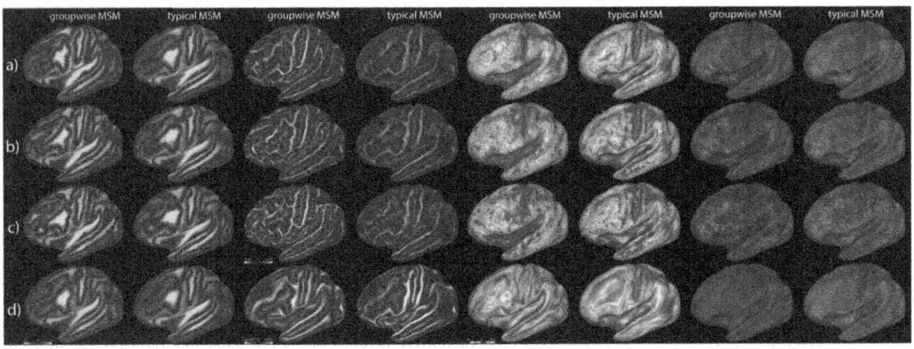

Fig. 4. Mean and variance maps for each types of registration for three groups (row a) NODE2117 (71 examples), b) NODE1988 (31), c) NODE2138 (14)) and the top of the hierarchy (row d) NODE2218 (2083)).

Table 2. Similarity metrics and distortion of selected groups NODE2117, NODE1988, NODE2138 and top level NODE2218.

Group	Method	Sulcal Depth		Curvature		Areal distortion	
		CC	DICE	CC	DICE	mean	95%
NODE2117	gw MSM	0.7897	0.6682	**0.4942**	**0.5471**	0.2598	0.5799
	typ MSM	0.7738	0.6506	0.2699	0.4215	0.2721	0.6164
NODE1988	gw MSM	0.8020	0.6725	**0.5385**	**0.5703**	0.2622	0.5884
	typ MSM	0.7861	0.6522	0.2929	0.4309	0.2677	0.622
NODE2138	gw MSM	0.8019	0.6748	**0.5613**	**0.5586**	0.2590	0.5861
	typ MSM	0.7739	0.6397	0.2864	0.4253	0.2584	0.5964
NODE2218 (Top node)	gw MSM	0.8059	0.6768	**0.3562**	**0.4441**	0.1607	0.3596
	typ MSM	0.7266	0.6103	0.2453	0.4063	0.1646	0.3655

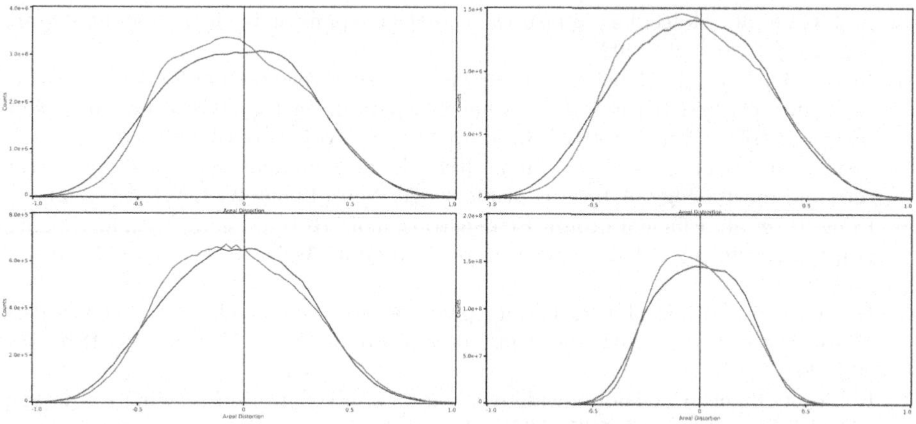

Fig. 5. Areal distortion in selected groups (NODE2117 top left, NODE1988 top right, NODE2138 bottom left) and top level (NODE2218 bottom right) comparing groupwise MSM (red) with typical registration (blue). (Color figure online)

5 Conclusions

In this paper, we propose a high performance cortical surface registration method combined with a hierarchical clustering of examples from the young adult HCP dataset. Our results show that the groupwise mode of newMSM significantly outperforms typical individual registration in terms of cross-correlation similarity and DICE overlap ratio of gyral crowns on the same level of distortion, while also preserves individual cortical signatures. In addition, we could improve the computational efficiency of MSM that made possible to implement groupwise registration and to run typical MSM registration much quicker. This work has important potential applications, including uncovering cortical folding patterns [17], studying cortical asymmetry [18], and broader implications for functional alignment and neuroimaging analyses. We should note, that the compared methods are different since typical MSM registration uses a blurry global average template. Spatial normalisation with templates may introduce bias that could remove details. Future work will involve further experiments for functional alignment and extensive downstream analysis using more subjects from the UK Biobank dataset. Also, we anticipate further performance improvements after replacing the optimiser libraries.

References

1. Ahmad, S., et al.: Deep learning deformation initialization for rapid groupwise registration of inhomogeneous image populations. Frontiers in Neuroinformatics **13**(May), 1–12 (2019)
2. Amunts, K., et al.: Brodmann's areas 17 and 18 brought into stereotaxic space-where and how variable? Neuroimage **11**(1), 66–84 (2000)

3. Ball, G., et al.: Molecular signatures of cortical expansion in the human fetal brain. bioRxiv pp. 2024–02 (2024)
4. Bhatia, K.K., et al.: Consistent groupwise non-rigid registration for atlas construction. In: 2004 2nd IEEE International Symposium on Biomedical Imaging: Nano to Macro (IEEE Cat No. 04EX821). pp. 908–911 Vol. 1. IEEE (2004)
5. Dimitrova, R., et al.: Preterm birth alters the development of cortical microstructure and morphology at term-equivalent age. NeuroImage **243**, 118488 (2021)
6. Dong, P., et al.: Efficient groupwise registration of MR brain images via hierarchical graph set shrinkage. Med. Image Comput. Comput. Assist. Interv. **11070**, 819–826 (Sep 2018)
7. Duan, D., et al.: Exploring folding patterns of infant cerebral cortex based on multi-view curvature features: Methods and applications. Neuroimage **185**, 575–592 (2019)
8. Elliott, L.T., et al.: Genome-wide association studies of brain imaging phenotypes in UK Biobank. Nature **562**(7726), 210–216 (2018)
9. Elseberg, J., et al.: Comparison of nearest-neighbor-search strategies and implementations for efficient shape registration. Journal of Software Engineering for Robotics **3**(1), 2–12 (2012)
10. Fischl, B., et al.: High-resolution intersubject averaging and a coordinate system for the cortical surface. Hum. Brain Mapp. **8**(4), 272–284 (1999)
11. Fischl, B., et al.: Cortical folding patterns and predicting cytoarchitecture. Cereb. Cortex **18**(8), 1973–1980 (2008)
12. Garrison, J.R., et al.: Paracingulate sulcus morphology is associated with hallucinations in the human brain. Nature communications **6**(1), 8956 (2015)
13. Glasser, M.F., et al.: The minimal preprocessing pipelines for the Human Connectome Project. Neuroimage **80**, 105–124 (2013)
14. Glasser, M.F., et al.: A multi-modal parcellation of human cerebral cortex. Nature **536**(7615), 171–178 (2016)
15. Gordon, E.M., et al.: Generation and evaluation of a cortical area parcellation from resting-state correlations. Cerebral cortex **26**(1), 288–303 (2016)
16. Guillon, L., et al.: Identification of rare cortical folding patterns using unsupervised deep learning. Imaging Neuroscience **2**, 1–27 (2024)
17. Guo, Y., Suliman, M.A., Williams, L., Glasser, M.F., O'Muircheartaigh, J., Robinson, E.: Uncovering common variants of cortical folding through hierarchical surface registration. OHBM (2023)
18. Guo, Y., Williams, L., Glasser, M.F., Suliman, M.A., Hammers, A., Van Essen, D.C., O'Muircheartaigh, J., Robinson, E.: Uncovering the asymmetry of common temporal lobe folding variants. OHBM (2024)
19. Haxby, J.V., et al.: Hyperalignment: Modeling shared information encoded in idiosyncratic cortical topographies. Elife **9** (Jun 2020)
20. Heckemann, R.A., et al.: Automatic anatomical brain MRI segmentation combining label propagation and decision fusion. NeuroImage **33**(1), 115–126 (2006)
21. Ishikawa, H.: Higher-order clique reduction without auxiliary variables. In: Proceedings of the IEEE Conference on Computer Vision and Pattern Recognition. pp. 1362–1369 (2014)
22. Jia, H., et al.: Absorb: Atlas building by self-organized registration and bundling. NeuroImage **51**(3), 1057–1070 (2010)
23. Joshi, S., et al.: Unbiased diffeomorphic atlas construction for computational anatomy. NeuroImage **23**, S151–S160 (2004)

24. Komodakis, N., et al.: Fast, approximately optimal solutions for single and dynamic MRFs. In: 2007 IEEE Conference on Computer Vision and Pattern Recognition. pp. 1–8. IEEE (2007)
25. Kong, X.Z., et al.: Mapping cortical brain asymmetry in 17,141 healthy individuals worldwide via the ENIGMA Consortium. Proceedings of the National Academy of Sciences **115**(22), E5154–E5163 (2018)
26. Langs, G., et al.: Functional geometry alignment and localization of brain areas. Advances in neural information processing systems **23** (2010)
27. Lombaert, H., et al.: Diffeomorphic spectral matching of cortical surfaces. In: Information Processing in Medical Imaging: 23rd International Conference, IPMI 2013, Asilomar, CA, USA, June 28–July 3, 2013. Proceedings 23. pp. 376–389. Springer (2013)
28. Lyu, I., et al.: Robust estimation of group-wise cortical correspondence with an application to macaque and human neuroimaging studies. Frontiers in neuroscience **9**, 210 (2015)
29. Margulies, D.S., et al.: Situating the default-mode network along a principal gradient of macroscale cortical organization. Proceedings of the National Academy of Sciences **113**(44), 12574–12579 (2016)
30. Plaze, M., et al.: "Where do auditory hallucinations come from?"-a brain morphometry study of schizophrenia patients with inner or outer space hallucinations. Schizophrenia Bulletin **37**(1), 212–221 (2011)
31. Ren, J., et al.: Sugar: Spherical ultrafast graph attention framework for cortical surface registration. Medical Image Analysis p. 103122 (2024)
32. Robinson, E.C., et al.: Multimodal surface matching with higher-order smoothness constraints. Neuroimage **167** (2018)
33. Robinson, E.C., et al.: MSM: A new flexible framework for multimodal surface matching. Neuroimage **100**, 414–426 (2014)
34. Robinson, E.C., et al.: Discrete optimisation for group-wise cortical surface atlasing. In: Proceedings of the IEEE Conference on Computer Vision and Pattern Recognition Workshops. pp. 2–8 (2016)
35. Suliman, M.A., et al.: A Deep-Discrete learning framework for spherical surface registration. In: Medical Image Computing and Computer Assisted Intervention – MICCAI 2022. pp. 119–129. Springer Nature Switzerland (2022)
36. Thual, A., et al.: Aligning individual brains with fused unbalanced Gromov Wasserstein. Advances in Neural Information Processing Systems **35**, 21792–21804 (2022)
37. Van Der Meer, D., et al.: The genetic architecture of human cortical folding. Science advances **7**(51), eabj9446 (2021)
38. Wang, Q.: Groupwise registration based on hierarchical image clustering and atlas synthesis. Human brain mapping **31**(8), 1128–1140 (2010)
39. Williams, L.Z., et al.: Structural and functional asymmetry of the neonatal cerebral cortex. Nature Human Behaviour pp. 1–14 (2023)
40. Wolz, R., et al.: Leap: learning embeddings for atlas propagation. NeuroImage **49**(2), 1316–1325 (2010)
41. Wright, R., et al.: Construction of a fetal spatio-temporal cortical surface atlas from in utero MRI: Application of spectral surface matching. NeuroImage **120**, 467–480 (2015)
42. Wu, G., et al.: SharpMean: groupwise registration guided by sharp mean image and tree-based registration. Neuroimage **56**(4), 1968–1981 (Jun 2011)
43. Yeo, B.T.T., et al.: Spherical demons: fast diffeomorphic landmark-free surface registration. IEEE Trans. Med. Imaging **29**(3), 650–668 (2010)

44. Ying, S., et al.: Hierarchical unbiased graph shrinkage (HUGS): a novel groupwise registration for large data set. NeuroImage **84**, 626–638 (2014)
45. Zhao, F., et al.: S3Reg: Superfast spherical surface registration based on deep learning. IEEE Trans. Med. Imaging **40**(8), 1964–1976 (Aug 2021)
46. Zhao, F., et al.: S3Reg: superfast spherical surface registration based on deep learning. IEEE transactions on medical imaging **40**(8), 1964–1976 (2021)

Optimising Region of Interest Registration for Multiple-Tissue Whole Slide Images

Alessio Fiorin[1,2,3], Laia Adalid Llansa[1,2], Elena Goyda[1,2], Vincenzo Della Mea[4], Anna Korzynska[5], Shrief Abdelazeez[5], Ramon Bosch Príncep[1,2], Alba Fischer Carles[1,2,6](✉), Noelia Gallardo Borràs[1,2,7], Marylène Lejeune[1,2,3], Daniel Mata Cano[1,2], Domenec Puig[3], Hatem A. Rashwan[3], Esther Sauras Colón[1,2,7], Mikel Relloso Ortiz de Uriarte[1,2], Laia Reverté Calvet[1,2], and Carlos López Pablo[1,2,3]

[1] Oncological Pathology and Bioinformatics Research Group, Institut d'Investigació Sanitària Pere Virgili, Tortosa, Spain
alessio.fiorin@estudiants.urv.cat

[2] Department of Pathology, Hospital de Tortosa Verge de la Cinta, Institut Català de la Salut, Tortosa, Spain

[3] Department of Computer Engineering and Mathematics, Universitat Rovira i Virgili, Tarragona, Spain

[4] Department of Mathematics, Computer Science and Physics, University of Udine, Udine, Italy

[5] Laboratory of Processing and Analysis of Microscopic Images, Nalecz Institute of Biocybernetics and Biomedical Engineering, Polish Academy of Sciences, Warsaw, Poland

[6] BCN MedTech, Department of Engineering, Universitat Pompeu Fabra, Barcelona, Spain
alba.fischer01@estudiant.upf.edu

[7] Campus Terres de l'Ebre, Universitat Rovira i Virgili, Tortosa, Spain

Abstract. Digital pathology transforms clinical workflows by enabling the analysis of whole slide images (WSIs) on computers. While most methods focus on haematoxylin and eosin (H&E) stained WSIs, immunohistochemistry (IHC) is crucial for biomarker analysis, particularly for assessing tumour-infiltrating lymphocyte (TIL) subtypes inside a region of interest (ROI) selected by a pathologist using Salgado's criteria. This study first investigates the naive approach, which verifies a state-of-the-art WSI registration method's robustness for registering single-tissue H&E and multiple-tissue IHC WSIs. Then, to simplify this first attempt by accomplishing registration between single tissues, the study proposes two approaches: splitting the multiple-tissue IHC WSIs, which is considered the baseline, and a virtualised splitting with an incremental resolution optimisation-based technique. ROI registration predictions for

Supplementary Information The online version contains supplementary material available at https://doi.org/10.1007/978-3-031-73480-9_26.

M. Modat et al. (Eds.): WBIR 2024, LNCS 15249, pp. 333–345, 2024.
https://doi.org/10.1007/978-3-031-73480-9_26

TIL assessment will be assessed on IHC WSIs using standard metrics and one derived from a standard landmark-based metric popular in the image registration field. Existing image detection-inspired metrics for evaluating ROI proposals will be proposed and tuned to consider the global viewpoint, where the ROI proposals lie. This study aims to establish a reliable and time-efficient ROI registration procedure for WSIs with multiple stained tissues. This method would enable efficient selection of the ROI from the H&E WSI and potentially reduce the need for pathologist intervention through automatic quality control.

Keywords: Region of interest · Whole slide image · Registration · Multiple tissues · Tumour-infiltrating lymphocyte

1 Introduction

Histological assessment is crucial to the disease's diagnosis, prognosis and treatment phases. As scientific research advances, computational pathology significantly impacts traditional clinical workflows within pathology laboratories [1]. Nowadays, thanks to advances in computer vision in digital pathology, pathologists can analyse digitised slide images, called whole slide images (WSIs), through the aid of computer-assisted pathology (CAP) tools [2].

Currently, most CAP tools focus on analysing WSIs of tissues stained with haematoxylin and eosin (H&E) [3]. However, in the diagnostic workflow, tissues stained with immunohistochemistry (IHC) are also analysed as they allow the scoring of specific biomarkers. Given these two types of staining, it is possible to unlock new fields of research and clinical applications by integrating information between several differently stained tissues [1].

This combination of information from multiple WSIs may be spatially registered to be combined at the tissue level, and the WSI registration technique aids in doing this [1]. This registration process is challenging as different technical factors come into play, mainly the manipulation of gigapixel-scale data, changes in the structure and shape of the other stained tissues, and the presence of artefacts and tears due to the tissue preparation phase [1]. Thus, WSI registration of slides with different staining techniques helps identify patterns and spatial relations within a given specimen [1].

Indeed, this WSI registration process is helpful for an in-depth assessment of tumour-infiltrating lymphocytes (TILs), as it is not possible to know the positiveness of specific biomarkers for these lymphocytes, which may promote tumour progression or inhibit it, by using H&E WSIs alone [4]. Therefore, pathologists, to have a comprehensive landscape of the immune environment for breast cancer (BC), should not only take into account the evaluation of TILs but also the patterns and spatial relationships of immune cells related to the world of TILs, such as macrophages CD163+ and T lymphocytes CD8+ [5,6].

This spatial alignment of cuts with different stains may help recognise a specific area called the region of interest (ROI) between various cuts, which is the area where TILs are assessed. It begins with the pathologist, who identifies the ROI for assessing TILs in the WSI of the H&E-stained tissue through Salgado's criteria provided in [4]. Once this is done, using the WSI registration process, the exact ROI is automatically found in the remaining sections stained with IHC procedures to recognise the different biomarkers. By doing that, pathologists can make assessments of biomarkers at the spatial level and within the ROI, providing additional insights into the biomarkers's distribution. However, artefacts, tears, and additional tissues within the WSIs can complicate this registration process. As mentioned earlier, artefacts can be introduced during tissue processing, cutting, staining (e.g., pen marks, ink, tissue folds, dust, air bubbles), or during the scanning phase (e.g., out-of-focus areas of the WSIs) [7–9]. Multiple tissues on a single slide are not considered artefacts since they are an intentional practice to reduce procedural costs. This choice is made during the tissue preparation phase rather than a mistake or oversight in the procedure.

Once the challenges of WSI registration between different stains and the issue of multiple tissues in the same WSI have been introduced, the first aim of this work is to test the robustness of a state-of-the-art (SOTA) procedure for registration under these complexities. The choice of the SOTA technique fell on the Virtual Alignment of pathoLogy Image Series (VALIS) method [10] as it is not only one of the best WSI registration algorithm proposals in the ACROBAT challenge [1,11], but also claims to be stain agnostic, scalable and robust. Therefore, the initial test determines whether VALIS is robust enough to accurately register a pair of H&E and IHC WSIs despite the presence of extra tissues.

This work proposes two approaches to facilitating registration for the VALIS technique to assess this test with a more straightforward task. The first approach consists of splitting the multiple-tissue IHC WSI into several single-tissue IHC WSIs, while the second approach splits the IHC WSI, but at a lower resolution and without making WSIs, so a virtual split.

The comparison of the three methods for the prediction of ROI will be conducted not only with standard metrics such as Intersection over Union (IoU) and Dice Score but also with a metric derived from relative Target Registration Error (rTRE), introduced in the ANHIR challenge [12], which assesses landmark-based registration error. In addition, metrics derived from the image detection field and then tweaked for this case study will be introduced [13,14].

By evaluating the naive test on the VALIS and then the two proposed approaches, this work points out if the multiple tissues on the IHC WSIs are a problem for this SOTA technique and if using the two proposed approaches can be mitigated. Thus, the aim is to select the correct tissue from the multiple-tissue IHC WSI, register it with the H&E image, and use it for ROI prediction. With the contribution of virtual splitting, this work aims to evaluate the possibility of reducing the time-consuming step of splitting multiple-tissue IHC WSIs. This novelty approach also seeks to incorporate an automated quality control function, significantly reducing the pathologist's participation in validation when needed.

2 Data Collection and Preparation

This work was carried out on twenty patients diagnosed with invasive breast carcinoma of no special type between 2007 and 2015. The samples used in this study were obtained from the Tumour Banks of the Hospital de Tortosa Verge de la Cinta (HTVC) and the Hospital Joan XXIII (HUJ23), Tarragona, Spain and from the Biobank of the Pere Virgili Research Institute (IISPV). The Ethics Committee of the IISPV approved this study (reference numbers 128/2022 & 218/2022). Each case consists of three WSIs: a WSI of breast cancer tissue stained with H&E, a WSI of breast cancer tissue with one or more lymphatic ganglion tissues stained using the CD163 marker, and a WSI of breast cancer tissue with one or more lymphatic ganglion tissues stained using the CD8 marker. Tissue positions may vary between IHC WSIs of the same specimen. The digitised H&E and IHC stained tissue slides are consecutive.

The tissue slides were digitised at 20× magnification using a PANNORAMIC 250 Flash 3DHistech scanner. To identify and separate the different tissues within each IHC WSI, an expert used QuPath, an open-source platform popular in digital pathology, to divide the tissue compartments with lines [15]. This manual annotation was necessary to have a ground truth of the tissue compartments, as they can be provided partially overlapping or with such tears and deformations that can lead to ambiguities in the accurate division of these within the IHC WSI. A pathologist, following Salgado's criteria [4], identified the ROI for evaluating TILs on the H&E image of each specimen. After finding the ROI, the pathologist annotated it using SlideViewer, proprietary software developed by 3DHistech. The ROIs annotated by the expert at the lowest level had a rectangular shape with dimensions of (4050 px $\pm$ 40 px) $\times$ (7831 px $\pm$ 20 px). ROIs are then converted into GeoJSON format, which is compatible with QuPath.

3 Method

The registration of the ROI from the H&E WSI onto the multiple-tissue IHC WSI is based on the VALIS technique [10], a SOTA method for WSI registration in histological images. If interested, it is possible to get into a schematic description of the main phases of VALIS registration implementation in [1] and, if needed, its implementation in https://github.com/MathOnco/valis. However, it is modified to perform only rigid registration, specifically a Euclidean transformation, to preserve the shape of the ROI, as all WSIs have identical resolutions. Therefore, non-rigid registration should be avoided, as it may alter the structural morphology of the tissue within the ROI, which could have consequences for the TIL assessment.

In this work, the H&E image is the reference image, while the multiple-tissue IHC WSI is the moving image. It can be seen as an example of a pair of reference and moving images from the same case study in the Supplementary Material. After registration, the coordinates of the ROI vertices are warped from the reference image to the moving image to reconstruct the ROI in the multiple-tissue IHC WSI.

This section will present three approaches: *Naive*, *Baseline*, and *VSIRO*. *Naive* uses VALIS directly for ROI registration, while *Baseline* simplifies the process by focusing on single-tissue alignment and introducing a quality control phase. *VSIRO* addresses the time-consuming preprocessing phase by implementing virtual splitting and using higher resolution when confidence is low.

3.1 Approaches

The *Naive* approach, depicted in Fig. 1, shows the initial registration test. It carries out the registration phase without preprocessing the H&E WSI as the reference image and the multiple-tissue IHC WSI as a moving image. By conducting this trial, it is possible to test the robustness of VALIS in registering the pair of WSIs even in the presence of extra tissue, providing an additional challenge to it.

To simplify the task, this work introduces the *Baseline* approach, depicted in Fig. 2. This method aims to facilitate the registration task by adapting it to the case study of the ACROBAT challenge. To make this fit, the approach splits the multiple-tissue IHC WSIs into different single-tissue IHC WSIs. To determine which single-tissue IHC WSIs are the ones with breast cancer tissue, the registration that minimises the median rTRE (mrTRE) of the features extracted from VALIS using Binary Robust Invariant Scalable Keypoints (BRISK) [16] and Oxford Visual Geometry Group (VGG) descriptor [17] is selected. These features are matched using brute force, with outliers removed using the Random Sample Consensus (RANSAC) algorithm [18].

However, physically adapting the original data to avoid the problem of extra tissue consumes a lot of time. This issue is because the single-tissue WSIs to be artificially created are of giga-pixel scale size, and different checks have to be performed compared to the compartmentalisation annotations previously carried out by experts. To optimise this initial step, this work proposes the ***V****irtual* ***S****plitting with* ***I****ncremental* ***R****esolution* ***O****ptimisation* (*VSIRO*) approach, as shown in Fig. 3. This method uses annotations from experts to identify tissue compartment areas in each multiple-tissue IHC WSI. It then creates multiple masks at a higher level for the multiple-tissue IHC WSIs. Thus, the method selects the same level for both WSIs since they share the exact resolution and number of levels. If there were differences in resolution or levels between the pair of WSIs, it may turn to similarity transformation registration. This approach allows VALIS to work with smaller images at lower resolution levels, speeding up the process.

Attention must be paid to the dimensions of the images as if they are too small; there is a risk of losing too much information to give input to the registration phase. Therefore, the *VSIRO* approach extracts the multiple-tissue IHC image from the lowest available level (level t) where the image width is larger but, at the same time, closer to the resize threshold than the VALIS width (850 px). After splitting the multiple-tissue image into n flat images containing a single tissue, *VSIRO* performs n registrations with the reference H&E image at the same level as each single-tissue IHC image. *VSIRO* computes the ratio between the second-lowest mrTRE and the lowest mrTRE among the n registrations. If this ratio is greater than or equal to a threshold r, set to 2, the second-lowest mrTRE is at least twice the lowest mrTRE, suggesting a reliable registration. If the ratio is less than r, it implies uncertainty in correctly identifying the tissue for registration. However, the set of r is a priori, and by deciding its value, if it is agreed to decrease r, there is a looser quality control. Otherwise, if it is increased, there will be stricter quality control. The procedure is repeated using the next lower level of the multiple-tissue IHC image to reach a more confident tissue selection. Registration is considered unreliable if it gets the minimum extractable level ($minLevel$) of the multiple-tissue IHC WSI without obtaining a reliable result (the ratio is less than r). Hardware limitations may also set this $minLevel$.

Suppose a reliable result is achieved. In that case, it warps the ROI of the reference image using the registration parameters of the IHC tissue with the lowest mrTRE among the n images.

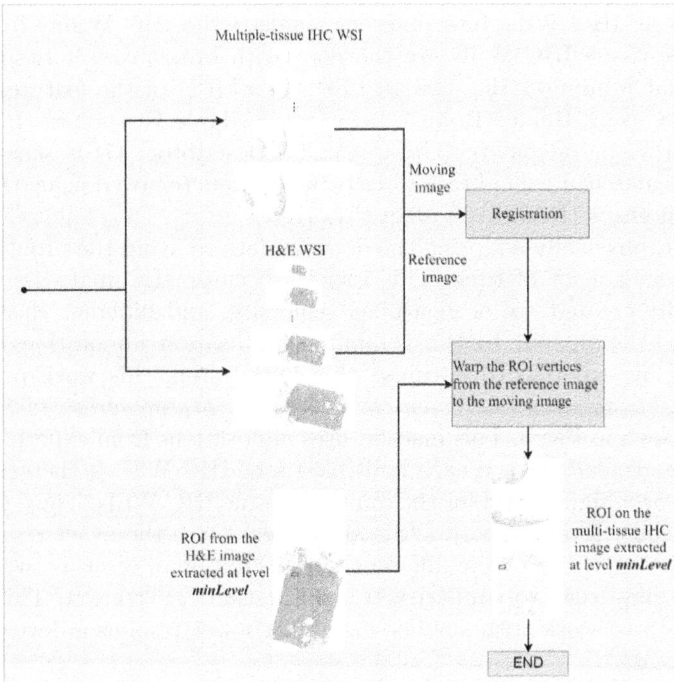

Fig. 1. Diagram illustrating the *Naive* approach for ROI registration. In this method, the H&E WSI is used as the reference image, and the multiple-tissue IHC WSI is the moving image.

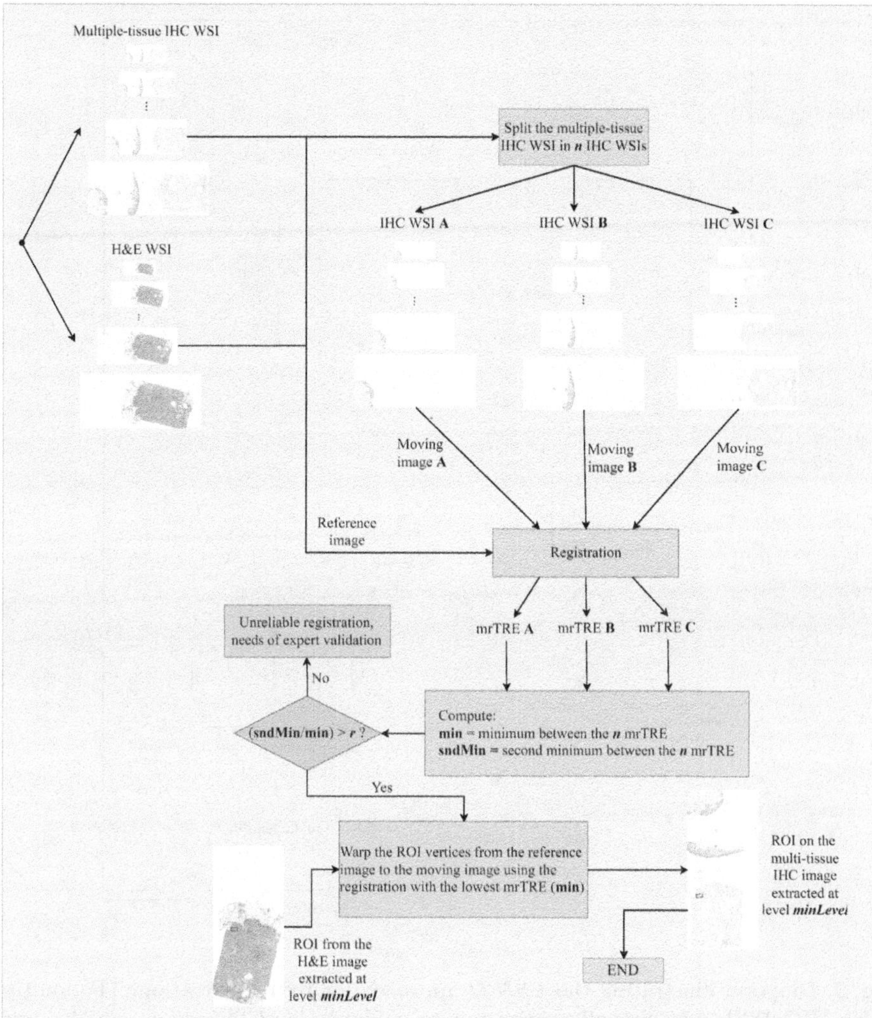

Fig. 2. Diagram depicting the *Baseline* method for ROI registration. The multiple-tissue IHC WSI is split into single-tissue IHC WSIs, each containing a single tissue type. Registration is performed separately on the H&E image using each IHC WSI. Subsequently, the ROI is determined using the registration parameters from the IHC WSI with the lowest mrTRE.

3.2 Metrics

To evaluate the three approaches and their likeness, this work first employs the popular IoU and Dice Score, which are adapted slightly for the case study (Eq. 1 and Eq. 2):

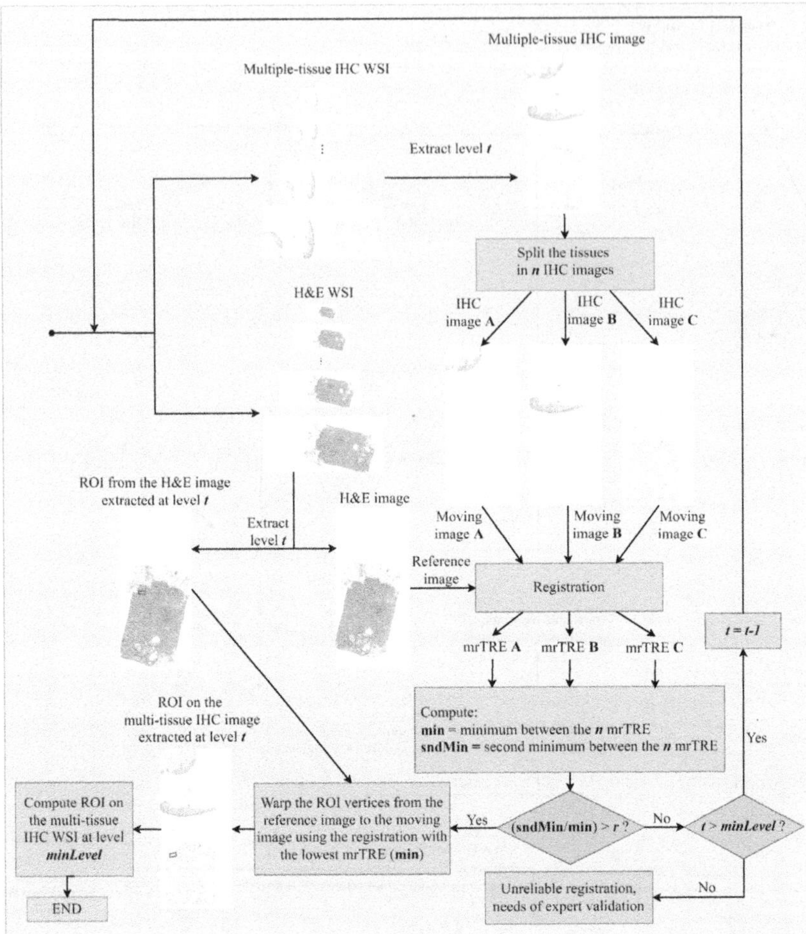

Fig. 3. Diagram illustrating the *VSIRO* approach for ROI registration. The multiple-tissue IHC WSIs are virtually separated at a higher level to extract smaller image compartments. Registration is then applied to these tissues using the corresponding H&E image extracted at the same level. The registration parameters chosen are from the IHC image with the lowest mrTRE. Quality control checks ensure accurate tissue selection; if discrepancies arise, it restarts by extracting the following lower-level images of both WSIs.

$$IoU_l^{ij} = \frac{|R_l^i \cap R_l^j|}{|R_l^i \cup R_l^j|} \tag{1}$$

$$DiceScore_l^{ij} = \frac{|2 \times (R_l^i \cap R_l^j)|}{|R_l^i| + |R_l^j|} \tag{2}$$

where R_l^i is the ROI proposal using the approach i on the WSI l, and R_l^j of the approach j on the WSI l. IoU and Dice Score are metrics for evaluating the overlap between two ROIs proposed by different approaches. IoU tends to penalize underestimation and overestimation more than Dice Score.

Additional metrics are then incorporated. First, it is introduced a metric called rTRE of the centroids (rTREc), presented in Eq. 3, which computes the rTRE between the centroids of paired ROIs from two compared approaches:

$$rTREc_l^{ij} = \frac{\|c_l^i - c_l^j\|_2}{d_l} \tag{3}$$

where c_l^i is defined as the centroid of the ROI of the WSI l computed using the approach i, c_j^i of the approach j and d_l is the length of the diagonal image l.

Next, it is redefined the Generalized IoU (GIoU) [13] and Distance IoU (DIoU) [14] metrics for this case study, respectively defined in Eq. 4 and Eq. 5, inspired from the bounding boxes used in image detection, as the ROIs are rectangular:

$$GIoU_l^{ij} = IoU_l^{ij} - \frac{|S_l^{ij} \setminus (R_l^i \cup R_l^j)|}{|S_l^{ij}|} \tag{4}$$

$$DIoU_l^{ij} = IoU_l^{ij} - \frac{\|c_l^i - c_l^j\|_2}{d_{S_l^{ij}}^2} \tag{5}$$

where S_l^{ij} is the smallest rectangle enclosing R_l^i and R_l^j in the image l. [13] suggests that S_l^{ij} can be a rectangle since it can have the same convex shape type as the enclosed shapes. After that, $d_{S_l^{ij}}^2$ is the length square of the diagonal of the S_l^{ij} rectangle.

Finally, these new metrics are proposed: Global GIoU (GGIoU) and Global GDIoU (GDIoU), respectively defined in Eq. 6 and Eq. 7, which take into consideration also the dimension of the entire image. The first one, as it can be seen from the derived name, takes inspiration from GIoU, while the latter takes from DIoU by doing a normalisation inspired from the rTRE metric:

$$GGIoU_l^{ij} = IoU_l^{ij} - \frac{|S_l^{ij} \setminus (R_l^i \cup R_l^j)|}{|l|} \tag{6}$$

$$GDIoU_l^{ij} = IoU_l^{ij} - \frac{\|c_l^i - c_l^j\|_2}{d_l} \tag{7}$$

4 Results

The following section presents the registration results by comparing the three approaches pairwise. This work provides two tables for each pair of biomarkers. The first table types (Table 1 and Table 3) show IoU, Dice Score, and rTREc outcomes, which are metrics close to the image registration field [12], while the other two tables (Table 2 and Table 4) propose outcomes metrics adapted for the case study, inspired by bounding boxes used in image detection.

Regarding the number of cycles for the *VSIRO* approach, it takes 1.1 ± 0.144 cycles for CD163 and 1.2 ± 0.244 cycles for CD8. This outcome implies that, in most cases, it only requires the images from the initial chosen level t, demonstrating the efficacy of the *VSIRO* approach. For the CD163 biomarker, the ROI registration times are 23 s ± 4 s for the *Naive* method, 5 h 11 m 44 s ± 34 m 22 s for the *Baseline* method, and 40 s ± 9 s for the *VSIRO* method. For the CD8 biomarker, the registration times are 15 s ± 2 s for the *Naive* method, 5h 25 m 13 s ± 35 m 11 s for the *Baseline* method, and 59 s ± 31 s for the *VSIRO* method.

Table 1. Approaches comparison for ROIs predicted on multiple-tissue CD163 WSIs from the H&E WSIs. Approaches are compared pairwise using IoU, Dice Score, and rTREc.

Approaches compared	IoU(↑)	DiceScore(↑)	rTREc(↓)
{*Naive*, *Baseline*}	0.461 ± 0.177	0.531 ± 0.192	0.093 ± 0.082
{*Naive*, *VSIRO*}	0.468 ± 0.181	0.534 ± 0.194	0.091 ± 0.082
{*Baseline*, *VSIRO*}	0.674 ± 0.137	0.756 ± 0.140	0.010 ± 0.009

Table 2. Approaches comparison for ROIs predicted on multiple-tissue CD163 WSIs from the H&E WSIs. Approaches are compared pairwise using GIoU, DIoU, GGIoU, and GDIoU.

Approaches compared	GIoU(↑)	DIoU(↑)	GGIoU(↑)	GDIoU(↑)
{*Naive*, *Baseline*}	0.163 ± 0.350	0.461 ± 0.177	0.453 ± 0.182	0.368 ± 0.239
{*Naive*, *VSIRO*}	0.184 ± 0.344	0.468 ± 0.181	0.459 ± 0.186	0.377 ± 0.242
{*Baseline*, *VSIRO*}	0.582 ± 0.225	0.674 ± 0.137	0.674 ± 0.138	0.664 ± 0.146

Table 3. Approaches comparison for ROIs predicted on multiple-tissue CD8 WSIs from the H&E WSIs. Approaches are compared pairwise using IoU, Dice Score, and rTREc.

Approaches compared	IoU(↑)	DiceScore(↑)	rTREc(↓)
{*Naive*, *Baseline*}	0.495 ± 0.187	0.553 ± 0.201	0.082 ± 0.067
{*Naive*, *VSIRO*}	0.497 ± 0.189	0.554 ± 0.202	0.062 ± 0.043
{*Baseline*, *VSIRO*}	0.650 ± 0.135	0.742 ± 0.131	0.034 ± 0.059

Table 4. Approaches comparison for ROIs predicted on multiple-tissue CD8 WSIs from the H&E WSIs. Approaches are compared pairwise using GIoU, DIoU, GGIoU, and GDIoU.

Approaches compared	GIoU(↑)	DIoU(↑)	GGIoU(↑)	GDIoU(↑)
{*Naive*, *Baseline*}	0.191 ± 0.367	0.495 ± 0.187	0.486 ± 0.192	0.413 ± 0.240
{*Naive*, *VSIRO*}	0.197 ± 0.366	0.497 ± 0.189	0.491 ± 0.193	0.435 ± 0.226
{*Baseline*, *VSIRO*}	0.548 ± 0.221	0.650 ± 0.135	0.647 ± 0.138	0.616 ± 0.176

5 Discussion

The *Naive* approach, which tries to register without preprocessing the H&E WSI with the multiple-tissue IHC WSI, cannot be used for this case study. This issue is because the task is too tricky for VALIS and is based only on its rigid registration phase. Therefore, the *Baseline* approach is recommended as a starting point, leading back to a more straightforward case study, so registration between single tissues with different stains. From an overall standpoint, the results between CD163 and CD8 are similar. These outcomes are due since they are biomarkers with the same colour stain from equal specimens, prepared from the same lab and digitised with the same WSI scanner [5]. The comparisons between the *Naive* method and the *Baseline* or *VSIRO* methods show a divergence in both approaches since ROI prediction using the *Naive* method requires intervention in 6/20 CD163 and 7/20 CD8 biomarker cases.

As can be noticed, there is a strong relation between the *Baseline* and *VSIRO* methods for both biomarkers. The failures encountered with CD163 occurred on two case studies with the *Baseline* method, which, even though it had chosen the correct single-tissue IHC WSI, could have been more successful in registering ROIs. For the CD8 biomarker, the quality control detected a failure in tissue selection on a case study with the *VSIRO* approach, which selected the wrong tissue. In this case, the ratio between the first and second minimum rTRE values did not pass the threshold r (ratio = 1.776), necessitating pathologist intervention. Note that even in the *Baseline* approach, it did not pass the r (ratio = 1.097) for the same case study. Therefore, it did not pass the quality control, but *Baseline* however took the correct tissue for the ROI. Based on a *VSIRO* split, this approach eliminates the need to physically split the multiple-tissue WSIs, providing a more time-efficient solution. *Baseline* and *VSIRO* workflows significantly reduce the need for pathologist intervention, limiting it to only the most complex cases when the quality control phase detects uncertainty in tissue selection.

The proposed metrics, GGIoU and GDIoU, are intended to enlarge the viewpoint of GIoU and DIoU metrics, which mainly focus on the local area comprising the two ROI proposals. The proposed metrics, GGIoU and GDIoU, consider the image size in which the ROI proposals lie. Unlike GIoU, the GGIoU metric is designed to minimise over-penalisation. In contrast, DIoU provides limited information because its penalty term is too minimal, as it proposes in the denominator the square length of the diagonal of the smallest rectangle surrounding the two ROIs.

6 Conclusion and Future Work

This work aimed to present the ROI registration case study between single-tissue WSIs and multiple-tissue WSIs. After the initial attempt to use a naive SOTA algorithm for WSI registration, this work found it not robust for this case study. The problem should be eased by returning to the single-tissue registration strategy to facilitate this task. However, using the *Baseline* approach by dividing the multiple-tissue WSIs into different single-tissue WSIs consumes too much time. Therefore, the *VSIRO* approach was presented to split the various tissue compartments without physically separating them into WSIs. In this way, it is more time-efficient and valuable in the clinical workflow. In addition, this work introduced a quality control assessment in both *Baseline* and *VSIRO* approaches to determine whether the methods are confident in computing the ROI prediction in the correct selected tissue. Furthermore, the *VSIRO* approach uses an incremental strategy that uses higher resolution images only if a smaller image size fails to pass the quality control phase.

The plan is to publish publicly the data and code for the research community. Pathologists will propose other ROIs on the H&E WSIs to evaluate their inter-observability and annotate the ROI on the IHC WSIs [4,6].

Future work involves taking into account multi-resolution ROI registration to tune the ROI, posing a challenge due to repetitive texture when working with a higher resolution [12]. Another future work will automate the entire pipeline, including automatically splitting the initial compartments without the experts' annotations. This will lead to a fully automatic end-to-end registration process, with pathologist intervention only in complex cases, such as when the quality control phase detects a possible failure.

Acknowledgments. This work was supported by BosomShield, a project that has received funding from Marie Skłodowska-Curie Doctoral Networks Actions (HORIZON-MSCA-2021-DN-01-01) under grant agreement 101073222. Further support was provided by SCARLET, a project funded by Proyectos Estratégicos Orientados a la Transición Ecológica y a la Transición Digital, from the 2021 call of the Ministerio de Ciencia e Innovación, with grant number TED2021-130081B-C22 and funding from NextGenerationEU.

Disclosure of Interests. The authors have no competing interests to declare that are relevant to the content of this article.

References

1. Weitz, P., Valkonen, M., Solorzano, L., Carr, C., Kartasalo, K., Boissin, C., Koivukoski, S., Kuusela, A., Rasic, D., Feng, Y., et al.: The ACROBAT 2022 challenge: automatic registration of breast cancer tissue. arXiv preprint arXiv:2305.18033 (2023)
2. Nam, S., Chong, Y., Jung, C.K., Kwak, T.Y., Lee, J.Y., Park, J., Rho, M.J., Go, H.: Introduction to digital pathology and computer-aided pathology. Journal of pathology and translational medicine **54**(2), 125 (2020)
3. Baxi, V., Edwards, R., Montalto, M., Saha, S.: Digital pathology and artificial intelligence in translational medicine and clinical practice. Modern Pathology **35**(1), 23–32 (2022)
4. Salgado, R., Denkert, C., Demaria, S., Sirtaine, N., Klauschen, F., Pruneri, G., Wienert, S., Van den Eynden, G., Baehner, F.L., Pénault-Llorca, F., et al.: The evaluation of tumor-infiltrating lymphocytes (TILs) in breast cancer: recommendations by an International TILs Working Group 2014. Annals of oncology **26**(2), 259–271 (2015)

5. Fortis, S.P., Sofopoulos, M., Sotiriadou, N.N., Haritos, C., Vaxevanis, C.K., Anastasopoulou, E.A., Janssen, N., Arnogiannaki, N., Ardavanis, A., Pawelec, G., et al.: Differential intratumoral distributions of CD8 and CD163 immune cells as prognostic biomarkers in breast cancer. Journal for immunotherapy of cancer **5**, 1–12 (2017)
6. Fiorin, A., López Pablo, C., Lejeune, M., Hamza Siraj, A., Della Mea, V.: Enhancing AI Research for Breast Cancer: A Comprehensive Review of Tumor-Infiltrating Lymphocyte Datasets. Journal of Imaging Informatics in Medicine pp. 1–13 (2024)
7. Smit, G., Ciompi, F., Cigéhn, M., Bodén, A., Van Der Laak, J., Mercan, C.: Quality control of whole-slide images through multi-class semantic segmentation of artifacts. In: Medical Imaging with Deep Learning (2021)
8. Schömig-Markiefka, B., Pryalukhin, A., Hulla, W., Bychkov, A., Fukuoka, J., Madabhushi, A., Achter, V., Nieroda, L., Büttner, R., Quaas, A., et al.: Quality control stress test for deep learning-based diagnostic model in digital pathology. Modern Pathology **34**(12), 2098–2108 (2021)
9. Schreiber, B., Denholm, J., Jaeckle, F., Arends, M.J., Branson, K., Schönlieb, C.B., Soilleux, E.: Rapid artefact removal and H&E-stained tissue segmentation. Scientific Reports **14**(1), 309 (2024)
10. Gatenbee, C.D., Baker, A.M., Prabhakaran, S., Swinyard, O., Slebos, R.J., Mandal, G., Mulholland, E., Andor, N., Marusyk, A., Leedham, S., et al.: Virtual alignment of pathology image series for multi-gigapixel whole slide images. Nature communications **14**(1), 4502 (2023)
11. Weitz, P., Valkonen, M., Solorzano, L., Carr, C., Kartasalo, K., Boissin, C., Koivukoski, S., Kuusela, A., Rasic, D., Feng, Y., et al.: ACROBAT–a multi-stain breast cancer histological whole-slide-image data set from routine diagnostics for computational pathology. arXiv preprint arXiv:2211.13621 (2022)
12. Borovec, J., Kybic, J., Arganda-Carreras, I., Sorokin, D.V., Bueno, G., Khvostikov, A.V., Bakas, S., Eric, I., Chang, C., Heldmann, S., et al.: ANHIR: automatic non-rigid histological image registration challenge. IEEE transactions on medical imaging **39**(10), 3042–3052 (2020)
13. Rezatofighi, H., Tsoi, N., Gwak, J., Sadeghian, A., Reid, I., Savarese, S.: Generalized intersection over union: A metric and a loss for bounding box regression. In: Proceedings of the IEEE/CVF conference on computer vision and pattern recognition. pp. 658–666 (2019)
14. Zheng, Z., Wang, P., Liu, W., Li, J., Ye, R., Ren, D.: Distance-IoU loss: Faster and better learning for bounding box regression. In: Proceedings of the AAAI conference on artificial intelligence. vol. 34, pp. 12993–13000 (2020)
15. Bankhead, P., Loughrey, M.B., Fernández, J.A., Dombrowski, Y., McArt, D.G., Dunne, P.D., McQuaid, S., Gray, R.T., Murray, L.J., Coleman, H.G., et al.: QuPath: Open source software for digital pathology image analysis. Scientific reports **7**(1), 1–7 (2017)
16. Leutenegger, S., Chli, M., Siegwart, R.Y.: Brisk: Binary robust invariant scalable keypoints. In: 2011 International Conference on Computer Vision. pp. 2548–2555 (2011). https://doi.org/10.1109/ICCV.2011.6126542
17. Simonyan, K., Vedaldi, A., Zisserman, A.: Learning local feature descriptors using convex optimisation. IEEE Transactions on Pattern Analysis and Machine Intelligence **36**(8), 1573–1585 (2014). https://doi.org/10.1109/TPAMI.2014.2301163
18. Cantzler, H.: Random sample consensus (ransac). Institute for Perception, Action and Behaviour, Division of Informatics, University of Edinburgh **3** (1981)

Automatic Registration of SHG and H&E Images with Feature-Based Initial Alignment and Intensity-Based Instance Optimization: Contribution to the COMULIS Challenge

Marek Wodzinski[1,2(✉)] and Henning Müller[2,3]

[1] Department of Measurement and Electronics, AGH University of Krakow, Krakow, Poland
wodzinski@agh.edu.pl

[2] University of Applied Sciences Western Switzerland (HES-SO Valais-Wallis) Information Systems Institute, Sierre, Switzerland

[3] Medical Faculty, University of Geneva, Geneva, Switzerland

Abstract. The automatic registration of noninvasive second-harmonic generation microscopy to hematoxylin and eosin slides is a highly desired, yet still unsolved problem. The task is challenging because the second-harmonic images contain only partial information, in contrast to the stained H&E slides that provide more information about the tissue morphology. Moreover, both imaging methods have different intensity distributions. Therefore, the task can be formulated as a multi-modal registration problem with missing data. In this work, we propose a method based on automatic keypoint matching followed by deformable registration based on instance optimization. The method does not require any training and is evaluated using the dataset provided in the Learn2Reg challenge by the COMULIS organization. The method achieved relatively good generalizability resulting in 88% of success rate in the initial alignment and average target registration error equal to 2.48 on the external validation set. We openly release the source code and incorporate it in the DeeperHistReg image registration framework.

Keywords: Image Registration · Deep Learning · SHG · H&E · Microscopy · Learn2Reg

1 Introduction

The automatic registration of microscopy images is still an active research area [4,21]. Recently, several notable contributions were proposed to automatic registration of whole slide images (WSIs) acquired using different stains, varying from contributions that focus on the quality of deformable registration [8,24–26], through methods that address the robustness of initial alignment [12,16,17,23],

M. Modat et al. (Eds.): WBIR 2024, LNCS 15249, pp. 346–357, 2024.
https://doi.org/10.1007/978-3-031-73480-9_27

ending with methods that propose ready-to-use software packages that can perform automatic registration without time-consuming parameter tuning or deep network retraining [7,14,22].

Nevertheless, the registration of microscopic images acquired using different imaging modalities is still an unsolved problem that requires further investigation. An example of a highly important task within the research area is the automatic registration of second-harmonic generation (SHG) microscopy images to bright-field images stained by hematoxylin and eosin (H&E) or immunohistochemistry (IHC) dyes. SHG microscopy is a non-invasive imaging technique, does not require exogenous labels, and allows one to easily visualize collagen fibers in the tissue microenvironment. This enables one to perform real-time observations without introducing the toxicity associated with staining agents or external artifacts. However, it only provides partial information while H&E or IHC present more information about the tissue morphology. Moreover, both imaging modalities have significantly different intensity distributions. Therefore, a dedicated solution that addresses both problems is required. The importance of the task motivated researchers from the COMULIS consortium to organize a dedicated sub-challenge during the well-recognized Learn2Reg challenge [9]. An exemplary pair of SHG and H&E images is shown in Fig. 1.

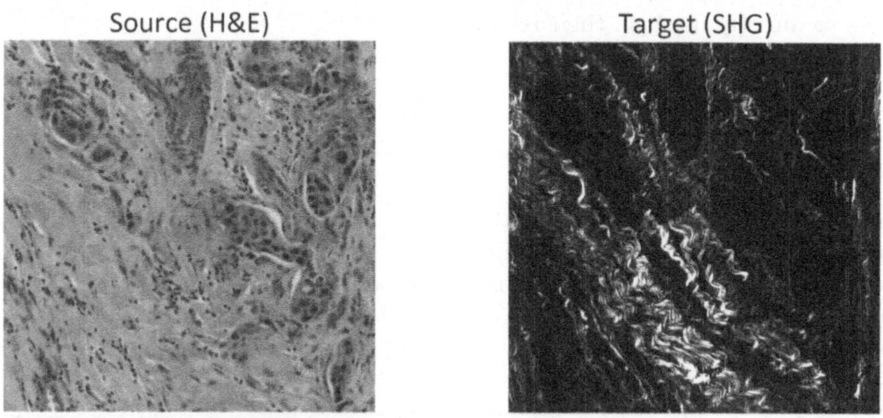

Fig. 1. Exemplary pair of H&E and SHG images. Note the significantly different intensity distributions and the amount of missing data in the SHG image.

So far, the work on SHG and H&E was rather limited. The researchers focused primarily on the registration of H&E to IHC slides [4,21] or the H&E to magnetic resonance images (MRIs) [1,2,20]. There are several different methods dedicated to the initial alignment, including both intensity-based methods [3,12,18], as well as the feature-based contributions [17]. The following deformable registration is still usually solved by iterative optimization due to the tremendous heterogeneity of WSIs [7,13,14,23,25], however several deep learning-based methods exist [8,24,26], unfortunately without exhaustive evaluation of the generalizability to previously unseen distributions. Methods often struggle with the following challenges related to the registration of WSIs: (i) large initial misalignment, (ii) tremendous size of input images, and (iii) missing data caused by different properties of staining agents. Nevertheless, there is an influential work presenting an automatic intensity-based method for the initial alignment of SHG and H&E images [11]. The method is based on an initial preprocessing that decreases the amount of information in the bright field images, followed by an iterative optimization of the initial alignment matrix. Even though the method is successful, it may struggle with recovering large rotational initial misalignment.

Contribution: In this work, we propose an automatic registration method dedicated to SHG and H&E images. The method consists of deep feature-based initial alignment allowing recovery of large deformations, followed by an optional deformable registration based on the instance optimization. We evaluate the method using the dataset released by the COMULIS organization during the Learn2Reg challenge. We show considerable generalizability and quality of the proposed initial alignment and argue whether the deformable registration is really necessary. We openly release the source code and incorporate the method in one of the open-source registration packages dedicated to WSIs, enabling the users to directly use it in their research.

2 Method

2.1 Overview

The proposed method consists of three steps: (i) preprocessing to make the features from both modalities as similar as possible, (ii) feature-based sparse initial alignment based on SuperPoint and SuperGlue, and (iii) dense deformable registration based on the instance optimization. The processing pipeline is presented in Fig. 2.

2.2 Preprocessing

Since the SHG and H&E images have substantially different intensity distributions and visualize distinct structures it is crucial to perform preprocessing that makes the general geometric features as similar as possible.

The H&E images are firstly converted to the HSV color space. Then, only the hue image is selected, normalized to [0–1] intensity range, and equalized by global histogram equalization. Then the image is filtered by a 5×5 median filter. The preprocessing of SHG images is similar with the difference that the images are equalized and filtered directly, without the conversion to HSV color space. The registration is performed using the original resolution of the images, however, for the feature-based initial alignment other resolution level are used as well (100, 200, 300, 400, and 500 pixels in each dimension).

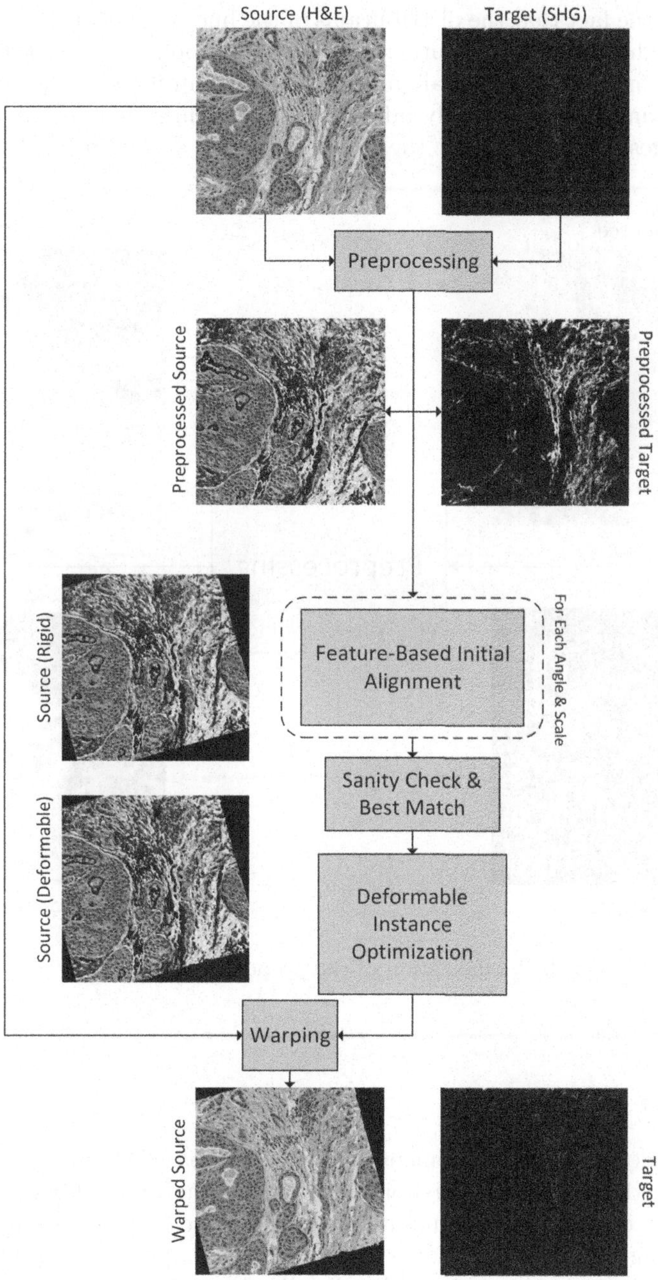

Fig. 2. Visualization of the registration pipeline presenting also the intermediate results. Best viewed zoomed in. The target SHG image is presented in the original intensity range to emphasize the task difficulty (zoom required to see the details).

The reason why we decided to perform the histogram equalization is connected with the fact that the distributions of the hue channel of H&E and the raw SHG images have significant outliers. Such distributions decrease the robustness of the following feature-based alignment. The median filtering was implemented because the images are strongly influenced by the impulse noise. An exemplary pair of images before and after the preprocessing is shown in Fig. 3.

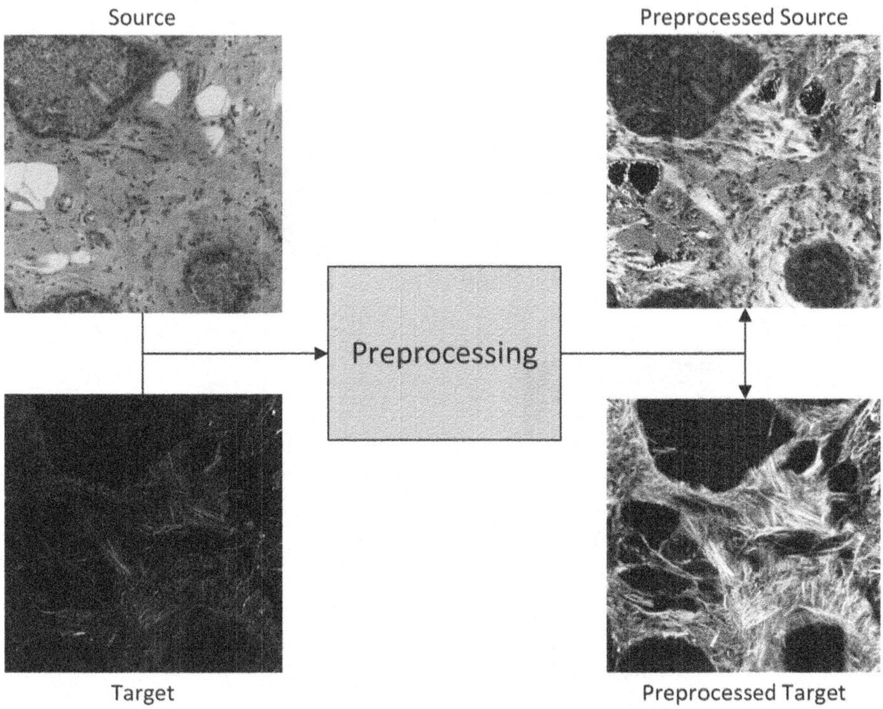

Fig. 3. Visualization of the proposed preprocessing.

2.3 Initial Alignment

The initial registration is implemented as a feature-based alignment of sparse key points using the SuperPoint keypoint extractor [5] and SuperGlue matcher [19]. They can be considered as foundation models dedicated to learning-based feature matching. Our previous works confirmed their superior generalizability into previously unseen cases and intensity distributions [23].

The SuperPoint and SuperGlue are not rotation and scale invariant. Therefore, we incorporate them in an exhaustive search that calculates the transformation and number of matches for each predefined resolution and initial rotation. Then, we filter the transformations that do not meet the predefined scale criteria (increase or decrease in scale more than 10%) and finally return the transform associated with the highest number of matched key points.

Finally, we compare the SuperPoint/SuperGlue pair to the classical SIFT + RANSAC method [15] and the recent OmniGlue model [10] showing its superiority. All the methods are evaluated in the same exhaustive framework.

2.4 Deformable Registration

The deformable registration is implemented as a multilevel instance optimization, an iterative optimization process. We decided to avoid training a deep network dedicated to the deformable registration because, with the available number of training samples, the method would not generalize into previously unseen cases, especially considering the fact that there are not openly available foundation models dedicated to deformable microscopy registration that could be used for fine-tuning the network in low data regimes.

The objective function is defined as a weighted sum of the local mutual information (MI) as the similarity metric, and the diffusive regularization as the regularizer, defined as:

$$O_{REG}(S, T, u) = MI(S_i \circ u_i, T_i) + \theta_i Reg(u_i), \quad (1)$$

where S_i, T_i are the source and target images at the i-th resolution level respectively, u_i is the calculated displacement field, θ_i denotes the regularization coefficient at i-th resolution level, MI denotes the local version of mutual information, Reg is the diffusive regularization, $\circ$ denotes the warping operation, and N is the number of resolution levels.

2.5 Dataset and Experimental Setup

The dataset consists of 166 pairs of SHD and H&E images among which 156 represent an unannotated training set and 10 are validation pairs with ground truth accessible through the Grand-Challenge platform. The images present human breast and pancreatic cancer tissue and were acquired at 40x magnification [6,11] at the Laboratory for Optical and Computational Instrumentation, Department of Biomedical Engineering, University of Wisconsin-Madison, Madison, WI 53706, USA. The dataset was released under a CC license and is used as a subtask in the Learn2Reg 2024 challenge.

All experiments were performed using a workstation equipped with a NVIDIA RTX A6000 GPU. We performed ablation studies to verify the robustness of different initial alignment methods. The source code and the hyperparameters used are openly available and incorporated in the *DeeperHistReg* registration framework [22].

3 Results

We evaluate the proposed approach using three methods: (i) target registration error (TRE) reported using the validation set with ground truth available through the Grand-Challenge platform, (ii) the ratio of successful initial alignments (evaluated both on the validation and the training set), (iii) visual analysis of the registered images. We also used the training set for the evaluation because the proposed method was not re-trained or fine-tuned using the samples available in the training set.

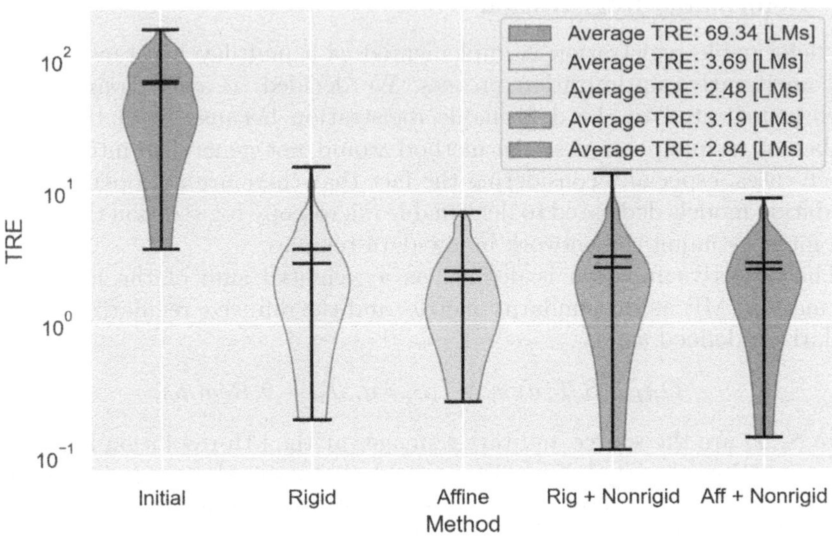

Fig. 4. The TRE calculated for the validation pairs using the Grand-Challenge platform.

The TRE is presented in Fig. 4. Due to a limited number of available submissions, we report only results presenting the influence of initial alignment and the following deformable registration. The percentage of successful initial alignments is presented in Table 1 and includes an ablation study related to the methods used in the feature-based registration. The success rate is the ratio of correctly registered pairs (visual inspection whether the images are meaningfully aligned) to obviously incorrect registrations (no improvement or worsening the results). To evaluate the success rate, we combined pairs from the training and validation subsets. Exemplary visualizations of the registration results are presented in Fig. 5.

4 Discussion

The proposed method achieved considerable robustness and generalizability (Fig. 4). The initial alignment method successfully registered all validation pairs and 88% of all the registration pairs combined together (Table 1). This is a significant achievement because: (i) part of the training pairs seem to be impossible to

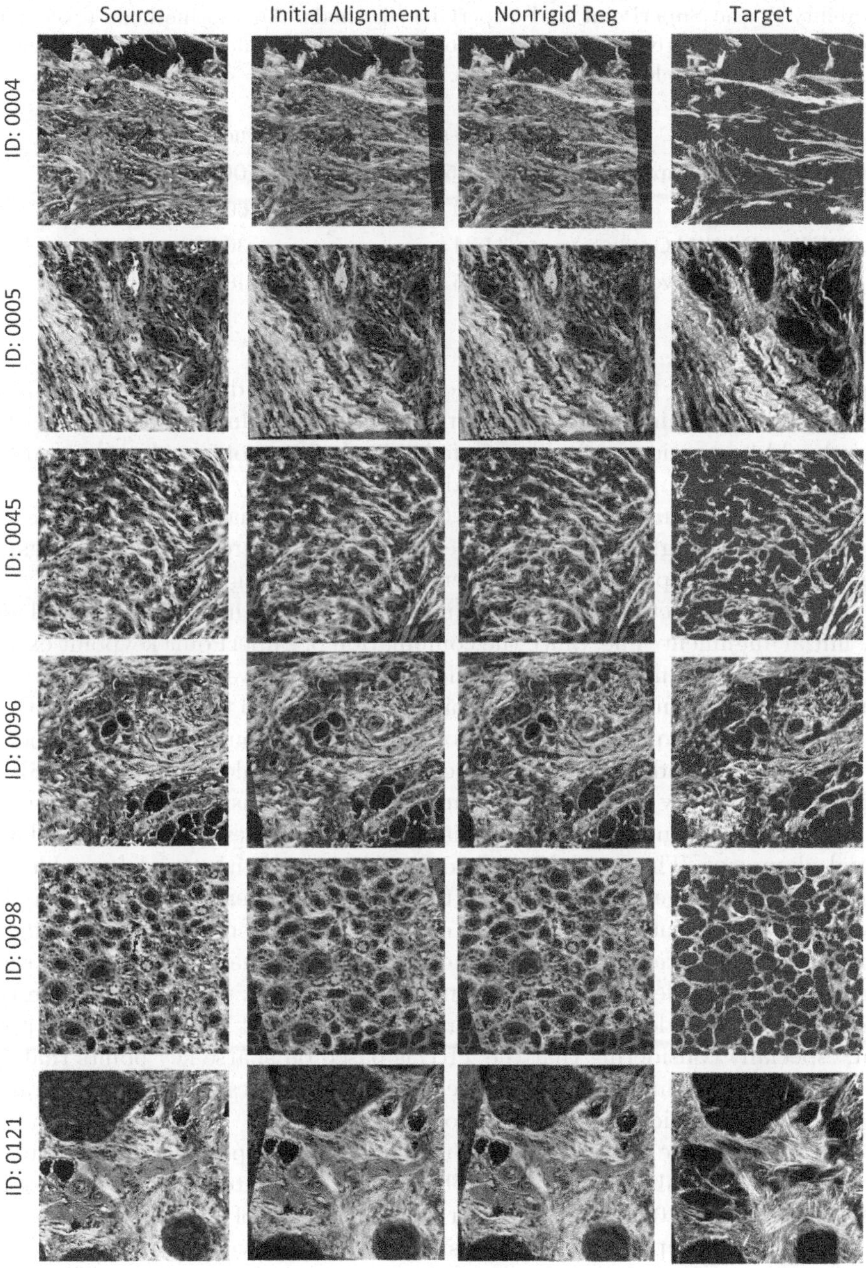

Fig. 5. Qualitative registration results using several samples from the validation subset. The source and target images are overlayed in different color channels to present the alignment quality. The preprocessed images are used for the presentation clarity. Note the small impact of the deformable registration.

Table 1. The success rate of several initial alignment algorithms. Note the generalizability of the SuperPoint and SuperGlue methods. All the methods (except the intensity-based iterative affine registration using mutual information) were incorporated in the multi-scale and multi-angle exhaustive search.

Method	Success Rate All [%] ↑	Success Rate Val [%] ↑
SuperPoint & SuperGlue	**87.95**	**100.00**
OmniGlue	48.79	60.00
SIFT + RANSAC	34.34	30.00
Iterative Multilevel Affine (MI)	37.95	40.00

register (Fig. 6), (ii) the images are so different that it is difficult to even capture the relations by humans. On the other hand, the following deformable registration should be considered as an optional step. It does not improve the registration quality for the majority of the slides (no statistically significant differences, Wilcoxon signed-rank test, p-value ¿ 0.05), however, may introduce undesired foldings (even though the proposed method does not introduce any foldings). It seems that the samples were not undergoing large nonrigid deformations.

The proposed method has several limitations. The main one is connected with the initial alignment time. Since the combination of SuperPoint keypoint extractor and SuperGlue matcher is not scale and rotation invariant, it is required to perform the matching using various initial rotations and scales. As a result, the initial alignment requires an average of 40 s to be computed (NVIDIA A6000). In fact, even methods that claim to be rotation and scale invariant (e.g. SIFT) express the property only in limited range of values. Thus, the exhaustive search of initial rotation and scale is beneficial for the robustness of the feature-based initial alignment. The second limitation is connected with the deformable registration. The images not only come from different intensity distributions but also present different structures (e.g. collagen fibers vs nuclei). This results in missing data that cannot be simply addressed by mutual information or other similarity metrics dedicated to multimodal registration. It is an open question whether the dense deformable registration is the correct approach to the problem, especially considering that the TRE is based on sparse key points that can be annotated only in regions with direct correspondences, thus the registration quality in other regions remains unknown.

Interestingly, the combination of SuperPoint and SuperGlue outperforms to more recent OmniGlue method. The observation is counterintuitive because the OmniGlue article [10] states a superior generalizability of the keypoint matcher. The claim is not supported by the results, and we show that, at least for multimodal microscopy images, the combination of SuperPoint and SuperGlue is superior to OmniGlue.

Finally, we point out an important observation related to the initial alignment and the available data. It would be significantly easier to register images for entire slides instead of cropped patches. For some of the cropped patches, it is

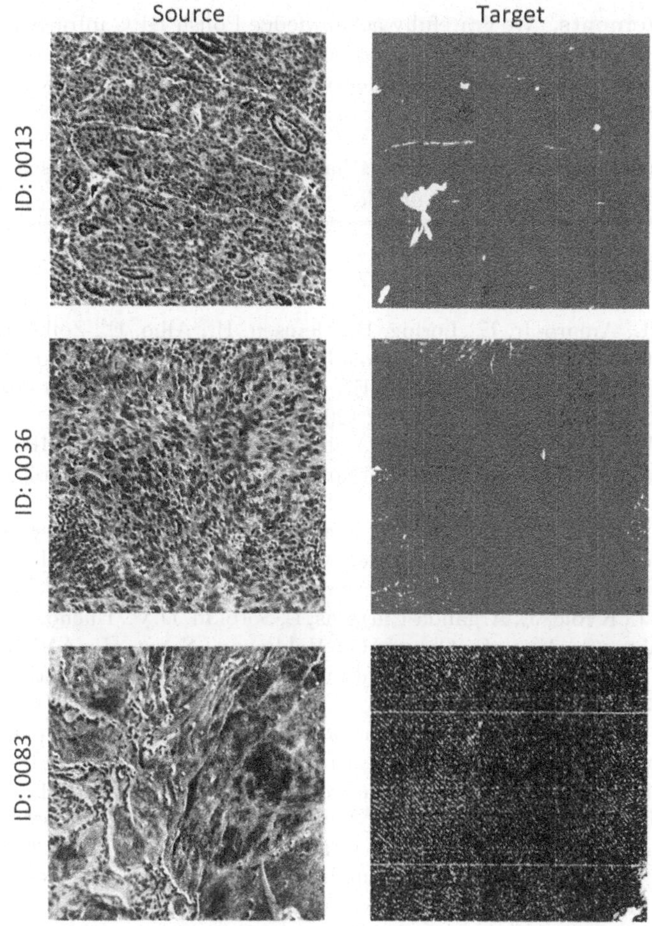

Fig. 6. Examples of several incorrectly registered pairs, however, according to the authors opinion, such pairs cannot be registered correctly. The images are presented after the preprocessing for the presentation clarity.

inherently difficult to find the sparse correspondences, however, larger contextual information could result in a significantly higher robustness of the feature-based registration. Maybe in future editions of the challenge the task could be split into two problems: (i) an initial alignment using resampled whole slide images, and (ii) the deformable registration of the cropped patches.

To conclude, we proposed an automatic registration method dedicated to SHD and H&E images. The method consists of dedicated preprocessing, robust feature-based initial alignment, and optional deformable registration. The method achieved considerable registration accuracy and robustness and is openly available to the scientific community.

Acknowledgements. We gratefully acknowledge Polish HPC infrastructure PLGrid support within computational grant no. PLG/2024/017079. The research was partially supported by the program "Excellence Initiative - Research University" for AGH University.

Disclosure of Interests. The authors have no competing interests to declare that are relevant to the content of this article.

References

1. Alegro, M., Amaro-Jr, E., Loring, B., Heinsen, H., Alho, E., Zollei, L., Ushizima, D., Grinberg, L.T.: Multimodal whole brain registration: MRI and high resolution histology. In: Proceedings of the IEEE conference on computer vision and pattern recognition workshops. pp. 194–202 (2016)
2. Alyami, W., Kyme, A., Bourne, R.: Histological validation of MRI: A review of challenges in registration of imaging and whole-mount histopathology. Journal of Magnetic Resonance Imaging **55**(1), 11–22 (2022)
3. Awan, R., Raza, S.E.A., Lotz, J., Weiss, N., Rajpoot, N.: Deep feature based cross-slide registration. Computerized Medical Imaging and Graphics **104**, 102162 (2023)
4. Borovec, J., Kybic, J., Arganda-Carreras, I., Sorokin, D.V., Bueno, G., Khvostikov, A.V., Bakas, S., Eric, I., Chang, C., Heldmann, S., et al.: ANHIR: automatic non-rigid histological image registration challenge. IEEE Transactions on Medical Imaging **39**(10), 3042–3052 (2020)
5. DeTone, D., Malisiewicz, T., Rabinovich, A.: Superpoint: Self-supervised interest point detection and description. In: Proceedings of the IEEE conference on computer vision and pattern recognition workshops. pp. 224–236 (2018)
6. Eliceiri, K., Li, B., Keikhosravi, A.: Multimodal biomedical dataset for evaluating registration methods (patches from tma cores). Zenodo, June (2020)
7. Gatenbee, C.D., Baker, A.M., Prabhakaran, S., Swinyard, O., Slebos, R.J., Mandal, G., Mulholland, E., Andor, N., Marusyk, A., Leedham, S., et al.: Virtual alignment of pathology image series for multi-gigapixel whole slide images. Nature Communications **14**(1), 4502 (2023)
8. Ge, L., Wei, X., Hao, Y., Luo, J., Xu, Y.: Unsupervised histological image registration using structural feature guided convolutional neural network. IEEE Transactions on Medical Imaging **41**(9), 2414–2431 (2022)
9. Hering, A., Hansen, L., Mok, T.C., Chung, A.C., Siebert, H., Häger, S., Lange, A., Kuckertz, S., Heldmann, S., Shao, W., et al.: Learn2Reg: comprehensive multi-task medical image registration challenge, dataset and evaluation in the era of deep learning. IEEE Transactions on Medical Imaging **42**(3), 697–712 (2022)
10. Jiang, H., Karpur, A., Cao, B., Huang, Q., Araujo, A.: OmniGlue: Generalizable Feature Matching with Foundation Model Guidance. In: Proceedings of the IEEE/CVF Conference on Computer Vision and Pattern Recognition. pp. 19865–19875 (2024)
11. Keikhosravi, A., Li, B., Liu, Y., Eliceiri, K.W.: Intensity-based registration of bright-field and second-harmonic generation images of histopathology tissue sections. Biomedical Optics Express **11**(1), 160–173 (2020)
12. Lin, Y., Liang, Z., He, Y., Huang, W., Guan, T.: End-to-end affine registration framework for histopathological images with weak annotations. Computer Methods and Programs in Biomedicine **241**, 107763 (2023)

13. Lotz, J., Olesch, J., Müller, B., Polzin, T., Galuschka, P., Lotz, J., Heldmann, S., Laue, H., González-Vallinas, M., Warth, A., et al.: Patch-based nonlinear image registration for gigapixel whole slide images. IEEE Transactions on Biomedical Engineering **63**(9), 1812–1819 (2015)
14. Lotz, J., Weiss, N., van der Laak, J., Heldmann, S.: Comparison of consecutive and restained sections for image registration in histopathology. Journal of Medical Imaging (2023)
15. Lowe, D.G.: Distinctive image features from scale-invariant keypoints. International Journal of Computer Vision **60**, 91–110 (2004)
16. Marzahl, C., Wilm, F., Tharun, L., Perner, S., Bertram, C.A., Kröger, C., Voigt, J., Klopfleisch, R., Maier, A., Aubreville, M., et al.: Robust quad-tree based registration on whole slide images. In: MICCAI Workshop on Computational Pathology. pp. 181–190. PMLR (2021)
17. Pyatov, V., Sorokin, D.: Affine registration of histological images using transformer-based feature matching. Pattern Recognition and Image Analysis **32**(3), 626–630 (2022)
18. Pyatov, V.A., Sorokin, D.V.: Tahir: Transformer-based affine histological image registration. In: International Conference on Pattern Recognition. pp. 541–552. Springer (2022)
19. Sarlin, P.E., DeTone, D., Malisiewicz, T., Rabinovich, A.: Superglue: Learning feature matching with graph neural networks. In: Proceedings of the IEEE/CVF conference on computer vision and pattern recognition. pp. 4938–4947 (2020)
20. Shao, W., Banh, L., Kunder, C.A., Fan, R.E., Soerensen, S.J., Wang, J.B., Teslovich, N.C., Madhuripan, N., Jawahar, A., Ghanouni, P., et al.: ProsRegNet: A deep learning framework for registration of MRI and histopathology images of the prostate. Medical Image Analysis **68**, 101919 (2021)
21. Weitz, P., Valkonen, M., Solorzano, L., Carr, C., Kartasalo, K., Boissin, C., Koivukoski, S., Kuusela, A., Rasic, D., Feng, Y., et al.: The ACROBAT 2022 Challenge: Automatic Registration Of Breast Cancer Tissue. arXiv 2305.18033 (2023)
22. Wodzinski, M., Marini, N., Atzori, M., Müller, H.: DeeperHistReg: Robust Whole Slide Images Registration Framework. arXiv:2404.14434 (2024)
23. Wodzinski, M., Marini, N., Atzori, M., Müller, H.: RegWSI: Whole slide image registration using combined deep feature-and intensity-based methods: Winner of the ACROBAT 2023 challenge. Computer Methods and Programs in Biomedicine **250**, 108187 (2024)
24. Wodzinski, M., Müller, H.: DeepHistReg: Unsupervised deep learning registration framework for differently stained histology samples. Computer Methods and Programs in Biomedicine **198**, 105799 (2021)
25. Wodzinski, M., Skalski, A.: Multistep, automatic and nonrigid image registration method for histology samples acquired using multiple stains. Physics in Medicine & Biology **66**(2), 025006 (2021)
26. Zhao, S., Lau, T., Luo, J., Eric, I., Chang, C., Xu, Y.: Unsupervised 3d end-to-end medical image registration with volume tweening network. IEEE Journal of Biomedical and Health Informatics **24**(5), 1394–1404 (2019)

Towards Fast and Accurate Non-rigid Liver Fusion

Stefano Fogarollo[1(✉)], Yannick Scharll[2], Reto Bale[2], and Matthias Harders[1]

[1] Department of Computer Science Interactive Graphics and Simulation Group (IGS), University of Innsbruck, Technikerstraße 21, Innsbruck, Austria
stefano.fogarollo@uibk.ac.at

[2] Interventional Oncology-Microinvasive Therapy (SIP), Department of Radiology, Medical University Innsbruck, Innsbruck, Austria

Abstract. Fast and accurate registration is crucial for minimally invasive liver interventions. Traditional methods, while effective at optimizing rigid and non-rigid deformations, often require minutes or even hours to complete, limiting their clinical applicability. As an alternative, deep learning methods have been proposed, which are capable of inference in seconds, after appropriate training. However, they do not generalize well. In this work, we propose a hybrid method that takes the best from both worlds, and consists of two steps. After a coarse initial deformation using conventional techniques, we learn to refine the displacements with a targeted neural network architecture. We evaluate the accuracy of fast conventional techniques, deep learning methods, and our hybrid approach on 30 challenging liver interventions at the Innsbruck University Hospital (20× CT-CT registration tasks, 10× CT-MR). Our method demonstrates superior robustness compared to deep learning techniques, while maintaining desired inference speed. We can significantly improve on the liver and vascular alignment when compared to a wide range of learning methods, while keeping the inference time below 3 seconds. To the best of our knowledge, this is the first instance where a deep learning method is used to refine initial displacements generated by a conventional method in the context of multi-modal deformable liver registration.

Keywords: liver registration · learning-based refinement · conventional techniques

1 Introduction

Minimally invasive liver tumor interventions require fast and accurate registration to ensure precise planning and evaluation of the treatment. This entails aligning two volumes such that the anatomical structures match correctly. Inference speed is also a critical factor, as fast results are necessary for effective

Supplementary Information The online version contains supplementary material available at https://doi.org/10.1007/978-3-031-73480-9_28.

M. Modat et al. (Eds.): WBIR 2024, LNCS 15249, pp. 358–369, 2024.
https://doi.org/10.1007/978-3-031-73480-9_28

decision-making during the intervention. Affine registration is not sufficient in the intra-procedural setting since liver deformations are highly non-linear [28]. Traditionally, the problem has been approached by iteratively optimizing the transformation based on an energy function balancing the image similarity cost and regularization terms. Given a moving I_m and a fixed I_f volume, we aim at predicting a dense displacement field ϕ aligning I_m to I_f, by optimizing:

$$\phi = \arg\min \mathcal{C}(I_m, I_f) \tag{1}$$

$$= \arg\min(\mathcal{C}_{\text{similarity}}(I_m \circ \phi, I_f) + \lambda \mathcal{C}_{\text{regularization}}(\phi)), \tag{2}$$

where $\circ$ is the warping operation, and λ is the regularization hyper-parameter. Popular choices for $\mathcal{C}_{\text{similarity}}$ include mean squared error [6], mutual information [25], and cross-correlation [2]; $\mathcal{C}_{\text{regularization}}$ enforces a spatially smooth deformation, and it is typically modeled as a function of the spatial gradients. Methods like SyN [2] and Elastix [14] are known to provide accurate and reliable results, but their computational intensity often results in impractical inference times [29]. Lately, conventional optimization techniques have been adapted to modern computational frameworks and hardware, achieving faster inference times [7].

Deep learning-based techniques have also been investigated to reduce the inference time. Since the work of [4], the deformable image registration problem has typically been approached as dense displacement field regression using neural networks, optimizing the loss function:

$$\mathcal{L}(I_m, I_f, \phi) = \mathcal{L}_{\text{similarity}}(I_m \circ \phi, I_f) + \lambda \mathcal{L}_{\text{regularization}}(\phi). \tag{3}$$

Typically $\mathcal{L}_{\text{similarity}}$ consists of mean squared error [6] or local normalized cross-correlation (LNCC); $\mathcal{L}_{\text{regularization}}$ usually incorporates isotropic diffusion [4] or bending energy [6]. This loss is often incorporating supervision from the available segmentation labels, with the DICE loss function [19]. Once the learnable parameters are optimized on a training dataset, the inference process is significantly faster compared to traditional techniques. Single-stage architectures like VoxelMorph [4], CycleMorph [12], and Transmorph [6] pioneered the usage of deep-learning for registration, but have difficulties in dealing with large and irregular deformations. In contrast, multi-stage pipelines incrementally refine the displacements, usually in a coarse-to-fine fashion, allowing for a more precise alignment. Recent works like LapIRN [20] PIViT [17], NICE-Trans [18], and ModeT [26], have achieved performance comparable to conventional methods. However, these techniques do not generalize well when the input data distribution differs from the training data distribution [22]. To deal with this distribution shift, instance optimization techniques like convexAdam [24] and NIO [22] learn instance-specific network parameters. However, the final optimized field can contain implausible displacements, as these are not learned from a training dataset but are optimized on a case-by-case basis. Learning modality-agnostic descriptors can also enhance the performance of these methods, but also requires hard-coded augmentation techniques and assumes sufficient training data [21]. Related to our work, the U-Net based network described in [30] targets CT-CT arterial to portal venous phase liver registration, but needs the liver mask as input, which is limiting in clinical scenarios. We explore also CT-MR fusion, and do not require liver segmentations during inference.

In this preliminary study, we focus on fast and accurate intra-patient liver fusion. The goal is to predict as fast as possible (within seconds) the deformation field aligning the liver in intra-patient abdominal scans, taken at different times and in different modalities (CT-MR). We propose to combine the advantages of both conventional and learning methods into a hybrid pipeline: first, we estimate an initial coarse deformation field ϕ_c with a fast conventional method iter_lbp [7], upsample to the original input resolution, and warp the moving image. Then, we use a SwinTransformer [16] encoder to create a feature pyramid for each input. We learn to refine the displacements with an attention-based neural network in a coarse-to-fine fashion. We predict the displacements from the coarsest scale of the pyramid, using the neighborhood attention map inspired by ModeT [26]. In this step, we do not predict another displacement field, but a refinement, with which we correct the initial ϕ_c, and then warp the original moving image. Our key contribution is a hybrid pipeline that combines the advantage of both worlds, speed and accuracy. To the best of our knowledge, this is the first instance of a deep learning method used to refine initial displacements generated by a conventional method for multi-modal deformable liver registration. This preliminary study represents a step towards practical and efficient liver registration in clinical settings. In the following, we introduce our method, report the results of the experiments on 30 procedures from the Innsbruck University Hospital, and perform ablation studies on the core components of the pipeline.

2 Method

Our hybrid pipeline consists of two steps: first, we obtain an initial coarse flow estimation ϕ_c with the non-learning method described in [7]. This step is carried out at a lower input resolution to save time during inference. Then, we learn a refinement field ϕ_Δ and resample ϕ_c to obtain the final displacement flow ϕ. Each step is detailed below and can be seen in Fig. 1.

Initial Coarse Estimation. We downsample the moving and fixed images I_m, I_f to half of their resolution, before applying iter_lbp, a non-learning method designed for inter- and intra-patient abdominal registration. First, MIND [8] feature descriptors are extracted at Förstner keypoints, then dynamic loopy belief propagation (LBP) is iterated until convergence to find the most probable displacements at keypoints. Following the original implementation, we use thin-plate splines extrapolation to produce a dense displacement field ϕ_c. We opt for this method since it shows robustness on abdominal registration tasks [7], it is fast ($\approx$ 2 seconds), and requires only minimal hyper-parameter tuning.

Next, we upsample the dense field to the original resolution, and warp the moving image to create I_w. Based on our experimental evaluation, this step already provides a decent alignment, but it is inherently limited by hand-crafted features and keypoints. Therefore, we develop a learning scheme to correct ϕ_c.

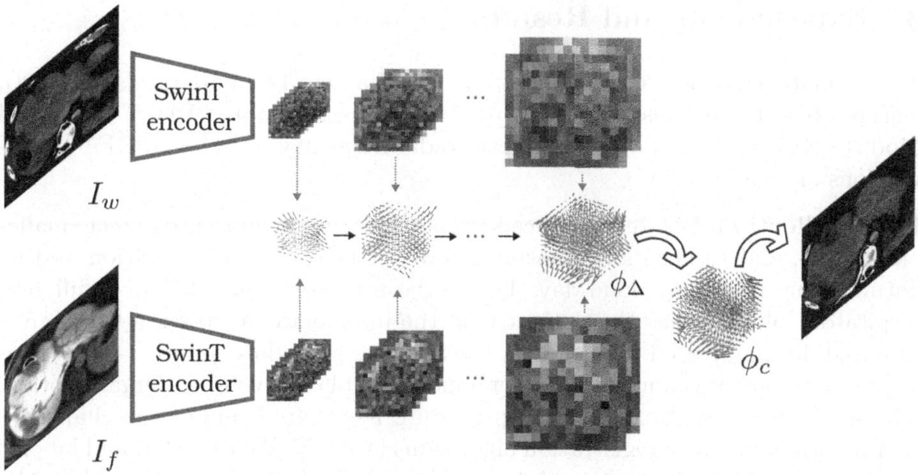

Fig. 1. Overview of our proposed refinement method, after an initial coarse alignment obtained with iter_lbp. We warp the moving image to produce I_w. Then, we learn to estimate a correction field ϕ_Δ with neighbourhood cross-attention (red dotted arrows) on the image pyramid, extracted with a SwinTransformer encoder. Finally, we resample the coarse initial field ϕ_c and warp the moving image. (Color figure online)

Displacements Correction. The proposed network adopts a dual-stream encoder-decoder architecture, with windowed self-attention in the encoding steps, and cross-attention in the decoding steps. First, we encode the inputs I_w, I_f with the same shared encoder, creating a feature pyramid for each input. We use a SwinTransformer architecture [16] to extract a 5-level pyramid. Each feature tensor in the pyramid, has half the spatial resolution of the previous, but double the feature channels.

Then, we proceed with the decoding steps: at each level l of the pyramid, we calculate the query and key tensors Q_l, K_l with a projection layer $Q_l = \mathcal{P}(F_l), K_l = \mathcal{P}(M_l)$, where F_l, M_l are the pyramidal feature tensors at level l from fixed and moving image, respectively. $\mathcal{P}$ is a shallow non-linear projection, consisting of a linear layer, followed by layer normalization [3]. Next, we compute the cross-attention neighborhood map as in ModeT [26], for each voxel p:

$$\mathcal{A}_l(p) = \text{softmax}(Q_l(p) \cdot K_l^T(\mathcal{N}_k(p)) + B), \tag{4}$$

where $\mathcal{N}_k(x)$ denotes the neighborhood of voxel x, a regular grid of size k. B is a learnable relative positional bias, initialized to all zeros. In case of multiple attention heads, we concatenate the output of all the heads. At each level l, we compute a refinement field with a regression head [4], and integrate it with the scaling-and-squaring approach [1]. Then, we resample the previous estimated field, and warp M_{l+1} accordingly. Finally, we resample the initial ϕ_c, with the finest-level correction, to produce ϕ. Like in related work, we use a spatial transformer network to warp the moving image with ϕ [11].

3 Experiments and Results

We evaluate traditional methods (Syn [2], Elastix [14], and deedsBCV [9]), learning-based methods (Voxelmorph [4], Transmorph [6], PIViT [17], and ModeT [26]), and ours on stereotactic radio-frequency ablation (SRFA) interventions on the liver [5].

Data Collection. Before the procedure, a MR scan is acquired to detect smaller metastases, and to get more information about the tumor composition and its nature. This is usually done days before the procedure, possibly in a different respiratory phase. Once the patient is in the interventional room, a CT scan is acquired during arterial phase. After the ablation procedure, another CT scan is acquired, in portal venous phase to enhance the ablation volume. Organ motion, different patient position, and the procedure itself, are some of the challenges that deformable image registration algorithms face [15]. We collected and labeled 30 procedures from the Innsbruck University Hospital, 10 of them including MR scans. We set up three representative registration tasks: 20 CT-CT *post-intra* (portal venous, arterial), 10 *pre-intra* (MR, arterial), and 10 *post-pre* (portal venous, MR). Image registration is first required to transfer the pre-procedure planning, onto the intra-procedure volume (*pre-intra*). In addition, registering the post-procedure scan onto the intra-procedure is necessary for the evaluation of the actual tumor removal (*post-intra*, *post-pre*) [15]. Particular attention is given to the alignment of hepatic vasculature, since it is a measure of how well the hepatic interior tissue is displaced. In each task, we rigidly pre-align the volumes (based on automatic segmentations [27]), resample to an isotropic resolution of 2.0 mm, crop to achieve voxel dimensions of $128 \times 128 \times 128$, and normalize the intensities in $[0, 1]$.

Evaluation Metrics. We measure the alignment of liver and vascular areas (hepatic vessels and portal vein), in terms of DICE, average surface distance (ASSD), and the 95th percentile of the Hausdorff distance (HD95). We also measure the percentage of voxels with non-positive Jacobian determinant $|J_s|_{\leq 0}$, and the inference time. ASSD and HD95 are measured in millimeters, while time is measured in seconds. In the following tables, arrows indicate whether higher or lower values are better, and text in bold represents the best results. We group the methods by category: traditional, instance optimization, learning methods, and ours.

Training. We carried out the experiments on a `NVIDIA A40`, using PyTorch [23] for the implementations, and Adam optimizer [13] (initial learning rate $1e-4$) for training the learning methods. We impose LNCC as $\mathcal{L}_{\text{similarity}}$ [4], and the diffusion regularizer as $\mathcal{L}_{\text{regularization}}$ [4]. We also use the DICE loss [19] on the liver segmentation masks [6]. For better generalization, we employ dropout and common data augmentation techniques: affine and deformable augmentations, Gaussian noise and blurring. We list the hyper-parameters in the Supplementary Information.

Post-intra Procedure Registration. For training the learning methods, we perform five cross-validation experiments (14 training, 2 validation, and 4 test cases) on the 20 CT-CT registration tasks. We visualize selected qualitative results in Fig. 2, and report quantitative results below in Table 1.

Table 1. Quantitative results on the *post-intra* procedure data.

Method	Liver DICE ↑	Liver ASSD ↓	Vasculature ASSD ↓	$\lvert J_s \rvert_{\leq 0}$ ↓	Time ↓
SyN	0.93±0.08	0.88±0.27	2.11±0.43	**0.0**	8.6±0.5
Elastix	0.96±0.05	0.70±0.18	1.88±0.41	**0.0**	32.1±0.4
deedsBCV	0.98±0.02	0.62±0.07	1.77±0.30	0.42±0.1	12.4±0.8
convexAdam	**0.98**±0.01	0.67±0.14	1.70±0.32	**0.0**	2.3±0.6
VoxelMorph	0.94±0.08	1.51±0.30	2.64±0.21	1.01±0.1	**0.1**±0.1
TransMorph	0.95±0.05	1.36±0.15	2.51±0.14	1.52±0.1	0.2±0.1
PIViT	0.97±0.04	1.25±0.18	2.34±0.12	0.41±0.1	**0.1**±0.1
ModeT	0.98±0.03	1.18±0.09	2.32±0.13	0.52±0.1	0.2±0.1
ours	0.98±0.02	**0.60**±0.05	**1.63**±0.18	**0.0**	2.1±0.3

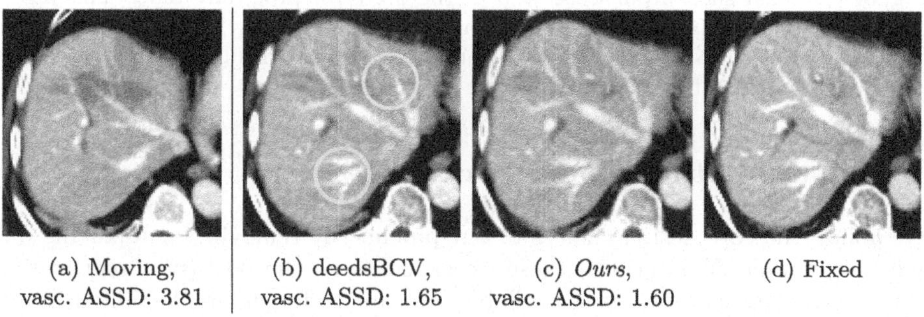

(a) Moving, vasc. ASSD: 3.81 (b) deedsBCV, vasc. ASSD: 1.65 (c) *Ours*, vasc. ASSD: 1.60 (d) Fixed

Fig. 2. CT axial slices from Case # 6 of the *post-intra* procedure data. Note how deedsBCV obtains a good but not perfect alignment on the vascular structures (yellow circles). (Color figure online)

Multi-modal Fusion. In this set of experiments, the number of available procedures is 10, so we perform 10 leave-one-out-cross validation experiments; in this dataset, there are no vasculature labels available, so we report metrics relative to the liver only. We report the results of the *pre-intra* fusion task in Table 2, and the *post-pre* fusion task in Table 3. Qualitative results are available in Fig. 3 and in Fig. 4, respectively.

Table 2. Quantitative results on the *pre-intra* procedure data.

Method	DICE ↑	ASSD ↓	HD95 ↓	$\lvert J_s \rvert_{\leq 0}$ ↓	Time ↓
SyN	0.90±0.09	3.86±1.31	14.1±5.3	**0.0**	7.5±0.3
Elastix	0.89±0.07	4.84±2.44	20.1±5.4	**0.0**	44.1±0.4
deedsBCV	**0.93±0.04**	2.12±0.52	**7.4±2.6**	**0.0**	10.9±0.4
convexAdam	0.90±0.05	3.24±1.43	12.6±8.5	**0.0**	2.2±0.4
VoxelMorph	0.85±0.09	7.18±2.16	19.2±5.7	0.63±0.1	**0.1**±0.1
TransMorph	0.86±0.11	6.84±1.93	17.6±5.1	0.31±0.1	0.2±0.1
PIViT	0.88±0.11	6.51±1.64	16.1±4.7	0.23±0.1	**0.1±0.1**
ModeT	0.87±0.05	6.22±1.85	20.3±6.1	**0.0**	0.2±0.1
ours	0.92±0.03	**2.01**±0.42	7.8±2.0	**0.0**	2.2±0.5

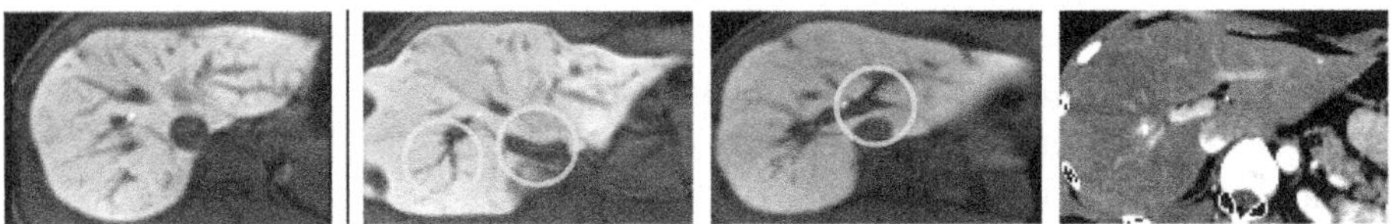

(a) Moving, liver ASSD: 8.19 (b) Elastix, liver ASSD: 4.62 (c) *Ours*, liver ASSD: 2.79 (d) Fixed

Fig. 3. CT axial slices from Case # 3 of the *pre-intra* procedure data. The yellow circles mark critical points for assessing the registration quality, in the context of the clinical application. Note how Elastix deformations are not regularized enough in this case (yellow circles). (Color figure online)

4 Ablation Studies and Discussion

We ablate the core design choices of our pipeline by removing or replacing our learning module. We perform these experiments on the *post-intra* registration task, since there is more data available. We ablate for different configurations of our pipeline, and report quantitative results below in Table 4 and qualitative results in Fig. 5. Then, we explore different hyper-parameters of iter_lbp, and report quantitative results in the Supplementary Information.

Discussion. In the case of complex deformations and with limited annotations like in liver SRFA, traditional or hybrid methods should be preferred over learning-based ones. Overall, we notice an increased difficulty in registering multi-modal scans (CT-MR, *pre-intra* and *post-pre*), compared to single-modality (CT-CT, *post-intra*). Our ablations reveal no single hyper-parameter setting is significantly better than the others for the initial liver alignment (see the Supplementary Information).

Regarding CT-CT registration, the liver is well aligned by most methods (ASSD < 1 mm), while in multi-modal registration, the average surface error is always above 2 mm. The average vasculature surface error is below the original

Table 3. Quantitative results on the *post-pre* procedure data.

Method	DICE ↑	ASSD ↓	HD95 ↓	$\|J_s\|_{\leq 0}$ ↓	Time ↓
SyN	0.88±0.11	3.75±1.41	13.2±7.3	**0.0**	7.4±0.3
Elastix	0.88±0.09	4.35±1.50	17.3±9.2	**0.0**	44.1±0.6
deedsBCV	0.90±0.09	**2.47±1.43**	10.4±4.1	0.38±0.1	11.0±0.4
convexAdam	0.89±0.07	3.52±1.84	15.3±8.1	**0.0**	2.2±0.5
VoxelMorph	0.84±0.11	6.28±1.32	16.9±6.1	0.52±0.1	**0.1±0.1**
TransMorph	0.85±0.08	5.92±1.51	20.1±6.4	0.23±0.1	0.2±0.1
PIViT	0.86±0.03	5.71±1.96	17.5±8.1	0.11±0.1	**0.1±0.1**
ModeT	0.86±0.05	5.65±1.63	14.6±7.5	0.12±0.1	0.2±0.1
ours	**0.91±0.06**	2.50±1.20	**9.7±3.2**	**0.0**	2.0±0.4

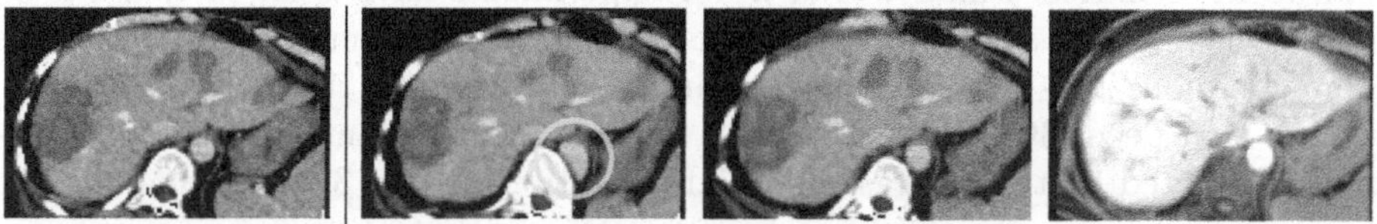

(a) Moving, liver ASSD: 3.69 (b) convexAdam, liver ASSD: 1.82 (c) *Ours*, liver ASSD: 1.51 (d) Fixed

Fig. 4. CT axial slices from Case # 4 of the *post-pre* procedure data. The yellow circle marks an area where the deformations are not plausible, since the aorta cannot exhibit such displacements. (Color figure online)

Table 4. Ablation studies on the proposed pipeline. Text in bold represents statistically significant difference in performance (paired t-tests, p-value < 0.05), with respect to the baseline. $-\mathcal{A}_l(p)$ denotes the replacement of the cross-attention neighborhood map (see Eq. 4) with a linear layer. We evaluate the combination of iter_lbp and ModeT, since it is performing better, and it is more similar to our approach than the other learning methods considered.

Method	Liver ASSD ↓	Vasculature ASSD ↓	$\|J_s\|_{\leq 0}$ ↓	Time ↓
baseline (ours)	0.60±0.05	1.63±0.18	0.0	2.1±0.3
iter_lbp	**0.81±0.11**	**1.84±0.35**	0.0	1.4±0.3
ModeT	**1.18±0.09**	**2.32±0.16**	**0.52±0.1**	**0.2±0.1**
$-\mathcal{A}_l(p)$	**0.78±0.13**	**1.79±0.31**	0.0	2.0±0.2
iter_lbp + ModeT	**0.72±0.01**	1.70±0.16	0.0	**2.4±0.2**

volume thickness in the CT-CT results, whereas in the multi-modal scenarios, there is still margin for improvement, at least qualitatively. The ablation studies support our design choices: conventional methods as well as learning-based methods appear to be not optimal for fast and accurate liver registration in

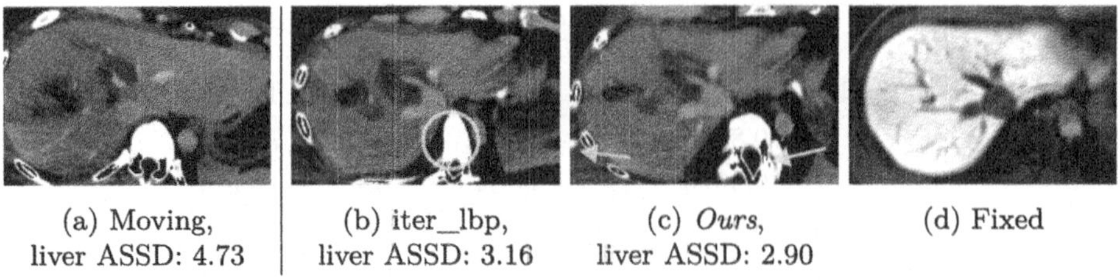

Fig. 5. CT axial slices from Case # 5 of the *post-pre* procedure data. The initial deformations obtained with iter_lbp do not necessarily exhibit plausibility, since they are not learned over a training data set. However, our refinement is capable of correcting most errors (yellow arrows). (Color figure online)

real-world scenarios. Finally, even if our experiments have been carried out on inputs with spatial shape 128 × 128, usually in clinical scenarios, 512 × 512 is more common. This impacts the running time of iterative traditional methods: we find that deedsBCV takes 186.2$\pm$3.1 seconds, while our method is only minimally impacted by the increased spatial resolution, taking 3.1$\pm$0.2 seconds.

Limitations. The refinement obtained with our method might not be sufficient to compensate for the errors of the initial displacements, as seen in Fig. 5. In particular, it would be beneficial to understand how much the initial coarse alignment impacts the final performance, after refinement. Additionally, since the initial alignment is done with a non-learning method, hyper-parameter tuning is necessary before applying it to different datasets. Our focus is on liver registration, but we note that the methods should be compared on bigger datasets [10].

5 Conclusions

Our hybrid pipeline demonstrates great potential for fast and accurate liver registration in challenging real-world scenarios. The ablation studies and empirical results on 30 real-world registration tasks support these findings. Our next steps include systematic benchmarking of additional methods for multi-modal liver fusion and improving accuracy, particularly on the hepatic vascular network. We also note that more work should be done towards encoding the images, or extracting information from the initial coarse displacements. The registration of non-contrast enhanced liver scans remains an open problem due to the high noise levels inherent in the image acquisition process. Plausibility of displacements is also another direction that should to be explored. Finally, we plan to extend our work on other organs and modalities [10]. This work significantly improves precision and speed over baseline methods, representing a promising advancement towards practical and efficient liver fusion in clinical settings.

Acknowledgements. The authors want to thank Nenad Radojicic for helping during the data annotation process. This project has been partially funded by Stiftung Propter

Homines, Vaduz, Liechtenstein, and it has been supported by the Austrian Science Foundation (FWF, project DOC 110). The computational results presented have been achieved (in part) using the Vienna Scientific Cluster (VSC).

Data, Code and/or Material Availability. Patients at Innsbruck University Hospital gave informed consent for anonymous use of their data. Code is available at https://git.uibk.ac.at/informatik/igs/open/etabeta_pub.

Conflict of Interest. The authors declare no conflict of interest.

References

1. Arsigny, V., Commowick, O., Pennec, X., Ayache, N.: A log-euclidean framework for statistics on diffeomorphisms. In: Larsen, R., Nielsen, M., Sporring, J. (eds.) MICCAI. pp. 924–931. Springer Berlin Heidelberg (2006)
2. Avants, B., Epstein, C., Grossman, M., Gee, J.: Symmetric diffeomorphic image registration with cross-correlation: Evaluating automated labeling of elderly and neurodegenerative brain. Medical Image Analysis **12**, 26–41 (2008). https://doi.org/10.1016/j.media.2007.06.004
3. Ba, J.L., Kiros, J.R., Hinton, G.E.: Layer normalization (2016), https://arxiv.org/abs/1607.06450
4. Balakrishnan, G., Zhao, A., Sabuncu, M.R., Guttag, J., Dalca, A.V.: Voxelmorph: A learning framework for deformable medical image registration. IEEE Transactions on Medical Imaging **38**, 1788–1800 (2019). https://doi.org/10.1109/TMI.2019.2897538
5. Bale, R., Widmann, G., Haidu, M.: Stereotactic radiofrequency ablation. CardioVascular and Interventional Radiology **34**, 852–856 (2011). https://doi.org/10.1007/s00270-010-9966-z
6. Chen, J., Frey, E.C., He, Y., Segars, W.P., Li, Y., Du, Y.: Transmorph: Transformer for unsupervised medical image registration. Medical Image Analysis **82**, 102615 (2022). https://doi.org/10.1016/j.media.2022.102615
7. Hansen, L., Heinrich, M.P.: Revisiting iterative highly efficient optimisation schemes in medical image registration. In: MICCAI. pp. 203–212. Springer (2021)
8. Heinrich, M.P., Jenkinson, M., Bhushan, M., Matin, T., Gleeson, F.V., Brady, S.M., Schnabel, J.A.: Mind: Modality independent neighbourhood descriptor for multimodal deformable registration. Medical Image Analysis **16**(7), 1423–1435 (2012). https://doi.org/10.1016/j.media.2012.05.008, special Issue on the 2011 Conference on Medical Image Computing and Computer Assisted Intervention
9. Heinrich, M.P., Jenkinson, M., Brady, M., Schnabel, J.A.: Mrf-based deformable registration and ventilation estimation of lung ct. IEEE Transactions on Medical Imaging **32**(7), 1239–1248 (2013). https://doi.org/10.1109/TMI.2013.2246577
10. Hering, A., Hansen, L., Mok, T.C.W., Chung, A.C.S., Siebert, H., Hager, S., Lange, A., Kuckertz, S., Heldmann, S., Shao, W., Vesal, S., Rusu, M., Sonn, G., Estienne, T., Vakalopoulou, M., Han, L., Huang, Y., Yap, P.T., Brudfors, M., Balbastre, Y., Joutard, S., Modat, M., Lifshitz, G., Raviv, D., Lv, J., Li, Q., Jaouen, V., Visvikis, D., Fourcade, C., Rubeaux, M., Pan, W., Xu, Z., Jian, B., Benetti, F.D., Wodzinski, M., Gunnarsson, N., Sjolund, J., Grzech, D., Qiu, H., Li, Z., Thorley, A., Duan, J., Grosbrohmer, C., Hoopes, A., Reinertsen, I., Xiao, Y., Landman, B., Huo, Y., Murphy, K., Lessmann, N., van Ginneken, B., Dalca, A.V., Heinrich,

M.P.: Learn2reg: Comprehensive multi-task medical image registration challenge, dataset and evaluation in the era of deep learning. IEEE Transactions on Medical Imaging **42**, 697–712 (2023). https://doi.org/10.1109/TMI.2022.3213983
11. Jaderberg, M., Simonyan, K., Zisserman, A., Kavukcuoglu, K.: Spatial transformer networks. In: Proceedings of the 28th International Conference on Neural Information Processing Systems - Volume 2. p. 2017-2025. NIPS'15 (2015)
12. Kim, B., Kim, D.H., Park, S.H., Kim, J., Lee, J.G., Ye, J.C.: Cyclemorph: Cycle consistent unsupervised deformable image registration. Medical Image Analysis **71**, 102036 (2021). https://doi.org/10.1016/j.media.2021.102036
13. Kingma, D.P., Ba, J.: Adam: a method for stochastic optimization. In: ICLR (2015)
14. Klein, S., Staring, M., Murphy, K., Viergever, M., Pluim, J.: elastix: A toolbox for intensity-based medical image registration. IEEE Transactions on Medical Imaging **29**, 196–205 (2010). https://doi.org/10.1109/TMI.2009.2035616
15. Laimer, G., Schullian, P., Jaschke, N., Putzer, D., Eberle, G., Alzaga, A., Odisio, B., Bale, R.: Minimal ablative margin (mam) assessment with image fusion: an independent predictor for local tumor progression in hepatocellular carcinoma after stereotactic radiofrequency ablation. European Radiology **30**(5), 2463–2472 (2020). https://doi.org/10.1007/s00330-019-06609-7
16. Liu, Z., Lin, Y., Cao, Y., Hu, H., Wei, Y., Zhang, Z., Lin, S., Guo, B.: Swin transformer: Hierarchical vision transformer using shifted windows. In: 2021 IEEE/CVF International Conference on Computer Vision (ICCV). pp. 9992–10002 (10 2021). https://doi.org/10.1109/ICCV48922.2021.00986
17. Ma, T., Dai, X., Zhang, S., Wen, Y.: Pivit: Large deformation image registration with pyramid-iterative vision transformer. In: MICCAI. pp. 602–612. Springer (2023)
18. Meng, M., Bi, L., Fulham, M., Feng, D., Kim, J.: Non-iterative coarse-to-fine transformer networks for joint affine and deformable image registration. In: MICCAI. pp. 750–760. Springer (2023)
19. Milletari, F., Navab, N., Ahmadi, S.A.: V-net: Fully convolutional neural networks for volumetric medical image segmentation. In: 2016 Fourth International Conference on 3D Vision (3DV). pp. 565–571 (2016). https://doi.org/10.1109/3DV.2016.79
20. Mok, T.C.W., Chung, A.C.S.: Large deformation diffeomorphic image registration with laplacian pyramid networks. In: MICCAI. pp. 211–221. Springer (2020)
21. Mok, T.C.W., Li, Z., Bai, Y., Zhang, J., Liu, W., Zhou, Y.J., Yan, K., Jin, D., Shi, Y., Yin, X., Lu, L., Zhang, L.: Modality-agnostic structural image representation learning for deformable multi-modality medical image registration (2024)
22. Mok, T.C.W., Li, Z., Xia, Y., Yao, J., Zhang, L., Zhou, J., Lu, L.: Deformable medical image registration under distribution shifts with neural instance optimization. In: Cao, X., Xu, X., Rekik, I., Cui, Z., Ouyang, X. (eds.) Machine Learning in Medical Imaging. pp. 126–136. Springer Nature Switzerland, Cham (2024)
23. Paszke, A., Gross, S., Massa, F., Lerer, A., Bradbury, J., Chanan, G., Killeen, T., Lin, Z., Gimelshein, N., Antiga, L., Desmaison, A., Köpf, A., Yang, E., DeVito, Z., Raison, M., Tejani, A., Chilamkurthy, S., Steiner, B., Fang, L., Bai, J., Chintala, S.: Pytorch: an imperative style, high-performance deep learning library. In: 33rd International Conference on Neural Information Processing Systems (2019)
24. Siebert, H., Hansen, L., Heinrich, M.P.: Fast 3D registration with accurate optimisation and little learning for learn2reg 2021. In: Biomedical Image Registration, Domain Generalisation and Out-of-Distribution Analysis. pp. 174–179 (2022)

25. Viola, P., Wells III, W.M.: Alignment by maximization of mutual information. International Journal of Computer Vision **24**(2), 137–154 (1997). https://doi.org/10.1023/A:1007958904918
26. Wang, H., Ni, D., Wang, Y.: Modet: Learning deformable image registration via motion decomposition transformer. In: MICCAI. pp. 740–749. Springer (2023)
27. Wasserthal, J., Breit, H.C., Meyer, M.T., Pradella, M., Hinck, D., Sauter, A.W., Heye, T., Boll, D.T., Cyriac, J., Yang, S., Bach, M., Segeroth, M.: Totalsegmentator: Robust segmentation of 104 anatomic structures in ct images. Radiology: Artificial Intelligence **5**(5), e230024 (2023). https://doi.org/10.1148/ryai.230024
28. Wise, P.A., Preukschas, A.A., Özmen, E., Bellemann, N., Norajitra, T., Sommer, C.M., Stock, C., Mehrabi, A., Müller-Stich, B.P., Kenngott, H.G., Nickel, F.: Intraoperative liver deformation and organ motion caused by ventilation, laparotomy, and pneumoperitoneum in a porcine model for image-guided liver surgery. Surgical Endoscopy **38**(3), 1379–1389 (2024). https://doi.org/10.1007/s00464-023-10612-x
29. Xu, Z., Lee, C.P., Heinrich, M.P., Modat, M., Rueckert, D., Ourselin, S., Abramson, R.G., Landman, B.A.: Evaluation of six registration methods for the human abdomen on clinically acquired ct. IEEE Transactions on Biomedical Engineering **63**(8), 1563–1572 (2016). https://doi.org/10.1109/TBME.2016.2574816
30. Xue, P., Zhang, J., Ma, L., Liu, M., Gu, Y., Huang, J., Liu, F., Pan, Y., Cao, X., Shen, D.: Structure-aware registration network for liver dce-ct images. IEEE Journal of Biomedical and Health Informatics **28**(4), 2163–2174 (2024). https://doi.org/10.1109/JBHI.2024.3350166

Author Index

M. Modat et al. (Eds.): WBIR 2024, LNCS 15249, pp. 371–373, 2024.
https://doi.org/10.1007/978-3-031-73480-9

SPRINGER NATURE

GPSR Compliance

The European Union's (EU) General Product Safety Regulation (GPSR) is a set of rules that requires consumer products to be safe and our obligations to ensure this.

If you have any concerns about our products, you can contact us on ProductSafety@springernature.com

In case Publisher is established outside the EU, the EU authorized representative is:

Springer Nature Customer Service Center GmbH
Europaplatz 3
69115 Heidelberg, Germany

The manufacturer's authorised representative in the EU is Springer Nature Customer Service Centre GmbH, Europaplatz 3, 69115 Heidelberg, Germany. If you have any concerns regarding our products, please contact ProductSafety@springernature.com

Printed and bound by CPI Group (UK) Ltd, Croydon, CR0 4YY
28/09/2026
02224064-0003